Student Solutions Manual

Charles W. Haines
Rochester Institute of Technology

Elementary Differential Equations and Boundary Value Problems

Ninth Edition

William E. Boyce
Richard C. DiPrima
Rensselaer Polytechnic Institute

WILEY

John Wiley & Sons, Inc.

COVER ART   Norm Christiansen

To order books or for customer service please, call 1-800-CALL WILEY (225-5945).

ISBN-13   978-0-470-38335-3

Printed in the United States of America

10 9 8 7 6 5 4 3 2

Printed and bound by Hamilton Printing Company

PREFACE

This supplement has been prepared for use in conjunction with the ninth editions of *Elementary Differential Equations and Boundary Value Problems* and *Elementary Differential Equations,* both by W.E. Boyce and R.C. DiPrima. The supplement contains a sampling of the problems from each section of the text. In most cases the complete details in determining the solutions are given while in the remainder of the problems helpful hints are provided. The problems chosen in each section represent, wherever possible, the variety of applications and types of examples that are covered in the written material of the text, thereby providing the student with more examples from which to learn.

Students should be aware that following these solutions is very different from designing and constructing one's own solution. Using this supplemental resource appropriately for learning differential equations is outlined as follows:

1. Make an honest attempt to solve the problem without using the guide.

2. If needed, glance at the beginning of the solution in the guide and then try again to generate the complete solution. Continue using the guide for hints when you reach an impasse.

3. Compare your final solution with the one provided to see whether yours is more or less efficient than the guide, since there is frequently more than one correct way to solve a problem.

4. Ask yourself why that particular problem was assigned.

The use of a symbolic computational software package, in many cases, would greatly simplify finding the solution to a given problem, but the details given in this solutions manual are important for the student's understanding of the underlying mathematical principles and applications. In other cases, these software packages are essential for completing the given problem, as the calculations and graphing would be overwhelming using analytical techniques. In these cases, some steps or hints are given and then reference is made to the use of an appropriate software package.

In order to simplify the text, the following abbreviations have been used:

D.E. differential equation(s)
O.D.E. ordinary differential equation(s)
P.D.E. partial differential equation(s)
I.C. initial condition(s)
I.V.P. initial value problem(s)
B.C. boundary condition(s)
B.V.P. boundary value problem(s)

I wish to express my appreciation to Mr. John Wellin and Dr. Josef Torok who have provided invaluable assistance with the preparation of the figures. Dr. Torok has also provided assistance with many of the solutions involving the use of symbolic computational software.

Charles W. Haines
Professor Emeritus
Mathematics and Mechanical Engineering
Rochester Institute of Technology
Rochester, New York
August 2008

Updated for the Ninth Edition by: Gabrielle Andries

Publisher's note:

Chapter Review Sheets are a new feature to this edition of the Student Solutions Manual. They correspond to Chapters 1-11, and are located at the back of the manual as noted on the Table of Contents.

CONTENTS

CHAPTER 1

Section 1.1, Page 7

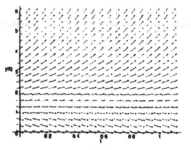

2. For y > 3/2 we see that y' > 0
    and thus y(t) is increasing there.
    For y < 3/2 we have y' < 0 and thus
    y(t) is decreasing there. Hence
    y(t) diverges from 3/2 as t→∞.

$y' = 2y - 3$

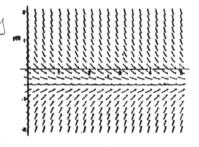

4. Observing the direction field,
    we see that for y>-1/2 we have
    y'<0, so the solution is
    decreasing. Likewise, for
    y<-1/2 we have y'>0 and
    thus y(t) is increasing. These
    results are consistent with the
    given D.E. Since the slopes get
    closer to zero as y gets closer
    to -1/2, we conclude that y→-1/2
    as t→∞.

$y' = -1 - 2y$

7. If all solutions approach 3, then 3 is the equilibrium
    solution and we want $\dfrac{dy}{dt}$ < 0 for y > 3 and $\dfrac{dy}{dt}$ > 0 for
    y < 3. Thus $\dfrac{dy}{dt}$ = 3-y, which is not unique as there are
    other possible answers, such as $\dfrac{dy}{dt}$ = 6-2y.

9. If solutions diverge from 2 then we want $\dfrac{dy}{dt}$ > 0 for y > 2
    and $\dfrac{dy}{dt}$ < 0 for y < 2. Thus $\dfrac{dy}{dt}$ = y - 2 is a possible D.E.

11. For y = 0 and y = 4 we have y' = 0 and thus y = 0 and
    y = 4 are equilibrium solutions. For y > 4, y' < 0 so if
    y(0) > 4 the solution approaches y = 4 from above. If
    0 < y(0) < 4, then y' > 0 and the solutions "grow" to y = 4
    as t→∞. For y(0) < 0 we see that y' < 0 and the solutions
    diverge from 0.

13. Since y' = y², y = 0 is the equilibrium solution and y' > 0
    for all y. Thus if y(0) > 0, solutions will diverge from
    0 and if y(0) < 0, solutions will aproach y = 0 as t→∞.

16. From Fig. 1.1.6 we see that y = 2 is an equilibrium
    solution and thus (c) and (g) are the only posible D.E. to

consider. Since $\frac{dy}{dt} > 0$ for $y > 2$, and $\frac{dy}{dt} < 0$ for $y < 2$ we conclude that (c) is the correct answer.

19. From Fig. 1.1.9 we see that $y = 0$ and $y = 3$ are equilibrium solutions, so (e) and (h) are the only possible D.E. Furthermore, in Fig. 1.1.9 we have $\frac{dy}{dt} < 0$ for $y > 3$ and for $y < 0$, and $\frac{dy}{dt} > 0$ for $0 < y < 3$. This tells us that (h) is the desired D.E.

21a. Let $q(t)$ be the number of grams of the chemical in the water at any time. Then $\frac{q(t)}{1,000,000}$ represents the concentration of the chemical in the pond at any time and hence $\frac{300q(t)}{1,000,000}$ is the rate at which the chemical leaves the pond per hour and $300(.01)$ represents the amount of the chemical coming into the pond per hour. Thus
$$\frac{dq}{dt} = 300(.01) - \frac{300q}{1,000,000} = 300(10^{-2} - 10^{-6}q).$$

21b. The equilibrium solution occurs when $q' = 0$, or $q = 10^4$ gm. Since $q' > 0$ for $q < 10^4$ gm and $q' < 0$ for $q > 10^4$ gm all solutions approach the equilibriium solution indpendent of the amount present at $t = 0$.

22. Let V be the volume, S the surface area and a the constant of proportionality. Then the D.E. expressing the evaporation is $\frac{dV}{dt} = -aS$, $a > 0$. Now $V = \frac{4}{3}\pi r^3$ and $S = 4\pi r^2$, so $S = 4\pi\left(\frac{3}{4\pi}\right)^{2/3} V^{2/3}$. Thus $\frac{dV}{dt} = -skV^{2/3}$, for $k > 0$.

25d.   28.

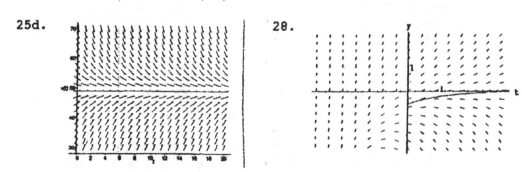

29. 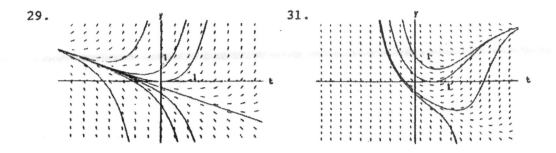 31.

Section 1.2 Page 15

1b. dy/dt = -2y+5 can be rewritten as $\dfrac{dy}{y-5/2}$ = -2dt. Thus

$\ln|y-5/2|$ = -2t+c_1, or y-5/2 = ce^{-2t}. y(0) = y_0 yields
c = y_0 - 5/2, so y = 5/2 + (y_0-5/2)e^{-2t}.
If y_0 > 5/2, the solution starts above the equilibrium
solution and decreases exponentially and approaches 5/2
as t→∞. Conversely, if y_0 < 5/2, the solution starts
below 5/2 and grows exponentially and approaches 5/2 from
below as t→∞.

2b. As in 1b., the D.E. can be rewritten as $\dfrac{dy}{y-5/2}$ = 2dt.

Integrating (as above) and solving yields y = 5/2 + ce^{2t},
so if y(0) = y_0 we find y_0 = 5/2 + c. Hence

y(t) = 5/2 + (y_0 - 5/2)e^{2t}. Again y = 5/2 is the equilbrium
solution, but all other solutions diverge from this due to
the positive exponential.

3a. Note that $\dfrac{dy}{dt}$ = -a(y - b/a) and thus $\dfrac{dy}{y-b/a}$ = -a.

Integration, as in the text, yields $\ln|y-b/a|$ = -at + c,
or y = $\dfrac{b}{a}$ + Ce^{-at}.

3c. (i) If a increases then b/a is smaller. Thus the
equilibrium solution is lower and it is reached sooner,
since e^{-at} decays faster for larger a. (ii) If b increases
then b/a (the equilibrium solution) is larger, but a is
constant, so there is no change in the rate of approach.
Similar analysis will yield the result for iii).

5a. Rewrite Eq.(ii) as $\dfrac{dy/dt}{y} = a$ and thus $\ln|y| = at+C$; or $y_1 = ce^{at}$.

5b. If $y = y_1(t) + k$, then $\dfrac{dy}{dt} = \dfrac{dy_1}{dt}$. Substituting both these into Eq.(i) we get $\dfrac{dy_1}{dt} = a(y_1+k) - b$. Since $\dfrac{dy_1}{dt} = ay_1$, this leaves $ak - b = 0$ and thus $k = b/a$. Hence $y = y_1(t) + b/a$ is the solution to Eq(i).

5c. Substitution of $y_1 = ce^{at}$ shows this is the same as that given in Eq.(17).

7b. From Eq.(11) we have $p = 900 + ce^{t/2}$. If $p(0) = p_0$, then $c = p_0 - 900$ and thus $p = 900 + (p_0 - 900)e^{t/2}$. If $p_0 < 900$, this decreases, so if we set $p = 0$ and solve for T (the time of extinction) we obtain $e^{T/2} = 900/(900-p_0)$, or $T = 2\ln[900/(900-p_0)]$ months.

9a. The solution to this problem is given by Eq.(26), which has a limiting velocity of 49 m/sec. Substituting $v = 48.02$ (which is 98% of 49) into Eq.(26) yields $e^{-t/5} = .02$. Solving for t we have $t = -5\ln(.02) = 19.56$ sec.

9b. Use Eq.(29) with $t = 19.56$.

11a. If the drag force is proportional to v^2 then $F = 98 - kv^2$ is the net force acting on the falling mass (m = 10kg.). Thus $10\dfrac{dv}{dt} = 98 - kv^2$, which has a limiting velocity of $v^2 = 98/k$. Setting $v^2 = 49^2$ gives $k = 98/49^2$ and hence $\dfrac{dv}{dt} = \dfrac{49^2 - v^2}{10/k} = (49^2-v^2)/245$.

11b. From part a we have $\dfrac{dv}{49^2-v^2} = \dfrac{dt}{245}$, which yields $\dfrac{1}{49}\tanh^{-1}\left(\dfrac{v}{49}\right) = \dfrac{t}{245} + C_0$. Setting t=0 and v=0, the initial point, we then have $0 = 0 + C_0$, or $C_0 = 0$. Thus $\tanh^{-1}\left(\dfrac{v}{49}\right) = t/5$, or $v(t) = 49\tanh(t/5)$ m/sec.

11c.

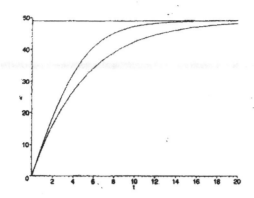

11d. In the graph of 11c, the solution for the linear drag force
lies below the solution for the quadratic drag force. This
latter solution approaches equilibrium faster since, as the
velociy increases, there is a larger drag force.

11e. Note that $\int \tanh(x)dx = \ln(\cosh(x)) + C$.

12a. $\dfrac{dQ}{dt} = -rQ$ yields $\dfrac{dQ/dt}{Q} = -r$, or $\ln|Q| = -rt + c_1$. Thus
$Q = ce^{-rt}$ and $Q(0) = 100$ yields $c = 100$. Hence $Q = 100e^{-rt}$.
Setting $t = 1$, we have $82.04 = 100e^{-r}$, which yields
$r = .1980/\text{wk}$ or $r = .02828/\text{day}$.

15a. Rewrite the D.E. as $\dfrac{du}{u-T} = -kdt$ and then integrate to find
$\ln|u-T| = -kt + c$. Thus $u - T = \pm Ce^{-kt}$. For $t = 0$ we have
$u_0 - T = \pm C$ and thus $u(t) = T + (u_0 - T)e^{-kt}$.

15b. Set $u(\tau) - T = \dfrac{u_0 - T}{2}$ when $t = \tau$ in the solution of part (a).

17a. Rewrite the D.E. as $\dfrac{dQ/dt}{Q-CV} = \dfrac{-1}{CR}$, thus, upon integrating and
simplifying, we get $Q = De^{-t/CR} + CV$. $Q(0) = 0 \Rightarrow D = -CV$ and
thus $Q(t) = CV(1 - e^{-t/CR})$.

17b. $\lim\limits_{t \to \infty} Q(t) = CV$ since $\lim\limits_{t \to \infty} e^{-t/CR} = 0$.

17c. In this case $R\dfrac{dQ}{dt} + \dfrac{Q}{C} = 0$, $Q(t_1) = CV$. The solution of this
D.E. is $Q(t) = Ee^{-t/CR}$, so $Q(t_1) = Ee^{-t_1/CR} = CV$, or
$E = CVe^{t_1/CR}$. Thus $Q(t) = CVe^{t_1/CR}e^{-t/CR} = CVe^{-(t-t_1)/CR}$ for $t \geq t_1$.

17a. CV = 20, CR = 2.5   17c. CV = 20, CR = 2.5, $t_1 = 10$

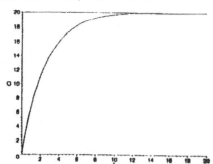 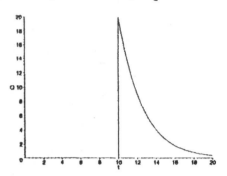

Section 1.3, Page 24

2. The D.E. is second order since there is a second
    derivative of y appearing in the equation. The equation
    is nonlinear due to the y^2 term (as well as due to the y^2
    term multiplying the y" term).

6. This is a third order D.E. since the highest derivative is
    y''' and it is linear since y and all its derivatives
    appear to the first power only. The terms t^3 and $\cos^2 t$ do
    not affect the linearity of the D.E.

8. For $y_1(t) = e^{-3t}$ we have $y'_1(t) = -3e^{-3t}$ and $y''_1(t) = 9e^{-3t}$.
    Substitution of these into the D.E. yields
    $9e^{-3t} + 2(-3e^{-3t}) - 3(e^{-3t}) = (9-6-3)e^{-3t} = 0.$

11. Substituting $y_1 = t^{1/2}$ into the D.E. we get
    $$2t^2\left(\frac{-1}{4}t^{-3/2}\right) + 3t\left(\frac{1}{2}t^{-1/2}\right) - t^{1/2} = \left(\frac{-1}{2}\right)t^{1/2} + \left(\frac{3}{2}\right)t^{1/2} - t^{1/2} = 0.$$

14. Recall that if $u(t) = \int_0^t f(s)ds$, then $u'(t) = f(t)$.

16. Differentiating e^{rt} twice and substituting into the D.E.
    yields $r^2 e^{rt} - e^{rt} = (r^2-1)e^{rt}$. If $y = e^{rt}$ is to be a
    solution of the D.E. then the last quantity must be zero
    for all t. Thus $r^2-1 = 0$ since e^{rt} is never zero.

19. Differentiating t^r twice and substituting into the D.E.
    yields $t^2[r(r-1)t^{r-2}] + 4t[rt^{r-1}] + 2t^r = [r^2+3r+2]t^r$. If
    $y = t^r$ is to be a solution of the D.E. the last term must
    be zero for all t and thus $r^2 + 3r + 2 = 0$.

22. The D.E. is second order since there are second partial derivatives of u(x,y). The D.E. is nonlinear due to the product of u(x,y) times u_x (or u_y).

26. Since $\dfrac{\partial u_1}{\partial t} = -\alpha^2 e^{-\alpha^2 t} \sin x$ and $\dfrac{\partial^2 u_1}{\partial x^2} = -e^{-\alpha^2 t} \sin x$ we have

$\alpha^2 [-e^{-\alpha^2 t} \sin x] = -\alpha^2 e^{-\alpha^2 t} \sin x$, which is true for all t and x.

29a. The free-body diagram is essentially shown in Fig. 1.3.1. The gravitational force is shown. The only other force is the tension, T, which acts towards the hinge along L.

29b. The component of the gravitational force (mg) along the tangent to the circular arc is given by $-mg\sin\theta$. The minus sign arises since θ is positive in the counterclockwise direction and the gravitational force is clockwise. Since T acts perpendicular to the tangent, there is no component of the tension in the tangential direction. Newton's Second Law states that F = ma. In this problem, since the motion is circular, it is appropriate to use polar coordinates and thus $a \neq \dfrac{dv}{dt}$ as we have used earlier. Since r = L is constant, the linear acceleration tangent to the circular motion is given by $L\dfrac{d^2\theta}{dt^2}$ and thus Newton's Second Law gives $-mg\sin\theta = mL\dfrac{d^2\theta}{dt^2}$.

29c. Dividing by mL and rearranging terms gives $\dfrac{d^2\theta}{dt^2} + \dfrac{g}{L}\sin\theta = 0$.

CHAPTER 2

Section 2.1, Page 39

1b. All solutions seem   1a.
to approach a line
in the region where
the negative and
positive slopes
meet each other.

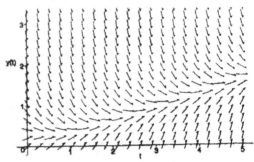

1c. $\mu(t) = \exp(\int 3dt) = e^{3t}$. Thus $e^{3t}(y'+3y) = e^{3t}(t+e^{-2t})$ or

$\dfrac{d}{dt}(ye^{3t}) = te^{3t} + e^t$. Integration of both sides yields

$ye^{3t} = \dfrac{1}{3}te^{3t} - \dfrac{1}{9}e^{3t} + e^t + c$, where integration by parts

is used on the right side, with u = t and dv = $e^{3t}dt$.
Division by e^{3t} gives $y(t) = ce^{-3t} +t/3 - 1/9$, so y
approaches t/3 - 1/9 as t → ∞. This is the line
identified in part b.

2c. $\mu(t) = e^{-2t}$.   3c. $\mu(t) = e^t$.

4c. $\mu(t) = \exp(\int\dfrac{dt}{t}) = e^{\ln t} = t$, so $(ty)' = 3t\cos 2t$, and

integration by parts yields the general solution.

6c. The equation must be divided by t so that it is in the
form of Eq.(3): y' + (2/t)y = (sint)/t. Thus
$\mu(t) = \exp(\int\dfrac{2dt}{t} = t^2$, and $(t^2y)' = t\sin t$. Integration
then yields $t^2y = -t\cos t + \sin t + c$.

7c. $\mu(t) = e^{t^2}$.   8c. $\mu(t) = \exp(\int\dfrac{4tdt}{1+t^2}) = (1+t^2)^2$.

11c. $\mu(t) = e^t$ so $(e^ty)' = 5e^t\sin 2t$. To integrate the right
side you can integrate by parts (twice), use an integral
table, or use a symbolic computational software program
to find $e^ty = e^t(\sin 2t - 2\cos 2t) + c$.

13. $\mu(t) = e^{-t}$ so that $(e^{-t}y)' = 2te^t$ and thus
$e^{-t}y = 2\int te^tdt + c = 2(te^t - \int e^tdt) + c = 2(te^t - e^t) + c$.
Thus $y(t) = 2(t-1)e^{2t} + ce^t$, so setting t = 0 we have
1 = -2 + c, or c = 3. Hence $y(t) = 2(t-1)e^{2t} + 3e^t$.

15. $\mu(t) = \exp(\int \frac{2dt}{t}) = t^2$ so that $(t^2y)' = t^3 - t^2 + t$.

   Integrating and dividing by t^2 gives
   $y = t^2/4 - t/3 + 1/2 + c/t^2$. Setting $t = 1$ and $y = 1/2$
   we have $c = 1/12$.

18. $\mu(t) = t^2$. Thus $(t^2y)' = t\sin t$ and
   $t^2y = -t\cos t + \sin t + c$. Setting $t = \pi/2$ and
   $y = 1$ yields $c = \pi^2/4 - 1$.

20. $\mu(t) = \exp\int \frac{t+1}{t} dt = \exp(t + \ln(t)) = te^t$.

21b. $\mu(t) = e^{-t/2}$ so $(e^{-t/2}y)' = 2e^{-t/2}\cos t$. Integrating (see
   comments in Prob.11) and dividing by $e^{-t/2}$ yields
   $y(t) = -\frac{4}{5}\cos t + \frac{8}{5}\sin t + ce^{t/2}$. Thus $y(0) = -\frac{4}{5} + c = a$,

   or $c = a + \frac{4}{5}$ and $y(t) = -\frac{4}{5}\cos t + \frac{8}{5}\sin t + (a + \frac{4}{5})e^{t/2}$.

21c. If $(a + \frac{4}{5}) = 0$, then the solution is oscillatory for all

   t, while if $(a + \frac{4}{5}) \neq 0$, the solution is unbounded as

   $t \to \infty$. Thus $a_0 = -\frac{4}{5}$.

21a.   25a.

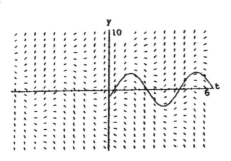   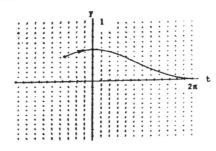

25b. $\mu(t) = \exp\int \frac{2dt}{t} = t^2$, so $(t^2y)' = \sin t$ and

   $y(t) = \frac{-\cos t}{t^2} + \frac{c}{t^2}$. Setting $t = -\frac{\pi}{2}$ yields

   $\frac{4c}{\pi^2} = a$ or $c = \frac{a\pi^2}{4}$ and hence $y(t) = \frac{a\pi^2/4 - \cos t}{t^2}$, which

   is unbounded as $t \to 0$ unless $a\pi^2/4 = 1$ or $a_0 = 4/\pi^2$.

25c. For $a = 4/\pi^2$ $y(t) = \dfrac{1 - \cos t}{t^2}$. To find the limit as

$t \to 0$ L'Hospital's Rule must be used:

$\lim\limits_{t \to 0} y(t) = \lim\limits_{t \to 0} \dfrac{\sin t}{2t} = \lim\limits_{t \to 0} \dfrac{\cos t}{2} = \dfrac{1}{2}$.

26b. $\mu(t) = \exp\!\int \dfrac{\cos(t)}{\sin(t)} dt = \exp(\ln(\sin t)) = \sin(t)$ and thus

$(\sin(t)y)' = e^t$. Hence $\sin(t)y = e^t + c$ or $y(t) = \dfrac{e^t + c}{\sin(t)}$.

Setting $t = 1$ and $y = a$ we get $c = a\sin 1 - e$ so
$y(t) = (e^t - e + a\sin 1)/\sin(t)$. If $y(t)$ is to remain
finite as $t \to 0$ the numerator, $e^t - e + a\sin 1$, must
approach 0 as $t \to 0$ and hence $a_0 = (e-1)/\sin 1$.

26c. Using a_0 we have $y(t) = (e^t - 1)/\sin(t)$, which approaches
1 as $t \to 0$, using l'Hospital's Rule.

30. $(e^{-t}y)' = e^{-t} + 3e^{-t}\sin t$ so

$e^{-t}y = -e^{-t} - 3e^{-t}\left(\dfrac{\sin t + \cos t}{2}\right) + c$ or

$y(t) = -1 - \left(\dfrac{3}{2}\right)e^{-t}(\sin t + \cos t) + ce^t$. Thus

$y(0) = -1 - \dfrac{3}{2} + c = y_0$ or $c = y_0 + \dfrac{5}{2}$. Now, if $y(t)$ is to

remain bounded as $t \to \infty$, we must have $c = 0$ so that
$y_0 = -5/2$.

32. Write the first term of Eq.(47) as $\dfrac{\int_0^t e^{s^2/4}ds}{e^{t^2/4}}$. In applying

L'Hospital's Rule, the derivative of the numerator term
is $e^{t^2/4}$ by the Fundamental Theorem of Calculus. The
derivative of the denominator is $(t/2)e^{t^2/4}$ and thus the
limit of both terms in Eq.(47) is 0 as $t \to \infty$.

33. $\mu(t) = e^{at}$ so the D.E. can be written as
$(e^{at}y)' = be^{at}e^{-\lambda t} = be^{(a-\lambda)t}$. If $a \neq \lambda$, then integration
and solution for y yields $y = [b/(a-\lambda)]e^{-\lambda t} + ce^{-at}$. Then
$\lim\limits_{t \to \infty} y$ is zero since both λ and a are positive numbers.

If $a = \lambda$, then the D.E. becomes $(e^{at}y)' = b$, which
yields $y = (bt+c)/e^{\lambda t}$ as the solution. L'Hospital's Rule
gives

$\lim\limits_{t \to \infty} y = \lim\limits_{t \to \infty} \dfrac{(bt+c)}{e^{\lambda t}} = \lim\limits_{t \to \infty} \dfrac{b}{\lambda e^{\lambda t}} = 0$.

35. There is no unique answer for this situation. One
possible answer is to assume $y(t) = ce^{-2t} + 3 - t$ (which
satisfies the given condition), then $y'(t) = -2ce^{-2t} - 1$.
Eliminating ce^{-2t} between the two equations yields
$y' + 2y = 5 - 2t$.

39. By Eq.(iii), Prob.38, $y(t) = A(t)\exp(-\int(-2)dt) = A(t)e^{2t}$.
Differentiating $y(t)$ and substituting into the D.E.
yields $A'(t) = t^2$ since the terms involving $A(t)$ add to
zero. Thus $A(t) = t^3/3 + c$, which substituted into $y(t)$
yields the solution.

42. Since $p(t) = \dfrac{1}{2}$, $y(t) = A(t)\exp(-\int\dfrac{dt}{2}) = A(t)e^{-t/2}$ and
$A'(t) = (3/2)t^2e^{t/2}$. Integration of A' and substituting
in $y(t)$ yields the desired solution.

Section 2.2, Page 47

Problems 1 through 20 follow the pattern of the examples
worked in this section. The first eight problems, however,
do not have I.C. so the integration constant, c, cannot be
found.

1. Write the equation in the form $ydy = x^2dx$. Integrating
the left side with respect to y and the right side with
respect to x yields
$$\frac{y^2}{2} = \frac{x^3}{3} + C, \text{ or } 3y^2 - 2x^3 = c.$$

4. For $y \neq -3/2$ multiply both sides of the equation by
$3 + 2y$ to get the separated equation
$(3+2y)dy = (3x^2-1)dx$. Integration then yields
$3y + y^2 = x^3 - x + c$.

6. We need $x \neq 0$ and $|y| < 1$ for this problem to be defined.
Separating the variables we get $(1-y^2)^{-1/2}dy = x^{-1}dx$.
Integrating each side yields $\arcsin y = \ln|x|+c$, so
$y = \sin[\ln|x|+c]$, $x \neq 0$ (note that $|y| < 1$). Also,
$y = \pm 1$ satisfy the D.E., since both sides are zero.

10a. Separating the variables we get $ydy = (1-2x)dx$, so

$\dfrac{y^2}{2} = x - x^2 + c$. Setting $x = 1$ and $y = -2$ we have $2 = c$

and thus $y^2 = 2x - 2x^2 + 4$ or $y = -\sqrt{2x - 2x^2 + 4}$. The negative square root must be used since $y(1) = -2$.

10b. The graph is the bottom half of the ellipse
$2x^2 - 2x + y^2 = 4$.

10c. Rewriting $y(x)$ as $-\sqrt{2(2-x)(x+1)}$, we see that y is defined for $-1 \le x \le 2$, However, since y' does not exist for $x = -1$ or $x = 2$, the solution is valid only for the open interval $-1 < x < 2$.

13. Separate variables by factoring the denominator of the right side to get $y\,dy = \dfrac{2x}{1+x^2}dx$. Integration yields $y^2/2 = \ln(1+x^2)+c$ and use of the I.C. gives $c = 2$. Thus $y = \pm\,[2\ln(1+x^2)+4]^{1/2}$, but we must discard the plus square root because of the I.C. Since $1 + x^2 > 0$, the solution is valid for all x.

15. Separating variables and integrating yields
$y + y^2 = x^2 + c$. Setting $y = 0$ when $x = 2$ yields $c = -4$ or $y^2 + y = x^2-4$. To solve for y complete the square on the left side by adding 1/4 to both sides. This yields
$y^2 + y + \dfrac{1}{4} = x^2 - 4 + \dfrac{1}{4}$ or $(y + \dfrac{1}{2})^2 = x^2 - 15/4$. Taking the square root of both sides yields
$y + \dfrac{1}{2} = \pm\sqrt{x^2 - 15/4}$, where the positive square root must be taken in order to satisfy the I.C. Thus
$y = -\dfrac{1}{2} + \sqrt{x^2 - 15/4}$, which is defined for $x^2 \ge 15/4$ or $x \ge \sqrt{15}/2$.

17a. Separating variables gives $(2y-5)dy = (3x^2-e^x)dx$ and integration then gives $y^2 - 5y = x^3 - e^x + c$. Setting $x = 0$ and $y = 1$ we have $1 - 5 = 0 - 1 + c$, or $c = -3$ and thus $y^2 - 5y - (x^3-e^x-3) = 0$. Using the quadratic formula then gives
$y(x) = \dfrac{5 \pm \sqrt{25+4(x^3-e^x-3)}}{2} = \dfrac{5}{2} - \sqrt{\dfrac{13}{4} + x^3 - e^x}$, where the negative square root is chosen so that $y(0) = 1$.

17c. The interval of definition for y must be found
numerically. Approximate values can be found by plotting
$y_1(x) = \dfrac{13}{4} + x^3$ and $y_2(x) = e^x$ and noting the values of x
where the two curves cross.

19a. We start with $\cos 3y\, dy = -\sin 2x\, dx$ and integrate to get
$\dfrac{1}{3}\sin 3y = \dfrac{1}{2}\cos 2x + c$. Setting $y = \pi/3$ when $x = \pi/2$
(from the I.C.) we find that $0 = -\dfrac{1}{2} + c$ or
$c = \dfrac{1}{2}$, so that $\dfrac{1}{3}\sin 3y = \dfrac{1}{2}\cos 2x + \dfrac{1}{2} = \cos^2 x$ (using the
appropriate trigonometric identity). To solve for y we
must choose the branch that passes through the point
$(\pi/2, \pi/3)$ and thus $3y = \pi - \arcsin(3\cos^2 x)$, or
$y = \dfrac{\pi}{3} - \dfrac{1}{3}\arcsin(3\cos^2 x)$.

19c. The solution in part a is defined only for
$0 \le 3\cos^2 x \le 1$, or $-\sqrt{1/3} \le \cos x \le \sqrt{1/3}$. Taking the
indicated square roots and then finding the inverse
cosine of each side yields $.9553 \le x \le 2.1863$, or
$|x-\pi/2| \le 0.6155$, as the approximate interval.

21. We have $(3y^2-6y)dy = (1+3x^2)dx$ so that $y^3-3y^2 = x + x^3 - 2$,
once the I.C. are used. From the D.E., the integral
curve will have a vertical tangent when
$3y^2 - 6y = 0$, or $y = 0,2$. For $y = 0$ we have
$x^3 + x - 2 = 0$, which is satisfied for $x = 1$, which is
the only zero of the function $w = x^3 + x - 2$. Likewise,
for $y = 2$, $x = -1$. Thus the solution is valid on $|x| < 1$.

23. Separating variables gives $y^{-2}dy = (2+x)dx$, so
$-y^{-1} = 2x + \dfrac{x^2}{2} + c$. $y(0) = 1$ yields $c = -1$ and thus
$y = \dfrac{-1}{\dfrac{x^2}{2} + 2x - 1} = \dfrac{2}{2 - 4x - x^2}$. This gives
$\dfrac{dy}{dx} = \dfrac{8 + 4x}{(2-4x-x^2)^2}$, so the minimum value is attained at
$x = -2$. Note that the solution is defined for
$-2 - \sqrt{6} < x < -2 + \sqrt{6}$ (by finding the zeros of the

denominator) and has vertical asymptotes at the end
points of the interval.

25. Separating variables and integrating yields
$3y + y^2 = \sin 2x + c$. $y(0) = -1$ gives $c = -2$ so that
$y^2 + 3y + (2-\sin 2x) = 0$. The quadratic formula, along
with the I.C., then gives $y = -\dfrac{3}{2} + \sqrt{\sin 2x + 1/4}$, which
is defined for $-.126 < x < 1.697$ (found by solving
$\sin 2x = -.25$ for x and noting $x = 0$ is the initial
point). Thus we have $\dfrac{dy}{dx} = \dfrac{\cos 2x}{(\sin 2x + \frac{1}{4})^{1/2}}$, which yields

$x = \pi/4$ as the only critical point in the above interval.
Using the second derivative test or graphing the solution
indicates the critical point is a maximum.

27a. By sketching the direction field or by using the D.E. we
note that $y' < 0$ for $y > 4$ and y' approaches zero as y
approaches 4. For $0 < y < 4$, $y' > 0$ and again approaches
zero as y approaches 4. Thus $\lim\limits_{t\to\infty} y = 4$ if $y_0 > 0$. For
$y_0 < 0$, $y' < 0$ for all y and hence y becomes negatively
unbounded $(-\infty)$ as t increases. If $y_0 = 0$, then $y' = 0$
for all t, so $y = 0$ for all t.

27b. Separating variables and using a partial fraction
expansion we have $(\dfrac{1}{y} - \dfrac{1}{y-4})dy = \dfrac{4}{3}t\,dt$. Hence
$\ln\left|\dfrac{y}{y-4}\right| = \dfrac{2}{3}t^2 + c_1$ and thus $\left|\dfrac{y}{y-4}\right| = e^{c_1}e^{2t^2/3} = ce^{2t^2/3}$,
where c is positive. For $y(0) = y_0 = .5$ this becomes
$\left|\dfrac{.5}{.5 - 4}\right| = c$ and thus $c = \dfrac{1}{7}$. Using this value for c
and solving for y yields $y(t) = \dfrac{4}{1 + 7e^{-2t^2/3}}$. Setting
this equal to 3.98 and solving for t yields
$t = 3.29527$.

29. Separating variables yields $\dfrac{cy+d}{ay+b}\,dy = dx$. If $a \neq 0$ and
$ay+b \neq 0$ then $dx = (\dfrac{c}{a} + \dfrac{ad-bc}{a(ay+b)})dy$. Integration then
yields the desired answer.

30a. Divide numerator and denomintor by $x \neq 0$.

30c. If $v = y/x$ then $y = vx$ and $\dfrac{dy}{dx} = v + x\dfrac{dv}{dx}$ and thus the

D.E. becomes $v + x\dfrac{dv}{dx} = \dfrac{v-4}{1-v}$. Subtracting v from both

sides yields $x\dfrac{dv}{dx} = \dfrac{v^2-4}{1-v}$.

30d. The last equation in (c) separates into $\dfrac{1-v}{v^2-4}dv = \dfrac{1}{x}dx$. To

integrate the left side use partial fractions to write

$\dfrac{1-v}{v-4} = \dfrac{A}{v-2} + \dfrac{B}{v+2}$, which yields $A = -1/4$ and $B = -3/4$.

Integration then gives $-\dfrac{1}{4}\ln|v-2| - \dfrac{3}{4}\ln|v+2| = \ln|x| - k$,

or $\ln|x^4||v-2||v+2|^3 = 4k$ after manipulations using

properties of the ln function. Thus $x^4|v-2||v+2|^3 = C$.

30e. Recalling that $v = y/x$ gives the desired solution.

31a. Simplifying the right side of the D.E. gives

$dy/dx = 1 + (y/x) + (y/x)^2$ so the equation is homogeneous.

31b. $y = vx$ gives $\dfrac{dy}{dx} = v + x\dfrac{dv}{dx}$, so substitution leads to

$v + x\dfrac{dv}{dx} = 1 + v + v^2$ or $\dfrac{dv}{1 + v^2} = \dfrac{dx}{x}$. Integrating, we

get $\arctan v = \ln|x| + c$ and substituting for v we obtain

$\arctan(y/x) - \ln|x| = c$.

33b. Dividing the numerator and denominator of the right side

by x and substituting $y = vx$ we get $v + x\dfrac{dv}{dx} = \dfrac{4v - 3}{2-v}$

which can be rewritten as $x\dfrac{dv}{dx} = \dfrac{v^2 + 2v - 3}{2 - v}$. Note that

$v = -3$ and $v = 1$ are solutions of this equation. For

$v \neq 1, -3$ separating variables gives

$\dfrac{2 - v}{(v+3)(v-1)} dv = \dfrac{1}{x}dx$. Applying a partial fraction

decomposition to the left side we obtain

$[\dfrac{1}{4}\dfrac{1}{v-1} - \dfrac{5}{4}\dfrac{1}{v+3}]dv = \dfrac{dx}{x}$, and upon integrating both sides

we find that $\dfrac{1}{4}\ln|v-1| - \dfrac{5}{4}\ln|v+3| = \ln|x| + c$.

Substituting for v and performing some algebraic manipulations we get the solution in the implicit form $|y-x| = c|y+3x|^5$. $v = 1$ and $v = -3$ yield $y = x$ and $y = -3x$, respectively, as solutions also.

35b. As in Prob. 33, substituting $y = vx$ into the D.E. we get
$v + x\dfrac{dv}{dx} = \dfrac{1+3v}{1-v}$, or $x\dfrac{dv}{dx} = \dfrac{(v+1)^2}{1-v}$. Note that $v = -1$ (or $y = -x$) satisfies this D.E. Separating variables yields $\dfrac{1-v}{(v+1)^2}dv = \dfrac{dx}{x}$. Integrating the left side by parts

(let $u = 1-v$ and $dw = \dfrac{dv}{(v+1)^2}$) we obtain

$\dfrac{v-1}{v+1} - \ln|v+1| = \ln|x| + c$. Letting $v = \dfrac{y}{x}$ then yields

$\dfrac{y-x}{y+x} - \ln\left|\dfrac{y+x}{x}\right| = \ln|x| + c$, or $\dfrac{y-x}{y+x} - \ln|y+x| = c$. The

answer in the text can be obtained by integrating the left side, above, using partial fractions. By differentiating both answers, it can be verified that indeed both forms satisfy the D.E.

Section 2.3, Page 59

2. Let $S(t)$ be the amount of salt that is present at any time t, then $S(0) = 0$ is the original amount of salt in the tank, 2γ is the amount of salt entering per minute, and $2(S/120)$ is the amount of salt leaving per minute (all amounts measured in grams). Thus $dS/dt = 2\gamma - 2S/120$, $S(0) = 0$. This is a linear equation, which has $e^{t/60}$ as its integrating factor. Thus the general solution is $S(t) = 120\gamma + ce^{-t/60}$. $S(0) = 0$ gives $c = -120\gamma$, so $S(t) = 120\gamma(1 - e^{-t/60})$ and hence $S(t) \to 120\gamma$ grams as $t \to \infty$.

3. We must first find the amount of salt that is present after 10 minutes. For the first 10 minutes (if we let $Q(t)$ be the amount of salt in the tank): $\dfrac{dQ}{dt} = \dfrac{1}{2}(2) - 2\dfrac{Q(t)}{100}$, $Q(0) = 0$. This is a linear equation which has the solution $Q(t) = 50(1 - e^{-t/50})$, as in Prob. 2, and thus $Q(10) = 50(1-e^{-.2}) = 9.063$ lbs. of salt in the tank after the first 10 minutes. At this point no more salt is allowed to enter, so the new I.V.P. (letting $P(t)$ be the amount of

salt in the tank after the first 10 minutes) is:

$$\frac{dP}{dt} = (0)(2) - 2\frac{P(t)}{100}, \quad P(0) = Q(10) = 9.063.$$ The solution

of this problem is $P(t) = 9.063e^{-.02t}$, which yields
$P(10) = 7.42$ lbs.

4.   Salt flows out of the tank at the rate of $\frac{Q(t)}{200+t}(2)$ lb/min.

since the volume of water in the tank at any time t is
200 + (1)(t) gallons (due to the fact that water flows into
the tank faster than it flows out). Thus the I.V.P. is

$$dQ/dt = (3)(1) - \frac{2}{200+t}Q(t), \quad Q(0) = 100,$$ which is a linear

equation with $(200+t)^2$ as its integrating factor.

8a. Set $S_0 = 0$ in Eq.(16) (or solve Eq.(15) with $S(0) = 0$).

8b. Set $r = .075$, $t = 40$ and $S(t) = \$1,000,000$ in the answer to
part (a) and then solve for k.

8c. Set $k = \$2,000$, $t = 40$ and $S(t) = \$1,000,000$ in the answer
to (a) and then solve numerically for r.

9.   Let S(t) be the amount of the loan remaining at time t, then
$dS/dt = .1S - k$, $S(0) = \$8,000$. Solving this for S(t) yields
$S(t) = 8000e^{.1t} - 10k(e^{.1t}-1)$. Setting $S = 0$ and substitution
of $t = 3$ gives $k = \$3,086.64$ per year. For 3 years this
totals \$9,259.92, so \$1,259.92 has been paid in interest.

10. Since we are assuming continuity, either convert the monthly
payment into an annual payment or convert the yearly
interest rate into a monthly interest rate for 240 months.
Then proceed as in Prob. 9.

11a. Using Eq. (15) we have $\dfrac{dS}{dt} = \dfrac{.09}{12}S - 800(1+\dfrac{t}{120})$ or

$$\frac{dS}{dt} - \frac{3}{400}S = -(800+\frac{20}{3}t), \quad S(0) = 100,000.$$ Using an

integrating factor and integration by parts (or using a D.E.

solver) we get $S(t) = \dfrac{6,080,000}{27} + \dfrac{8000}{9}t + ce^{3t/400}$. Using

the I.C. yields $c = \dfrac{-3,380,000}{27}$. Substituting this value

into S, setting $S(t) = 0$, and solving numerically for t
yields $t \cong 135.36$ months.

14a. We have $\dfrac{dy}{y}$ = (.1+.2sint)dt, by separating variables, and

thus y(t) = cexp(.1t-.2cost). y(0) = 1 gives c = $e^{.2}$, so
y(t) = exp(.2+.1t-.2cost). Setting y = 2 yields
ln2 = .2 + .1t - .2cosτ, which can be solved numerically to
give τ = 2.9632. If y(0) = y_0 , then as above,
y(t) = $y_0$exp(.2+.1t-.2cost). Thus if we set y = $2y_0$ we get
the same numerical equation for τ and hence the doubling
time has not changed.

16. If T is the temperature of the coffee at any time t, then
$\dfrac{dT}{dt}$ = -k(T - 70°); T(0) = 200°, T(1) = 190°. The solution of
this linear equation will involve k (the cooling rate) and
the integration constant c. Use T(0) = 200 to find c and
then use T(1) = 190 to evaluate k.

18a. Eq.(i) is a linear equation with the integrating factor e^{kt}.
Thus $(e^{kt}u)' = k(T_0 + T_1\cos\omega t)e^{kt}$ and hence
$e^{kt}u = T_0e^{kt} + kT_1\int\cos\omega te^{kt}dt + c$. Evaluating the integral
(by parts or by a symbolic software package) and dividing by
e^{kt} yields $u(t) = T_0 + kT_1\dfrac{k\cos\omega t + \omega\sin\omega t}{k^2 + \omega^2} + ce^{-kt}$. Note
that the last term approaches zero as t → ∞ for any I.C.,
and that the rest of the solution oscillates about u(t) = T_0

18c. Recall that Rcos[ω(t-τ)] = Rcosωtcosωτ + Rsinωtsinωτ.
Comparing this with the oscillatory portion of the above
solution we have $R\cos\omega\tau = \dfrac{k^2T_1}{k^2+\omega^2}$ and $R\sin\omega\tau = \dfrac{k\omega T_1}{k^2+\omega^2}$ since
these are the coefficients of cosωt and sinωt respectively.
By squaring and adding we find $R^2 = \dfrac{k^2T_1^2}{k^2+\omega^2}$ and by dividing
we find tanωτ = ω/k.

19a. The required D.E. is dQ/dt = kr + P - $\dfrac{Q(t)}{V}$r, since kr is
the rate of water pollutant entering the lake, P is the rate
of pollutant entering directly and Q(t)r/V is the rate at
which the pollutant leaves the lake. The I.C. is Q(0) = Vc_0.
Since c = Q(t)/V, the I.V.P. may be rewritten
Vc'(t) = kr + P - rc, c(0) = c_0, which has the solution
$c(t) = k + \dfrac{P}{r} + (c_0 - k - \dfrac{P}{r})e^{-rt/V}$.

19b. Set k = 0, P = 0, t = T and c(T) = $.5c_0$ in the solution
found in (a).

20a. If we measure x positively upward from the ground, then

Eq.(4) of Section 1.1 becomes $m\frac{dv}{dt} = -mg$, since there is no

air resistance. Thus the I.V.P. for v(t) is dv/dt = -g,

v(0) = 20, which gives v(t) = 20 - gt. Since $\frac{dx}{dt}$ = v(t) we

get x(t) = 20t - (g/2)t^2 + c. Then x(0) = 30 gives c = 30

and thus x(t) = 20t - (g/2)t^2 + 30. At the maximum height

v(t$_m$) = 0 and thus t$_m$ = 20/9.8 = 2.04 sec., which when

substituted in the equation for x(t) yields the maximum

height.

21. If v is positive in the upward direction then the drag

force $-\frac{1}{30}$v is downward when v is positive and upward

when v is negative. The I.V.P. in this case is

$m\frac{dv}{dt} = -\frac{1}{30}v - mg$, v(0) = 20.

23a. The I.V.P. is $m\frac{dv}{dt} = mg - .75v$, v(0) = 0 and v is

measured positively downward. Since m = 180/32, the D.E.

becomes $\frac{dv}{dt} = 32 - \frac{2}{15}v$ and thus v(t) = 240(1-e$^{-2t/15}$) so

that v(10) = 176.7 ft/sec.

23b. Integration of v(t) as found in (a) yields

x(t) = 240t + 1800(e$^{-2t/15}$-1), x is measured positively

down from the altitude of 5000 feet. Set t = 10 to find

the distance traveled when the parachute opens.

23c. After the parachute opens the I.V.P. is $m\frac{dv}{dt} = mg-12v$,

v(0) = 176.7, which has the solution

v(t) = 161.7e$^{-32t/15}$ + 15 and where t = 0 now represents

the time the parachute opens. Letting t→∞ yields the

limiting velocity of 15 ft/sec.

23d. Integrate v(t) as found in (c) to find

x(t) = 15t - 75.8e$^{-32t/15}$ + C$_2$. C$_2$ = 75.8 since x(0) = 0,

x now being measured from the point where the parachute

opens. Setting x = 3925.5 will then yield the length of

time the skydiver is in the air after the parachute

opens.

26a. As in Prob.21, $m\dfrac{dv}{dt} = -mg - kv$, $v(0) = v_0$.

26b. From part (a) $v(t) = -\dfrac{mg}{k} + [v_0 + \dfrac{mg}{k}]e^{-kt/m}$. As $k \to 0$

this has the indeterminant form of $-\infty + \infty$. Thus rewrite

$v(t)$ as $v(t) = [-mg + (v_0k + mg)e^{-kt/m}]/k$ which has the

indeterminant form of $0/0$, as $k \to 0$ and hence

L'Hospital's Rule may be applied with k as the variable.

27a. The equation of motion is $m(dv/dt) = w-R-B$ which, in this

problem, is $\dfrac{4}{3}\pi a^3\rho(dv/dt) = \dfrac{4}{3}\pi a^3\rho g - 6\pi\mu a v - \dfrac{4}{3}\pi a^3\rho'g$. The

limiting velocity occurs when $dv/dt = 0$.

27b. Since the droplet is motionless, $v = dv/dt = 0$, we have

the equation of motion $0 = (\dfrac{4}{3})\pi a^3\rho g - Ee - (\dfrac{4}{3})\pi a^3\rho'g$,

where ρ is the density of the oil and ρ' is the density

of air. Solving for e yields the answer.

28. All three parts can be answered from one solution if k
represents the resistance and if the method of solution
of Example 4 is used. Thus we have

$m\dfrac{dv}{dt} = mv\dfrac{dv}{dx} = mg - kv$, $v(0) = 0$, where we have assumed

the velocity is a function of x. The solution of this
I.V.P. involves a logarithmic term, and thus the answers
to parts (a) and (c) must be found using a numerical
procedure.

29a. Use Eq.(30)

29b. Note that 32 $ft/sec^2 = 78,545$ m/hr^2.

30b. From part a) $\dfrac{dx}{dt} = v = u\cos A$ and hence

$x(t) = (u\cos A)t + d_1$. Since $x(0) = 0$, we have $d_1 = 0$ and

$x(t) = (u\cos A)t$. Likewise $\dfrac{dy}{dt} = -gt + u\sin A$ and

therefore $y(t) = -gt^2/2 + (u\sin A)t + d_2$. Since $y(0) = h$

we have $d_2 = h$ and $y(t) = -gt^2/2 + (u\sin A)t + h$.

30d. Let t_w be the time the ball reaches the wall. Then

$x(t_w) = L = (ucosA)t_w$ and thus $t_w = \dfrac{L}{ucosA}$. For the ball

to clear the wall $y(t_w) \geq H$ and thus (setting

$t_w = \dfrac{L}{ucosA}$, $g = 32$ and $h = 3$ in y) we get

$\dfrac{-16L^2}{u^2cos^2A} + LtanA + 3 \geq H.$

30e. Setting $L = 350$ and $H = 10$ we get $\dfrac{-161.98}{cos^2A} + 350\dfrac{sinA}{cosA} \geq 7$

or $7cos^2A - 350cosAsinA + 161.98 \leq 0$. This can be solved
numerically or by plotting the left side as a function of
A and finding where the zero crossings are.

30f. Setting $L = 350$, and $H = 10$ in the answer to part d

yields $\dfrac{-16(350)^2}{u^2cos^2A} + 350tanA = 7$, where we have chosen the

equality sign since we want to just clear the wall.

Solving for u^2 we get $u^2 = \dfrac{1,960,000}{175sin2A-7cos^2A}$. Now u will

have a minimum when the denominator has a maximum. Thus
$350cos2A + 7sin2A = 0$, or $tan2A = -50$, which yields
$A = .7954$ rad. and $u = 106.89$ ft./sec.

Section 2.4, Page 75

1. If the equation is written in the form of Eq.(1), then
   $p(t) = (lnt)/(t-3)$ and $g(t) = 2t/(t-3)$. These are defined
   and continuous on the intervals $(0,3)$ and $(3,\infty)$, but
   since the initial point is $t = 1$, the solution will be
   continuous on $0 < t < 3$.

4. $p(t) = 2t/(2-t)(2+t)$ and $g(t) = 3t^2/(2-t)(2+t)$, which
   have discontinuities at $t = \pm2$. Since $y_0 = -3$, the
   solution will be continuous on $-\infty < t < -2$.

8. Theorem 2.4.2 guarantees a unique solution to the D.E.
   through any point (t_0,y_0) such that $t_0^2 + y_0^2 < 1$ since
   $\dfrac{\partial f}{\partial y} = -y/(1-t^2-y^2)^{1/2}$ is defined and continuous only for
   $1-t^2-y^2 > 0$. Note also that $f = (1-t^2-y^2)^{1/2}$ is defined

and continuous in this region as well as on the boundary $t^2+y^2 = 1$. The boundary can't be included in the final region due to the discontinuity of $\frac{\partial f}{\partial y}$ there.

11. In this case $f = \dfrac{1+t^2}{y(3-y)}$ and $\dfrac{\partial f}{\partial y} = \dfrac{1+t^2}{y(3-y)^2} - \dfrac{1+t^2}{y^2(3-y)}$, which are both continuous everywhere except for $y = 0$ and $y = 3$.

13. The D.E. may be written as $ydy = -4tdt$ so that $\dfrac{y^2}{2} = -2t^2+c$, or $y^2 = C-4t^2$. The I.C. then yields $y_0^2 = C$, so that $y^2 = y_0^2 - 4t^2$ or $y = \pm\sqrt{y_0^2-4t^2}$, which is defined for $4t^2 < y_0^2$ or $|t| < |y_0|/2$. Note that $y_0 \neq 0$ since Theorem 2.4.2 does not hold there.

17. From the direction field (or the given D.E.) it is noted that for $t > 0$ and $y < 0$ that $y' < 0$, so $y \to -\infty$ for $y_0 < 0$. Likewise, for $0 < y_0 < 3$, $y' > 0$ and $y' \to 0$ as $y \to 3$, so $y \to 3$ for $0 < y_0 < 3$ and for $y_0 > 3$, $y' < 0$ and again $y' \to 0$ as $y \to 3$, so $y \to 3$ for $y_0 > 3$. For $y_0 = 3$, $y' = 0$ and $y = 3$ for all t and for $y_0 = 0$, $y' = 0$ and $y = 0$ for all t.

22a. For $y_1 = 1-t$, $y_1' = -1$, so substitution into the D.E.
gives $\quad -1 = \dfrac{-t+[t^2+4(1-t)]^{1/2}}{2}$

$\qquad = \dfrac{-t+[(t-2)^2]^{1/2}}{2}$

$\qquad = \dfrac{-t+|t-2|}{2}$. By the definition of the absolute value, the right side is -1 if $(t-2) \geq 0$. Setting $t = 2$ in y_1 we get $y_1(2) = -1$, as required by the I.C.

22b. By Theorem 2.4.2 we are guaranteed a unique solution only where $f(t,y) = \dfrac{-t+(t^2+4y)^{1/2}}{2}$ and $f_y(t,y) = (t^2+4y)^{-1/2}$ are continuous. In this case the initial point $(2,-1)$ lies in the region $t^2 + 4y \leq 0$, so $\dfrac{\partial f}{\partial y}$ is not continuous and hence the theorem is not applicable and there is no contradiction.

22c. For $y = ct + c^2$ follow the steps of Prob. 22a. If
$y = y_2(t)$ then we must have $ct + c^2 = -t^2/4$ for all t,
which is not possible since c is a constant.

23b. $\phi(t) = t^{-1}$ gives $\phi'(t) = -t^{-2}$ so $\phi' + \phi^2 = 0$. $\phi(t) = ct^{-1}$
gives $\phi'(t) = -ct^{-2}$, so $\phi' + \phi^2 \neq 0$ unless $c = 0$ or $c = 1$.

25. $[y_1(t) + y_2(t)]' + p(t)[y_1(t) + y_2(t)] =$
$y_1'(t) + p(t)y_1(t) + y_2'(t) + p(t)y_2(t) = 0 + g(t)$.

27a. For $n = 1$, we have $y' + [p(t)-q(t)]y = 0$, which is
linear. Thus Eq.(3) gives
$y(t) = c\mu^{-1}(t) = ce^{-\int[p(t)-q(t)]dt}$, since $g(t) = 0$.

27b. Let $v = y^{1-n}$ then $\dfrac{dv}{dt} = (1-n)y^{-n}\dfrac{dy}{dt}$ so $\dfrac{dy}{dt} = \dfrac{1}{1-n}y^n\dfrac{dv}{dt}$, for
$n \neq 1$. Substituting into the D.E. yields
$\dfrac{y^n}{1-n}\dfrac{dv}{dt} + p(t)y = q(t)y^n$ or
$v' + (1-n)p(t)y^{1-n} = (1-n)q(t)$, or
$v' + (1-n)p(t)v = (1-n)q(t)$, which is a linear D.E. for v.

28. $n = 3$ so $v = y^{-2}$ and $\dfrac{dv}{dt} = -2y^{-3}\dfrac{dy}{dt}$ or $\dfrac{dy}{dt} = -\dfrac{1}{2}y^3\dfrac{dv}{dt}$.
Substituting this into the D.E. gives
$-\dfrac{1}{2}y^3\dfrac{dv}{dt} + \dfrac{2}{t}y = \dfrac{1}{t^2}y^3$. Simplifying and using
$y^{-2} = v$ then gives the linear D.E. $v' - \dfrac{4}{t}v = -\dfrac{2}{t^2}$. Thus
$\mu(t) = \dfrac{1}{t^4}$ and $v(t) = ct^4 + \dfrac{2}{5t} = \dfrac{2+5ct^5}{5t}$. Solving for y
gives $y = \pm[5t/(2+5ct^5)]^{1/2}$.

29. $n = 2$ so $v = y^{-1}$ and $\dfrac{dv}{dt} = -y^{-2}\dfrac{dy}{dt}$. Thus the D.E.
becomes $-y^2\dfrac{dv}{dt} - ry = -ky^2$ or $\dfrac{dv}{dt} + rv = k$. Hence
$\mu(t) = e^{rt}$ and $v = k/r + ce^{-rt}$. Setting $v = 1/y$ then
yields the solution.

32. Since $g(t)$ is continuous on the interval
$0 \leq t \leq 1$ and hence we solve the I.V.P.
$y_1' + 2y_1 = 1$, $y_1(0) = 0$ on that interval to obtain

$y_1 = 1/2 - (1/2)e^{-2t}$, $0 \le t \le 1$. For $1 < t$, $g(t) = 0$; and
hence we solve $y_2' + 2y_2 = 0$ to obtain $y_2 = ce^{-2t}$, $1 < t$.
The solution y of the original I.V.P. must be continuous
at $t = 1$ (since its derivative must exist) and hence we
need c in y_2 so that y_2 at 1 has the same value as y_1 at
1. Thus

$ce^{-2} = 1/2 - e^{-2}/2$ or $c = (1/2)(e^2-1)$ and we obtain

$$y = \begin{cases} 1/2 - (1/2)e^{-2t} & 0 \le t \le 1 \\ 1/2(e^2-1)e^{-2t} & 1 \le t \end{cases} \qquad \text{and}$$

$$y' = \begin{cases} e^{-2t} & 0 \le t \le 1 \\ (1-e^2)e^{-2t} & 1 < t. \end{cases}$$

Evaluating the two parts of y' at $t_0 = 1$ we see that they
are different, and hence y' is not continuous at $t_0 = 1$.

Section 2.5, Page 88

3. From the graph, or by setting $\dfrac{dy}{dt} = y(y-1)(y-2) = 0$, we
find that $y = 0,1,2$ are the critical points. The graph of
$y(y-1)(y-2)$ is positive for $0 < y < 1$ and $2 < y$ and
negative for $1 < y < 2$. Thus $y(t)$ is increasing
$(\dfrac{dy}{dt} > 0)$ for $0 < y < 1$ and $2 < y$ and decreasing
$(\dfrac{dy}{dt} < 0)$ for $1 < y < 2$. Therefore 0 and 2 are unstable
critical points while 1 is an asymptotically stable
critical point.

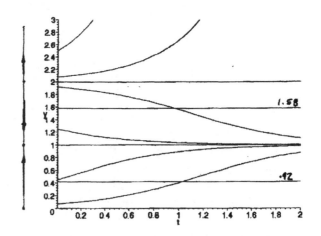

The phase line and several solutions are shown. The
inflection points of the solutions (1.58 and .42) are
found by determining where y(y-1)(y-2) has its relative
max and min points. These lines are are also shown.

5. $\dfrac{dy}{dt}$ is zero only when

   $e^{-y}-1=0$, or $y = 0$.

   Since $\dfrac{dy}{dt}>0$ for $y<0$

   and $\dfrac{dy}{dt}<0$ for $y>0$

   we conclude that $y=0$
   is an asymptotically
   stable critical point.

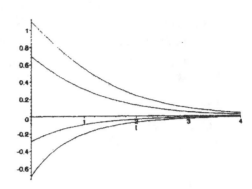

7c. Separate variables to get $\dfrac{dy}{(1-y)^2}$ = kdt. Integration

   yields $\dfrac{1}{1-y}$ = kt + c, or $y = 1 - \dfrac{1}{kt + c} = \dfrac{kt + c - 1}{kt + c}$.

   Setting t = 0 and $y(0) = y_0$ yields $y_0 = \dfrac{c-1}{c}$ or

   $c = \dfrac{1}{1-y_0}$. Hence $y(t) = \dfrac{(1-y_0)kt + y_0}{(1-y_0)kt + 1}$. For $y_0 < 1$

   we have $y \rightarrow (1-y_0)k/(1-y_0)k = 1$ as $t \rightarrow \infty$. For $y_0 > 1$
   the denominator will have a zero for some value of t,
   depending on the values chosen for y_0 and k. Thus the
   solution has a discontinuity at that point.

9.   Setting $\dfrac{dy}{dt}$ = 0 we find y = 0, ± 1 are the critical

   points. We have $\dfrac{dy}{dt} > 0$ for $|y| > 1$ while $\dfrac{dy}{dt} < 0$ for

   $|y| < 1$ we conclude that y = -1 is asymptotically stable,
   y = 0 is semistable, and y = 1 is unstable.

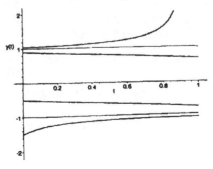

11. $y = b^2/a^2$ and $y = 0$ are the only critical points. For
     $0 < y < b^2/a^2$, $\dfrac{dy}{dt} < 0$ and thus $y = 0$ is asymptotically
     stable. For $y > b^2/a^2$, $dy/dt > 0$ and thus $y = b^2/a^2$ is
     unstable. All solutions that start above $y = b^2/a^2$ will
     continue to increase and all solutions that start for y
     between 0 and b^2/a^2 will decay to zero. If $y(0) = b^2/a^2$
     then $y(t) = b^2/a^2$ for all t, since this is an equilibrium
     point.

14. If $f'(y_1) < 0$ then the slope of f is negative at y_1 and
     thus $f(y) > 0$ for $y < y_1$ and $f(y) < 0$ for $y > y_1$ since
     $f(y_1) = 0$. Hence y_1 is an asymptotically stable critical
     point. A similar argument will yield the result for
     $f'(y_1) > 0$.

16a. Setting $\dfrac{dy}{dt} = 0$ we have $ry\ln(K/y) = 0$, so $y = 0$ and
     $y = K$ are the critical points. The graph for $r = 2$ and
     $K = 4$ is shown, and we see
     that $\dfrac{dy}{dt} > 0$ for $0 < t < K = 4$
     and $\dfrac{dy}{dt} < 0$ for $t > K = 4$. Thus
     $y = 0$ is unstable and $y = K$
     is asymptotically stable.

16b. The derivative of $y\ln(K/y)$ is $\ln(K/y) - 1$, so the graph of
     $\dfrac{dy}{dt}$ vs y has a maximum point at $y = K/e$. Thus $\dfrac{dy}{dt}$ is
     positive and increasing for $0 < y < K/e$ and hence $y(t)$ is
     concave up for that interval. Similarly $\dfrac{dy}{dt}$ is positive and
     decreasing for $K/e < y < K$ and thus $y(t)$ is concave down for
     that interval.

16c. $\ln(K/y)$ is very large for small values of y and thus
     $(ry)\ln(K/y) > ry(1 - y/K)$ for small y. Since $\ln(K/y)$ and
     $(1 - y/K)$ are both strictly decreasing functions of y and
     since $\ln(K/y) = (1 - y/K)$ only for $y = K$, we may conclude
     that $\dfrac{dy}{dt} = (ry)\ln(K/y)$ is never less than $\dfrac{dy}{dt} = ry(1-y/K)$.

17a. If $u = \ln(y/K)$ then $y = Ke^u$ and $\dfrac{dy}{dt} = Ke^u\dfrac{du}{dt}$ so that the
     D.E. becomes $du/dt = -ru$.

18a. The D.E. is $dV/dt = k - \alpha\pi r^2$. The volume of a cone of
height L and radius r is given by $V = \pi r^2 L/3$ where $L = hr/a$
from symmetry. Solving for r yields the desired solution.

18b. Equilibrium is given by $k - \alpha\pi r^2 = 0$.

18c. The equilibrium height must be less than h.

20b. Use the results of Problem 14.

20c. Y is defined to be Ey_2, where y_2 was found in part a.

20d. Differentiate Y with respect to E.

21a. Set $\dfrac{dy}{dt} = 0$ and solve for y using the quadratic formula.

21b. Use the results of Prob. 14.

21d. If $h > rK/4$ there are no critical points (see part a) and
$\dfrac{dy}{dt} < 0$ for all t.

24a. If $z = x/n$ then $dz/dt = \dfrac{1}{n}\dfrac{dx}{dt} - \dfrac{x}{n^2}\dfrac{dn}{dt}$. Use of
Equations (i) and (ii) then gives the D.E. (iii) and the
I.C. is $z(0) = 1$ since $n(0) = x(0)$.

24b. Separate variables to get $\dfrac{dz}{z(1-\nu z)} = -\beta dt$. Using
partial fractions this becomes $\dfrac{dz}{z} + \dfrac{\nu dz}{1-\nu z} = -\beta dt$.
Integration and solving for z yields the answer.

24c. From part b, find $z(20)$ when $\beta = \nu = 1/8$.

25b. For $a = 0$, $\dfrac{dy}{dt}$ is always negative, so $y = 0$ is
semistable. For $a > 0$ we have $\dfrac{dy}{dt} < 0$ for $|y| > \sqrt{a}$ and
$\dfrac{dy}{dt} > 0$ for $|y| < -\sqrt{a}$, so $y = \sqrt{a}$ is asymptotically
stable and $y = \sqrt{a}$ is unstable.

25c. The graphs are shown (on the next page) for $a = 4$. Note
that $\dfrac{d}{dy}(a-y^2) = 0$ for $y = 0$ and thus $y = 0$ is an inflection
point for the solution that crosses the t-axis.

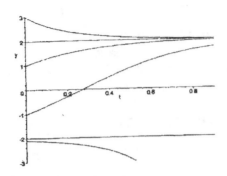

28a. Observe that x = p and x = q are critical points. Also
     note that dx/dt > 0 for
     x < min(p,q) and x > max(p,q) while dx/dt < 0 for x
     between min(p,q) and max(p,q). Thus x = min(p,q) is an
     asymptotically stable point while x = max(p,q) is
     unstable. To solve the D.E., separate variables and use
     partial fractions to obtain $\dfrac{1}{q-p}[\dfrac{dx}{q-x} - \dfrac{dx}{p-x}] = \alpha dt$.
     Integration and solving for x yields the solution.

28b. x = p is a semistable critical point and since $\dfrac{dx}{dt} > 0$,
     x(t) is an increasing function. Thus for x(0) = 0, x(t)
     approaches p as t → ∞. To solve the D.E., separate
     variables and integrate.

Section 2.6, Page 99

3.   M(x,y) = $3x^2-2xy+2$ and N(x,y) = $6y^2-x^2+3$, so $M_y = -2x = N_x$
     and thus the D.E. is exact. Integrating M(x,y) with
     respect to x we get $\psi(x,y) = x^3 - x^2y + 2x + h(y)$.
     Taking the partial derivative of this with respect to y
     and setting it equal to N(x,y) yields $-x^2+h'(y) =$
     $6y^2-x^2+3$, so that $h'(y) = 6y^2 + 3$ and $h(y) = 2y^3 + 3y$.
     Substitute this h(y) into $\psi(x,y)$ and recall that the
     equation which defines y(x) implicitly is $\psi(x,y) = c$.
     Thus $x^3 - x^2y + 2x + 2y^3 + 3y = c$ is the equation that
     implicitly defines the solution.

5.   Writing the equation in the form M(x,y)dx + N(x,y)dy = 0
     gives M(x,y) = ax + by and N(x,y) = bx + cy. Thus
     $M_y = b = N_x$ and the equation is exact. Integrating M(x,y)
     with respect to x yields $\psi(x,y) = (a/2)x^2 + bxy + h(y)$.
     Differentiating ψ with respect to y (x constant) and
     setting $\psi_y(x,y) = N(x,y)$ we find that h'(y) = cy and thus

$h(y) = (c/2)y^2$. Hence the solution is given by
$(a/2)x^2 + bxy + (c/2)y^2 = C$ or $ax^2 + 2bx + cy^2 = k$

7. $M_y(x,y) = e^x\cos y - 2\sin x = N_x(x,y)$ and thus the D.E. is
     exact. Integrating $M(x,y)$ with respect to x gives
     $\psi(x,y) = e^x\sin y + 2y\cos x + h(y)$. Finding $\psi_y(x,y)$ from
     this and setting that equal to $N(x,y)$ yields $h'(y) = 0$
     and thus $h(y)$ is a constant. Hence an implicit solution
     of the D.E. is $e^x\sin y + 2y\cos x = c$. The solution $y = 0$
     is also valid since it satisfies the D.E. for all x.

9. $M_y = N_x$ so the D.E. is exact. If you try to find $\psi(x,y)$
     by integrating $M(x,y)$ with respect to x you must
     integrate by parts. Instead find $\psi(x,y)$ by integrating
     $N(x,y)$ with respect to y to obtain $\psi(x,y) = e^{xy}\cos 2x - 3y$
     $+ g(x)$. Now find $g(x)$ by differentiating $\psi(x,y)$ with
     respect to x and set that equal to $M(x,y)$, which yields
     $g'(x) = 2x$ or $g(x) = x^2$. As before the implicit solution
     is $\psi(x,y) = c$

12. As long as $x^2 + y^2 \neq 0$, we can simplify the equation by
     multiplying both sides by $(x^2 + y^2)^{3/2}$. This gives the
     exact equation $xdx + ydy = 0$. The solution to this
     equation is given implicitly by $x^2 + y^2 = c$. If you
     apply Theorem 2.6.1 and its construction without the
     simplification, you get $(x^2 + y^2)^{-1/2} = C$ which can be
     written as $x^2 + y^2 = c$ under the same assumption required
     for the simplification.

14. $M_y = 1$ and $N_x = 1$, so the D.E. is exact. Integrating
     $M(x,y)$ with respect to x yields
     $\psi(x,y) = 3x^3 + xy - x + h(y)$. Differentiating this with
     respect to y and setting $\psi_y(x,y) = N(x,y)$ yields
     $h'(y) = -4y$ or $h(y) = -2y^2$. Thus the implicit solution
     is $3x^3 + xy - x - 2y^2 = c$. Setting $x = 1$ and $y = 0$ gives
     $c = 2$ so that $2y^2 - xy + (2+x-3x^3) = 0$ is the implicit
     solution satisfying the given I.C. Use the quadratic
     formula to find $y(x)$, where the negative square root is
     used in order to satisfy the I.C. The solution will be
     valid for $24x^3 + x^2 - 8x - 16 > 0$. Using a numerical
     procedure (or graphically) this cubic equation has one
     zero for $x \approx .9846$.

15. We want $M_y(x,y) = 2xy + bx^2$ to be equal to
    $N_x(x,y) = 3x^2 + 2xy$. Thus we must have $b = 3$. This
    gives $\psi(x,y) = \dfrac{1}{2}x^2y^2 + x^3y + h(y)$ and consequently
    $h'(y) = 0$. After multiplying through by 2, the solution
    is given implicitly by $x^2y^2 + 2x^3y = c$.

19. $M_y(x,y) = 3x^2y^2$ and $N_x(x,y) = 1 + y^2$ so the equation is
    not exact by Theorem 2.6.1. Multiplying by the
    integrating factor $\mu(x,y) = 1/xy^3$ we get
    $x + \dfrac{(1+y^2)}{y^3}y' = 0$, which is an exact equation since
    $M_y = N_x = 0$ (it is also separable). In this case
    $\psi = \dfrac{1}{2}x^2 + h(y)$ and $h'(y) = y^{-3} + y^{-1}$ so that
    $x^2 - y^{-2} + 2\ln|y| = c$ gives the solution implicitly. Note
    that $y(x) = 0$ also satisfies the given D.E.

22. Multiplication of the given D.E. (which is not exact) by
    $\mu(x,y) = xe^x$ yields $(x^2 + 2x)e^x\sin y\, dx + x^2e^x\cos y\, dy$,
    which is exact since $M_y(x,y) = N_x(x,y) = (x^2+2x)e^x\cos y$.
    To solve this exact equation it's easiest to integrate
    $N(x,y) = x^2e^x\cos y$ with respect to y to get
    $\psi(x,y) = x^2e^x\sin y + g(x)$. Finding ψ_x and setting that
    equal to $(x^2+2x)e^x\sin y$ yields $g'(x) = 0$.

23. **This problem is similar to the derivation leading up to
    Eq.(26). Assuming that μ depends only on y, we find from
    Eq.(25) that $\mu' = Q\mu$, where $Q = (N_x - M_y)/M$ must depend
    on y alone. Solving this last D.E. yields $\mu(y)$ as given.
    This method provides an alternative approach to Problems
    25 through 31.**

25. The equation is not exact so we must attempt to find an
    integrating factor. Since $\dfrac{1}{N}(M_y-N_x) = \dfrac{3x^2 + 2x + 3y^2 - 2x}{x^2 + y^2} = 3$
    is a function of x alone there is an integrating factor
    depending only on x, as shown in Eq.(26). Then $d\mu/dx =$
    3μ, and the integrating factor is $\mu(x) = e^{3x}$. Multiplying
    all terms in the given D.E. by e^{3x} will then yield an
    exact D.E.

26. An integrating factor can be found which is a function of x only, yielding $\mu(x) = e^{-x}$. Alternatively, you might recognize that $y' - y = e^{2x} - 1$ is a linear first order equation which can be solved as in Section 2.1.

27. Using the results of Prob. 23, it can be shown that $\mu(y) = y$ is an integrating factor. Thus multiplying the D.E. by y gives $ydx + (x - ysiny)dy = 0$, which can be identified as an exact equation. Alternatively, one can rewrite the last equation as $(ydx + xdy) - ysiny\,dy = 0$. The first term is $d(xy)$ and the last can be integrated by parts. Thus we have $xy + ycosy - siny = c$.

29. Simplify the D.E. by multiplying by siny (which is really an integrating factor) to obtain $e^{x}sinydx + e^{x}cosydy + 2ydy = 0$, which is exact. The first two terms are just $d(e^{x}siny)$ and thus, $e^{x}siny + y^{2} = c$.

31. Using the results of Prob. 24, it can be shown that $\mu(xy) = xy$ is an integrating factor. Thus, multiplying by xy we have $(3x^{2}y + 6x)dx + (x^{3} + 3y^{2})dy = 0$, which can be identified as an exact equation. Alternatively, we can observe that the above equation can be written as $d(x^{3}y) + d(3x^{2}) + d(y^{3}) = 0$, so that $x^{3}y + 3x^{2} + y^{3} = c$.

Section 2.7, Page 109

1d. The exact solution to this I.V.P. is $y = \phi(t) = t + 2 -e^{-t}$.

3a. The Euler formula is $y_{n+1} = y_{n} + h(2y_{n} - t_{n} + 1/2)$ for $n = 0,1,2,3$ and with $t_{0} = 0$ and $y_{0} = 1$. Thus, for $h = .1$,
$y_{1} = y_{0} + .1(2y_{0} - t_{0} + 1/2) = 1.25$,
$y_{2} = 1.25 + .1[2(1.25) - (.1) + 1/2] = 1.54$,
$y_{3} = 1.54 + .1[2(1.54) - (.2) + 1/2] = 1.878$, and
$y_{4} = 1.878 + .1[2(1.878) - (.3) + 1/2] = 2.2736$.

3b. Use the same formula as in Prob. 3a, except now $h = .05$ and $n = 0,1...7$. Notice that only results for $n = 1,3,5$ and 7 are needed to compare with part a.

3c. Again, use the same formula as above with $h = .025$ and $n = 0,1...15$. Notice that only results for $n = 3,7,11$ and 15 are needed to compare with parts a and b.

3d. $y' = 1/2 - t + 2y$ is a first order linear D.E. Rewrite
the equation in the form $y' - 2y = 1/2 - t$ and multiply
both sides by the integrating factor e^{-2t} to obtain
$(e^{-2t}y)' = (1/2 - t)e^{-2t}$. Integrating the right side by
parts and multiplying by e^{2t} we obtain $y = ce^{2t} + t/2$.
The I.C. $y(0) = 1 \rightarrow c = 1$ and hence the solution of the
I.V.P. is $y = \phi(t) = e^{2t} + t/2$. Thus $\phi(0.1) = 1.2714$,
$\phi(0.2) = 1.59182$, $\phi(0.3) = 1.97212$, and $\phi(0.4) = 2.42554$.

4d. The exact solution to this I.V.P. is
$y = \phi(t) = (6\cos t + 3\sin t - 6e^{-2t})/5$.

6.   For $y(0) > 0$ the
solutions appear to
converge to a number
between 0 and 2. Note
that $y=0$ is an
equilibrium solution
For $y(0)<0$ the
solutions diverge.

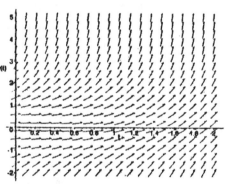

9.   All solutions seem
to diverge.

13a. The Euler formula is
$y_{n+1} = y_n + h(\dfrac{4-t_ny_n}{1 + y_n^2})$, where $t_0 = 0$ and
$y_0 = y(0) = -2$. Thus, for $h = .1$, we get
$y_1 = -2 + .1(4/5) = -1.92$
$y_2 = -1.92 + .1(\dfrac{4-.1(-1.92)}{1 + (1.92)^2}) = -1.83055$
$y_3 = -1.83055 + .1(\dfrac{4-.2(-1.83055)}{1+(1.83055)^2}) = -1.7302$

$$y_4 = -1.7302 + .1(\frac{4-.3(-1.7302)}{1 + (1.7302)^2}) = -1.617043$$

$$y_5 = -1.617043 + .1(\frac{4 - .4(-1.617043)}{1 + (1.617043)^2}) = -1.488494.$$

Thus, $y(.5) \cong -1.488494$.

15a. The Euler formula is

$$y_{n+1} = y_n + .1 (\frac{3t_n^2}{3y_n^2-4}), \text{ where } t_0 = 1 \text{ and } y_0 = 0. \text{ Thus}$$

$$y_1 = 0 + .1 (\frac{3}{-4}) = -.075 \text{ and}$$

$$y_2 = -.075 + .1 (\frac{3(1.1)^2}{3(-.075)^2-4}) = -.166134.$$

15c. There are two factors that explain the large differences. From the D.E., the slope of y, y', becomes very "large" for values of y near -1.155. Also, the slope changes sign at y = -1.155. Thus for part a, $y(1.7) \cong y_7 = -1.178$, which is close to -1.155 and the slope y' here is large and positive, creating the large change in $y_8 \cong y(1.8)$. For part b, $y(1.65) \cong -1.125$, resulting in a large negative slope, which yields $y(1.70) \cong -3.133$. The slope at this point is now positive and the remainder of the solutions "grow" to -3.098 for the approximation to y(1.8).

16. For the four step sizes given, the approximte values for y(.8) are 3.5078, 4.2013, 4.8004 and 5.3428. Thus, since these changes are still rather "large", it is hard to give an estimate other than y(.8) is at least 5.3428. By using h = .005, .0025 and .001, we find further approximate values of y(.8) to be 5.576, 5.707 and 5.790. Thus a better estimate now is for y(.8) to be between 5.8 and 6. No reliable estimate is obtainable for y(1), which is consistent with the direction field of Prob.9.

18. It is helpful, in understanding this problem, to also calculate $y'(t_n) = y_n(.1y_n^2 - t_n)$. For $\alpha = 2.38$ this term remains positive and grows very large for $t_n > 2$. On the other hand, for $\alpha = 2.37$ this term decreases and eventually becomes negative for $t_n \cong 1.6$ (for h = .01). For $\alpha = 2.37$ and h = .1, .05 and .01, y(2.00) has the approximations of 4.48, 4.01 and 3.50 respectively. A small step size must be used, due to the sensitivety of the slope field, given by $y_n(.1y_n^2 - t_n)$.

22. Using Eq.(8) we have $y_{n+1} = y_n + h(2y_n - 1) = (1+2h)y_n - h$.
Setting $n + 1 = k$ (and hence $n = k-1$) this becomes
$y_k = (1 + 2h)y_{k-1} - h$, for $k = 1,2,\ldots$. Since $y_0 = 1$,
we have $y_1 = 1 + 2h - h = 1 + h = (1 + 2h)/2 + 1/2$, and
hence $y_2 = (1 + 2h)y_1 - h = (1 + 2h)^2/2 + (1 + 2h)/2 - h$
$= (1 + 2h)^2/2 + 1/2$;
$y_3 = (1 + 2h)y_2 - h = (1 + 2h)^3/2 + (1 + 2h)/2 - h$
$= (1 + 2h)^3/2 + 1/2$. Continuing in this fashion (or
using induction) we obtain $y_k = (1 + 2h)^k/2 + 1/2$. For
fixed $t > 0$ choose $h = t/k$. Then substitute for h in the
last formula to obtain $y_k = (1 + 2t/k)^k/2 + 1/2$. Letting
$k \to \infty$ we find (See hint for Prob 20d.)
$y(t) = \lim_{k \to \infty} y_k = e^{2t}/2 + 1/2$, which is the exact solution.

Section 2.8, Page 118

1. Let $s = t-1$ and $w(s) = y(t(s)) - 2$, then when $t = 1$ and
$y = 2$ we have $s = 0$ and $w(0) = 0$. Also,
$$\frac{dw}{ds} = \frac{dw}{dt} \cdot \frac{dt}{ds} = \frac{d}{dt}(y-2)\frac{dt}{ds} = \frac{dy}{dt}$$ (since $t = s+1$) and hence
$$\frac{dw}{ds} = (s+1)^2 + (w+2)^2,$$ upon substitution into the given
D.E.

4a. Following Ex. 1 of the text, from Eq.(7) we have
$$\phi_{n+1}(t) = \int_0^t f(s,\phi_n(s))ds,$$ where $f(t,\phi) = -1 - \phi$. Thus if
$\phi_0(t) = 0$, then $\phi_1(t) = -\int_0^t ds = -t$;
$$\phi_2(t) = -\int_0^t (1-s)ds = -t + \frac{t^2}{2};$$
$$\phi_3(t) = -\int_0^t (1-s + \frac{s^2}{2})ds = -t + \frac{t^2}{2} - \frac{t^3}{2\cdot 3};$$
$$\phi_4(t) = -\int_0^t (1 - s + \frac{s^2}{2} - \frac{s^3}{3!})ds = -t + \frac{t^2}{2} - \frac{t^3}{3!} + \frac{t^4}{4!}.$$

Based upon these we hypothesize that $\phi_n(t) = \sum_{k=1}^{n} \frac{(-1)^k t^k}{k!}$

and use mathematical induction to verify this form for
$\phi_n(t)$. First, let $n = 1$, then $\phi_1(t) = -t$, so it is
certainly true for $n = 1$. Then, using Eq.(7) again we
have:

$$\phi_{n+1}(t) = -\int_0^t [1 + \phi_n(s)]\, ds = -t - \sum_{k=1}^{n} \frac{(-1)^k t^{k+1}}{(k+1)!}$$

$$= \sum_{k=0}^{n} \frac{(-1)^{k+1} t^{k+1}}{(k+1)!} = \sum_{i=1}^{n+1} \frac{(-1)^i t^i}{i!},\ \text{where } i = k+1. \ \text{Since this}$$

is the same form for $\phi_{n+1}(t)$ as derived from $\phi_n(t)$ above, we have verified by mathematical induction that $\phi_n(t)$ is as given.

4c. From part a, let $\phi(t) = \lim_{n \to \infty} \phi_n(t) = \sum_{k=1}^{\infty} \frac{(-1)^k t^k}{k!}$

$$= -t + \frac{t^2}{2} - \frac{t^3}{3!} + \dots\ .$$

Since this is a power series, recall from calculus that:

$$e^{at} = \sum_{k=0}^{\infty} \frac{a^k t^k}{k!} = 1 + at + \frac{a^2 t^2}{2} + \frac{a^3 t^3}{3!} + \dots\ . \ \text{If we let}$$

$a = -1$, then we have $e^{-t} = 1 - t + \frac{t^2}{2} - \frac{t^3}{3!} + \dots = 1 + \phi(t)$.

Hence $\phi(t) = e^{-t} - 1$.

4b.

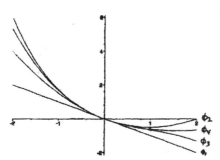

4d.

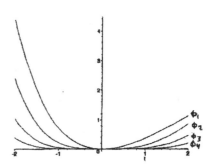

From the plot in 4d, it appears that ϕ_4 is a very good estimate for $|t| < 1$.

7a. As in Prob.4,

$$\phi_1(t) = \int_0^t (s\phi_0(s) + 1)\, ds = s \Big|_0^t = t$$

$$\phi_2(t) = \int_0^t (s^2 + 1)\, ds = \left(\frac{s^3}{3} + s\right)\Big|_0^t = t + \frac{t^3}{3}$$

$$\phi_3(t) = \int_0^t \left(s^2 + \frac{s^4}{3} + 1\right) ds = \left(\frac{s^3}{3} + \frac{s^5}{3\cdot 5} + s\right)\Big|_0^t = t + \frac{t^3}{3} + \frac{t^5}{3\cdot 5}.$$

Based upon these we hypothesize that:

$$\phi_n(t) = \sum_{k=1}^{n} \frac{t^{2k-1}}{1 \cdot 3 \cdot 5 \cdots (2k-1)}$$ and use mathematical induction

to verify this form for $\phi_n(t)$, which is clearly true for
$n = 1$. Using Eq.(7) again we have:

$$\phi_{n+1}(t) = \int_0^t \left(\sum_{k=1}^{n} \frac{s^{2k}}{1 \cdot 3 \cdot 5 \cdots (2k-1)} + 1 \right) ds$$

$$= \sum_{k=1}^{n} \frac{t^{2k+1}}{1 \cdot 3 \cdot 5 \cdots (2k+1)} + t$$

$$= \sum_{k=0}^{n} \frac{t^{2k+1}}{1 \cdot 3 \cdot 5 \cdots (2k+1)}$$

$$= \sum_{i=1}^{n+1} \frac{t^{2i-1}}{1 \cdot 3 \cdot 5 \cdots (2i-1)}$$, where $i = k+1$. Since this is

the same form for $\phi_{n+1}(t)$ as derived from $\phi_n(t)$ above, we
have verified by mathematical induction that $\phi_n(t)$ is as
given.

7b. Your plot should show that the estimates appear to be
     converging.

10a. From Eq.(7) we have $\phi_{n+1} = \int_0^t [1-\phi_n^3(s)]ds$. Setting
     $\phi_0(t) = 0$ yields the desired iterates.

10b. The iterates
     appear
     to diverge.

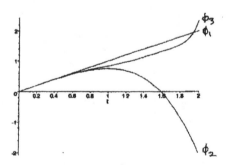

11. First, recall that $\sin x = x - \frac{x^3}{3!} + \frac{x^5}{5!} + O(x^7)$. Now, for

     this problem $\phi_1(t) \int_0^t [1-\sin\phi_0(s)]ds = t$ and hence

$$\phi_2(t) = \int_0^t [1-\sin(s)]ds = \int_0^t [1 - (s - \frac{s^3}{3!} + \frac{s^5}{5!} - O(s^7)]ds$$

$$= t - \frac{t^2}{2!} + \frac{t^4}{4!} - \frac{t^6}{6!} + O(t^8).$$ For ϕ_3 we need to

find $\sin[\phi_2(t)]$, which is given by

$$\sin[\phi_2(t)] = \phi_2(t) - \phi_2^3(t)/3! + \phi_2^5/5! + O(t^7)$$

$$= (t - \frac{t^2}{2!} + \frac{t^4}{4!} - \frac{t^6}{6!}) - \frac{(t - \frac{t^2}{2!})^3}{3!} + \frac{t^5}{5!} + O(t^7),$$

where we have retained only the terms less than $O(t^7)$.
Now use this in $\phi_3(t) = \int_0^t [1-\sin(\phi_2(s))]ds$, which gives
the desired answer up to $O(t^8)$.

13. If x=0 then $\phi_n(x) = 0$ for all n and $\lim_{n\to\infty}\phi_n(x) = 0$.

   If 0<x<1 let $x = \frac{1}{r}$, so r>1 and $\lim_{n\to\infty}\phi_n(x) = \lim_{n\to\infty}\phi_n(\frac{1}{r^n}) = 0$.

   If x=1 then $\phi_n(x) = 1$ for all n and $\lim_{n\to\infty}\phi_n(x) = 1$, so
   that indeed $\phi_n(x)$ converges to a discontinuous functiom.

Section 2.9, Page 130

2. Using the given difference equation we have for n=0,
    $y_1 = y_0/2$; for n=1, $y_2 = 2y_1/3 = y_0/3$; and for n=2,
    $y_3 = 3y_2/4 = y_0/4$. Thus we guess that $y_n = y_0/(n+1)$, and
    the given equation then gives $y_{n+1} = \frac{n+1}{n+2}y_n = y_0/(n+2)$,
    which, by mathematical induction, verifies $y_n = y_0/(n+1)$
    as the solution for all n. $\lim_{n\to\infty} y_n = 0$, as y_0 is constant.

5. From the given equation we have $y_1 = .5y_0+6$.

    $y_2 = .5y_1 + 6 = (.5)^2 y_0 + 6(1 + \frac{1}{2})$ and

    $y_3 = .5y_2 + 6 = (.5)^3 y_0 + 6(1 + \frac{1}{2} + \frac{1}{4})$. In general, then

    $y_n = (.5)^n y_0 + 6(1 + \frac{1}{2} + \cdots + \frac{1}{2^{n-1}})$

    $= (.5)^n y_0 + 6(\frac{1 - (1/2)^n}{1 - 1/2})$

$$= (.5)^n y_0 + 12 - (.5)^n 12$$

$$= (.5)^n (y_0 - 12) + 12.$$ Mathematical induction can now be used to prove that this is the correct solution. Note that $y_n \to 12$ as $n \to \infty$ and thus $y_n = 12$ is an equilibrium solution.

7. From Eq.(12) we have $y_{n+1} = (1+r)y_n$ since $b=0$ and $r = \dfrac{.07}{365}$ is the daily interest rate. Thus $y_1 = (1+r)y_0$, $y_2 = (1+r)^2 y_0, \ldots, y_n = (1+r)^n y_0$. Setting $n = 365$ we have $(1+r)^{365} = 1.0725$, so the effective annual yield is 7.25%.

10. As in Ex.(1), the governing equation is $y_{n+1} = \rho y_n - b$, which has the solution $y_n = \rho^n y_0 - \dfrac{1-\rho^n}{1-\rho} b$ (Eq.(14) with a negative b). Setting $y_{360} = 0$ and solving for b we obtain

$$b = \frac{(1-\rho)\rho^{360} y_0}{1-\rho^{360}},$$ where $\rho = 1.0075$ for part a.

13. You must solve Eq.(14) numerically for ρ when $n = 240$, $y_{240} = 0$, $b = -\$900$ and $y_0 = \$95,000$.

14. Substituting Eq.(25), $u_n = \dfrac{\rho-1}{\rho} + v_n$, into Eq.(21) we get

$$\frac{\rho-1}{\rho} + v_{n+1} = \rho(\frac{\rho-1}{\rho} + v_n)(1 - \frac{\rho-1}{\rho} - v_n) \text{ or}$$

$$v_{n+1} = -\frac{\rho-1}{\rho} + (\rho-1 + \rho v_n)(\frac{1}{\rho} - v_n)$$

$$= \frac{1-\rho}{\rho} + \frac{\rho-1}{\rho} - (\rho-1)v_n + v_n - \rho v_n^2 = (2-\rho)v_n - \rho v_n^2$$

15a. For $u_0 = .2$ we have $u_1 = 3.2u_0(1\to) = .512$ and $u_2 = 3.2u_1(1-u_1) = .7995392$. Likewise $u_3 = .51288406$, $u_4 = .7994688$, $u_5 = .51301899$, $u_6 = .7994576$ and $u_7 = .5130404$. Continuing in this fashion, $u_{14} = u_{16} = .79945549$ and $u_{15} = u_{17} = .51304451$.

16. To plot the stairsteps extend the graph of Fig.2.9.2(c) outside the interval [0,1]. For part b) choose any $u_0 > 1$. Then the first step goes from the x-axis down to the parabola (in the fourth quadrant) then in the

negative direction to the line y = x in the third
quadrant. The steps after that are similar to those in
Fig. 2.9.2.

17.   For both parts of this problem a computer spreadsheet
      was used and an initial value of u_0 = .2 was chosen.
      Different initial values or different computer programs
      may need a slightly different number of iterations to
      reach the limiting value.

17a. The limiting value of .65517 (to 5 decimal places) is
     reached after approximately 100 iterations for ρ = 2.9.
     The limiting value of .66102 (to 5 decimal places) is
     reached after approximately 200 iterations for ρ = 2.95.
     The limiting value of .66555 (to 5 decimal places) is
     reached after approximately 910 iterations for ρ = 2.99.

17b. The solution oscillates between .63285 and .69938 after
     approximately 400 iterations for ρ = 3.01. The solution
     oscillates between .59016 and .73770 after approximately
     130 iterations for ρ = 3.05. The solution oscillates
     between .55801 and .76457 after approximately 30
     iterations for ρ = 3.1. For each of these cases
     additional iterations verified the oscillations were
     correct to five decimal places.

18.   For an initial value of .2 and ρ = 3.448 we have the
      solution oscillating between .4403086 and .8497146.
      After approximately 3570 iterations the eighth decimal
      place is still not fixed, though. For the same initial
      value and ρ = 3.45 the solution oscillates between the
      four values: .43399155, .84746795, .44596778 and
      .85242779 after 3700 iterations.. For ρ = 3.449, the
      solution is still varying in the fourth decimal place
      after 3570 iterations, but there appear to be four
      values.

Miscellaneous Problems, Page 132

Before trying to find the solution of a D.E. it is necessary
to know its type. The student should first classify the D.E.
before looking at this section, which indentifies the type of
each D.E. in Problems 1 through 32.

      1. Linear   2. Separable

      3. Exact

4. Rewrite the D.E. as $\dfrac{dy}{dx} + (2x-1) = -3(2x-1)$, which is linear, with integrating factor $e^{(x^2-x)}$.

5. Exact   6. Linear

7. Separable   8. Linear

9. Exact   10. Separable

11. Exact   12. Linear

13. Separable   14. Exact

15. Separable   16. Exact

17. Linear   18. Linear with $\mu(x) = e^{2x}$

19. Exact   20. Separable

21. Exact   22. Separable

23. linear, $\mu(t) = te^t$ 24. Exact.

25. Exact. Your calculations might be simplified if you write the D.E. as $\dfrac{2xdx}{y} - \dfrac{x^2dy}{y^2} + \dfrac{-ydx}{x^2+y^2} + \dfrac{xdy}{x^2+y^2} = 0$ and show that the first two terms are exact and the last two terms are exact. Integrating each pair is then straight forward.

26. Homogeneous   28. Linear or Homogenous
    (See Sect 2.2, Prob.30)

27. Letting $u = x^2$ yields $\dfrac{dy}{dx} = 2x\dfrac{dy}{du}$ and thus $\dfrac{dy}{du} - 2yu = 2y^3$ which is linear in $u(y)$.

29. Homogeneous   30. Integrating factor, depends on x only

31. Integrating factor, depends on y only

32. From Prob.23, Sect. 2.4, this is a Bernoulli equation with $n = 2$. Let $v(x) = y$, then the linear equation $xv' - v = -e^{2x}$ is obtained.

34b. For $y_1(t) = 1/t$ $y_1' = -1/t^2$ and substitution into the D.E. shows that $y_1(t)$ is indeed a solution. Comparing the D.E. with the equation in Prob.33 we see that $q_1(t) = -1/t^2$, $q_2(t) = -1/t$ and $q_3(t) = 1$. Hence, using the method suggested in Prob.33, we set $y = y_1(t) + \dfrac{1}{v(t)}$ in the D.E. to obtain $\dfrac{dv}{dt} = -(-1/t + 2y_1)v - 1$, or

$\dfrac{dv}{dt} + \dfrac{1}{t}v = -1$. Thus $v(t) = \dfrac{c-t^2}{2t}$ and the second solution is $y_2(t) = \dfrac{1}{t} + \dfrac{2t}{c-t^2}$.

36. Let $v = y'$, then $v' = y''$ and thus the D.E. becomes $t^2v' + 2tv - 1 = 0$ or $t^2v' + 2tv = 1$. The left side is recognized as $(t^2 v)'$ and thus we may integrate to obtain $t^2 v = t + c$ (otherwise, divide both sides of the D.E. by t^2 and find the integrating factor, which is just t^2 in this case). Solving for $v = \dfrac{dy}{dt}$ we find

$\dfrac{dy}{dt} = 1/t + c/t^2$ so that $y = \ln t + c_1/t + c_2$.

38. If $v = y'$, the D.E. becomes $v' + tv^2 = 0$. This equation is separable and has the solution $-v^{-1} + t^2/2 = c$ or $v = y' = -2/(c_1 - t^2)$ where $c_1 = 2c$. We must consider separately the cases $c_1 = 0$, $c_1 > 0$ and $c_1 < 0$. If $c_1 = 0$, then $y' = 2/t^2$ or $y = -2/t + c_2$. If $c_1 > 0$, let $c_1 = k^2$. Then $y' = -2/(k^2-t^2) = -(1/k)[1/(k-t) + 1/(k+t)]$, so that $y = (1/k)\ln|(k-t)/(k+t)|+c_2$. If $c_1 < 0$, let $c_1 = -k^2$. Then $y' = 2/(k^2+t^2)$ so that $y = (2/k)\tan^{-1}(t/k) + c_2$. Finally, we note that $\dfrac{dy}{dt} = 0$ satisfies the D.E. and thus $y = $ constant is also a solution.

42. Following the procedure outlined, let $v = dy/dt$, then $y'' = \dfrac{dv}{dt} = v\dfrac{dv}{dy}$. Thus the D.E. becomes $yv\dfrac{dv}{dy} + v^2 = 0$ so that $v = 0$ and $y\dfrac{dy}{dy} + v = 0$. The D.E.is separable with

the solution $v = c/y$ (which includes $v = 0$ for $c = 0$).
Since $v = \dfrac{dy}{dt} = \dfrac{c}{y}$, we conclude that $y^2 = c_1 t + c_2$, by
separating variables and integrating.

45. Again let $v = y'$ and $v' = v\,dv/dy$ to obtain
$2y^2 v\dfrac{dv}{dy} + 2yv^2 = 1$, or $2y^2 v\,dv + 2yv^2\,dy = dy$. The left side
can be written as $d(y^2 v^2)$ and thus integration gives
$y^2 v^2 = y + c_1$ and thus $v = \pm y^{-1}(y + c_1)^{1/2}$. Setting $v = y'$
and separating variables gives $\pm y\,dy/(y+c_1)^{1/2} = dt$. On
observing that the left side of the equation can be
written as $\pm[(y+c_1) - c_1]dy/(y+c_1)^{1/2}$ we integrate and
find $\pm(2/3)(y-2c_1)(y+c_1)^{1/2} = t + c_2$.

47. If $v = y'$, then $v' = v\dfrac{dv}{dy}$ and the D.E. becomes

$v\dfrac{dv}{dy} + v^2 = 2e^{-y}$. Dividing by $v \neq 0$ we obtain

$\dfrac{dv}{dy} + v = 2v^{-1}e^{-y}$, which is a Bernoulli equation with

$n = -1$ (see Prob.27, Sect. 2.4). Let $w(y) = v^2$, then
$\dfrac{dw}{dy} = 2v\dfrac{dv}{dy}$ and the D.E. then becomes
$\dfrac{dw}{dy} + 2w = 4e^{-y}$, which is linear in w. Its solution is
$w = v^2 = ce^{-2y} + 4e^{-y}$. Setting $v = v'$ and separating

variables gives $\dfrac{dy}{\pm\sqrt{4e^{-y}+ce^{-2y}}} = \dfrac{e^y dy}{\pm\sqrt{4e^y+c}} = dt$.

Integrating and solving for e^y yields $e^y = (t+c_2)^2 + c_1$.

48. Since both t and y are missing, either approach used
above will work. In this case it's easier to use the
approach of Problems 36–41, so let $v = y'$ and thus $v' = y''$
and the D.E. becomes $v\dfrac{dv}{dt} = 2$.

51. The variable y is missing. Let $v = y'$, then $v' = y''$ and
the D.E. becomes $vv' - t = 0$. The solution of this
separable equation is $v^2 = t^2 + c_1$. Substituting $v = y'$
and applying the I.C. $y'(1) = 1$, we obtain $y' = t$. The
positive square root was chosen because $y' > 0$ at $t = 1$.
Solving this last equation and applying the I.C. $y(1) = 2$,
we obtain $y = t^2/2 + 3/2$.

CHAPTER 3

Section 3.1, Page 144

3. Assume $y = e^{rt}$, which gives $y' = re^{rt}$ and $y'' = r^2 e^{rt}$.
    Substitution into the D.E. yields $(6r^2 - r - 1)e^{rt} = 0$.
    Since $e^{rt} \neq 0$, we have the characteristic equation
    $6r^2 - r - 1 = 0$, or $(3r+1)(2r-1) = 0$. Thus
    $r = -1/3,\ 1/2$ and $y = c_1 e^{t/2} + c_2 e^{-t/3}$.

5. The characteristic equation is $r^2 + 5r = 0$, so the roots
    are $r_1 = 0$, and $r_2 = -5$. Thus
    $$y = c_1 e^{0t} + c_2 e^{-5t} = c_1 + c_2 e^{-5t}.$$

7. The characteristic equation is $r^2 - 9r + 9 = 0$ so the
    quadratic formula gives $r = (9 \pm \sqrt{81-36})/2 = (9 \pm 3\sqrt{5})/2$.
    Hence $y = c_1 \exp[(9+3\sqrt{5})t/2] + c_2 \exp[(9-3\sqrt{5})t/2]$.

10. Substituting $y = e^{rt}$ in the
    D.E. we obtain the
    characteristic equation
    $r^2 + 4r + 3 = 0$, which has
    the roots $r_1 = -1$, $r_2 = -3$.
    Thus $y = c_1 e^{-t} + c_2 e^{-3t}$ and
    $y' = -c_1 e^{-t} - 3c_2 e^{-3t}$.

    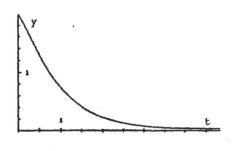

    Substituting $t = 0$ we then have $c_1 + c_2 = 2$ and
    $-c_1 - 3c_2 = -1$, yielding $c_1 = 5/2$ and $c_2 = -1/2$. Thus
    $y = \dfrac{5}{2}e^{-t} - \dfrac{1}{2}e^{-3t}$ and hence $y \to 0$ as $t \to \infty$.

15. The characteristic equation is $r^2 + 8r - 9 = 0$, so that
    $r_1 = 1$ and $r_2 = -9$ and the general solution is
    $y = c_1 e^{t} + c_2 e^{-9t}$. Since the I.C. are given at $t = 1$, it
    is convenient to write the general solution in the form
    $y = k_1 e^{(t-1)} + k_2 e^{-9(t-1)}$. Note that
    $c_1 = k_1 e^{-1}$ and $c_2 = k_2 e^{9}$. The advantage of the latter
    form of the general solution becomes clear when we apply
    the I.C. $y(1) = 1$ and $y'(1) = 0$. This latter form of y

gives $y' = k_1 e^{(t-1)} - 9k_2 e^{-9(t-1)}$ and thus setting $t = 1$ in y and y' yields the equations $k_1 + k_2 = 1$ and $k_1 - 9k_2 = 0$. Solving for k_1 and k_2 we find that $y = (9e^{(t-1)} + e^{-9(t-1)})/10$. The graph starts at $t = 1$ with zero slope and since $e^{(t-1)}$ has a positive exponent for $t > 1$, $y \to \infty$ as $t \to \infty$.

17. Comparing the given solution to Eq(17), we see that $r_1 = 2$ and $r_2 = -3$ are the two roots of the characteristic equation. Thus we have $(r-2)(r+3) = 0$, or $r^2 + r - 6 = 0$ as the characteristic equation. Hence the given solution is for the D.E. $y'' + y' - 6y = 0$.

19. The roots of the characteristic equation are $r = 1, -1$ and thus the general solution is $y(t) = c_1 e^t + c_2 e^{-t}$.

$y(0) = c_1 + c_2 = \dfrac{5}{4}$ and $y'(0) = c_1 - c_2 = -\dfrac{3}{4}$, yielding $y(t) = \dfrac{1}{4} e^t + e^{-t}$. From this $y'(t) = \dfrac{1}{4} e^t - e^{-t} = 0$ or $e^{2t} = 4$ or $t = \ln2$ and $y(\ln2) = \dfrac{1}{4}(2) + \dfrac{1}{2} = 1$. Since $y''(t) = y(t)$ is positive at $t = \ln2$, this is a minimum point. Note that $y(t) \to \infty$ as $t \to \infty$.

21. The general solution is $y = c_1 e^{-t} + c_2 e^{2t}$. Using the I.C. we obtain $c_1 + c_2 = \alpha$ and $-c_1 + 2c_2 = 2$, so adding the two equations we find $3c_2 = \alpha + 2$. If y is to approach zero as $t \to \infty$, c_2 must be zero. Thus $\alpha = -2$.

24. The roots of the characteristic equation are given by $r_1 = -2$, $r_2 = \alpha - 1$ and thus $y(t) = c_1 e^{-2t} + c_2 e^{(\alpha-1)t}$. Hence, for $\alpha < 1$, all solutions tend to zero as $t \to \infty$. For $\alpha > 1$, the second term becomes unbounded, but not the first, so there are no values of α for which all solutions become unbounded.

25a. The characteristic equation is $2r^2 + 3r - 2 = 0$, so $r_1 = -2$ and $r_2 = 1/2$ and $y = c_1 e^{-2t} + c_2 e^{t/2}$. The I.C. yield $c_1 + c_2 = 1$ and $-2c_1 + \dfrac{1}{2}c_2 = -\beta$ so that $c_1 = (1 + 2\beta)/5$ and $c_2 = (4 - 2\beta)/5$.

25b. Setting $\beta = 1$ and differentiating we obtain
$y' = (-6e^{-2t} + e^{t/2})/5$. Setting this equal to zero and
solving for t yields $t = \dfrac{2}{5}\ln 6$.

25c. From part (a), if $\beta = 2$ then $y(t) = e^{-2t}$ and the solution
simply decays to zero. For $\beta > 2$, the solution becomes
unbounded negatively, and again there is no minimum
point. For $0 < \beta < 2$ there is always a minimum point, as
found in part (b).

28a. The roots of the characteristic equation are given by
$r = \dfrac{-b \pm \sqrt{b^2 - 4ac}}{2a}$. For the roots to be real and different
we must have $b^2 - 4ac > 0$. If they are to be negative then
we must have $b > 0$ (since we are given $a > 0$) and $c > 0$. This
latter condition comes from the fact that if $c \le 0$ then
$\sqrt{b^2 - 4ac} \ge b$ and hence the numerator of r would give both
positive and negative values, or a zero if $c = 0$.

Section 3.2, Page 155

2. $W(\cos t, \sin t) = \begin{vmatrix} \cos t & \sin t \\ -\sin t & \cos t \end{vmatrix} = \cos^2 t + \sin^2 t = 1.$

4. $W(x, xe^x) = \begin{vmatrix} x & xe^x \\ 1 & e^x + xe^x \end{vmatrix} = xe^x + x^2e^x - xe^x = x^2e^x.$

8. Dividing by $(t-1)$ we have $p(t) = -3t/(t-1)$, $q(t) = 4/(t-1)$
and $g(t) = \sin t/(t-1)$, so the only point of discontinuity is
$t = 1$. By Theorem 3.2.1, the largest interval is $-\infty < t < 1$,
since the initial point is $t_0 = -2$.

12. $p(x) = 1/(x-2)$, $q(x) = \tan x$ and $g(x) = 0$, so $x = \pi/2$, 2,
$3\pi/2$, $5\pi/2 \ldots$ are points of discontinuity. Since $t_0 = 3$,
the interval specified by Theorem 3.2.1 is $2 < x < 3\pi/2$.

14. For $y = t^{1/2}$, $y' = \dfrac{1}{2}t^{-1/2}$ and $y'' = -\dfrac{1}{4}t^{-3/2}$. Thus
$yy'' + (y')^2 = -\dfrac{1}{4}t^{-1} + \dfrac{1}{4}t^{-1} = 0$. $y = 1$ is also a solution
since $y' = y'' = 0$. If $y = c_1(1) + c_2 t^{1/2}$ is substituted

in the D.E. you will get

$(c_1 + c_2 t^{1/2})(-\dfrac{c_2}{4} t^{-3/2}) + (\dfrac{c_2}{2} t^{-1/2})^2 = -\dfrac{c_1 c_2}{4} t^{-3/2}$, which is

zero only if $c_1 = 0$ or $c_2 = 0$. Thus the linear combination of two solutions is not, in general, a solution. Theorem 3.2.2 is not contradicted however, since the D.E. is not linear.

15. $y = \phi(t)$ is a solution of the D.E. so $L[\phi](t) = g(t)$. Since L is a linear operator, $L[c\phi](t) = cL[\phi](t) = cg(t)$. But, since $g(t) \neq 0$, $cg(t) = g(t)$ if and only if $c = 1$. This is not a contradiction of Theorem 3.2.2 since the linear D.E. is not homogeneous.

18. $W(f,g) = W(t,g) = \begin{vmatrix} t & g \\ 1 & g' \end{vmatrix} = tg' - g = t^2 e^t$, or $g' - \dfrac{1}{t} g = t e^t$.

This has an integrating factor of $\dfrac{1}{t}$ and thus

$\dfrac{1}{t} g' - \dfrac{1}{t^2} g = e^t$ or $(\dfrac{1}{t} g)' = e^t$. Integrating and

multiplying by t we obtain $g(t) = t e^t + ct$.

22. From Section 3.1, e^t and e^{-2t} are two solutions, and

since $W(e^t, e^{-2t}) = \begin{vmatrix} e^t & e^{-2t} \\ e^t & -2e^{-2t} \end{vmatrix} = -3e^{-t} \neq 0$ they form a

fundamental set of solutions. To find the fundamental set specified by Theorem 3.2.4, let $y(t) = c_1 e^t + c_2 e^{-2t}$, where c_1 and c_2 satisfy

$c_1 + c_2 = 1$ and $c_1 - 2c_2 = 0$ for y_1. Solving, we find

$y_1 = \dfrac{2}{3} e^t + \dfrac{1}{3} e^{-2t}$. Likewise, c_1 and c_2 satisfy

$c_1 + c_2 = 0$ and $c_1 - 2c_2 = 1$ for y_2, so that

$y_2 = \dfrac{1}{3} e^t - \dfrac{1}{3} e^{-2t}$.

26. For $y_1 = x$, we have $x^2(0) - x(x+2)(1) + (x+2)(x) = 0$ and

for $y_2 = xe^x$ we have $x^2(x+2)e^x - x(x+2)(x+1)e^x + (x+2)xe^x = 0$.

From Prob. 4, $W(x, xe^x) = x^2 e^x \neq 0$ for $x > 0$, so y_1 and y_2

form a fundamental set of solutions for $x > 0$.

28a. From Sect. 3.1 the characteristic eq. is

$r^2-r-2 = (r-2)(r+1) = 0$ and thus e^{-t} and e^{2t} are

solutions. Since $\begin{vmatrix} e^{-t} & e^{2t} \\ -e^{-t} & 2e^{2t} \end{vmatrix} = 3e^t \neq 0$, e^{-t} and e^{2t} are

fundamental solutions.

28b. For particular choices of c_1 and c_2, all three are
solutions by Theorem 3.2.2.

28c. $W(y_2,y_3) = \begin{vmatrix} e^{2t} & -2e^{2t} \\ 2e^{2t} & -4e^{2t} \end{vmatrix} = -4e^{4t} + 4e^{4t} = 0$ and thus y_2 and

y_3 do not form a fundamental set of solutions. Similar
calculations show that $[y_1,y_3]$ and $[y_1,y_4]$ are
fundamental solutions but y_4 and y_5 do not form a
fundamental set of solutions.

29. Writing the D.E. in the form of Eq.(21), we have
$p(t) = -(t+2)/t$. Thus Eq.(22) yields

$W(t) = cexp[-\int \frac{-(t+2)}{t} dt] = ct^2e^t$.

34. From Eq.(22) we have $W(y_1,y_2) = cexp[-\int p(t)dt]$, where
$p(t) = 2/t$ from the D.E. Thus $W(y_1,y_2) = c/t^2$. Since
$W(y_1,y_2)(1) = 2$ we find $c = 2$ and thus $W(y_1,y_2)(5) = 2/25$.

38. Let c be the point in I at which both y_1 and y_2 vanish.
Then $W(y_1,y_2)(c) = y_1(c)y_2'(c) - y_1'(c)y_2(c) = 0$. Since the
Wronskian is zero the functions y_1 and y_2 cannot form a
fundamental set.

40. Suppose that y_1 and y_2 have a point of inflection at t_0
and either $p(t_0) \neq 0$ or $q(t_0) \neq 0$. Since $y_1''(t_0) = 0$ and
$y_2''(t_0) = 0$ it follows from the D.E. that
$p(t_0)y_1'(t_0) + q(t_0)y_1(t_0) = 0$ and
$p(t_0)y_2'(t_0) + q(t_0)y_2(t_0) = 0$. If $p(t_0) = 0$ and $q(t_0) \neq 0$
then $y_1(t_0) = y_2(t_0) = 0$, and $W(y_1,y_2)(t_0) = 0$ so the
solutions cannot form a fundamental set. If $p(t_0) \neq 0$
and $q(t_0) = 0$ then $y_1'(t_0) = y_2'(t_0) = 0$ and $W(y_1,y_2)(t_0) = 0$,

so again the solutions cannot form a fundamental set. If
$p(t_0) \neq 0$ and $q(t_0) \neq 0$ then $y_1'(t_0) = -q(t_0)y_1(t_0)/p(t_0)$
and $y_2'(t_0) = -q(t_0)y_2(t_0)/p(t_0)$ and thus

$W(y_1,y_2)(t_0) = y_1(t_0)y_2'(t_0) - y_1'(t_0)y_2(t_0)$

$\quad = y_1(t_0)[-q(t_0)y_2(t_0)/p(t_0)] - [-q(t_0)y_1(t_0)/p(t_0)]y_2(t_0)$

$\quad = 0.$

41. Suppose that
$P(x)y'' + Q(x)y' + R(x)y = [P(x)y']' + [f(x)y]'$. On
expanding the right side and equating coefficients, we
find $f'(x) = R(x)$ and $P'(x) + f(x) = Q(x)$. These two
conditions on f can be satisfied if
$R(x) = Q'(x) - P''(x)$ which gives the necessary condition
$P''(x) - Q'(x) + R(x) = 0$.

44. We have $P(x) = x$, $Q(x) = -\cos x$, and $R(x) = \sin x$ and the
condition for exactness from Prob. 41 is satisfied. Also,
from Prob. 41, $f(x) = Q(x) - P'(x) = -\cos x - 1$, so the D.E.
becomes $(xy')' - [(1 + \cos x)y]' = 0$. Hence
$xy' - (1 + \cos x)y = c_1$. This is a first order linear
D.E. and the integrating factor (after dividing by x) is
$\mu(x) = \exp[-\int x^{-1}(1 + \cos x)dx]$. The general solution is
$y = [\mu(x)]^{-1}[c_1\int_{x_0}^{x} t^{-1}\mu(t)dt + c_2]$.

46. We want to choose $\mu(x)$ and $f(x)$ so that $\mu(x)P(x)y'' +$
$\mu(x)Q(x)y' + \mu(x)R(x)y = [\mu(x)P(x)y']' + [f(x)y]'$.
Expand the right side and equate coefficients of y'', y'
and y. This gives $\mu'(x)P(x) + \mu(x)P'(x) + f(x) =$
$\mu(x)Q(x)$ and $f'(x) = \mu(x)R(x)$. Differentiate the first
equation and then eliminate $f'(x)$ to obtain the adjoint
equation $P\mu'' + (2P' - Q)\mu' + (P'' - Q' + R)\mu = 0$.

48. $P = 1-x^2$, $Q = -2x$ and $R = \alpha(\alpha+1)$. Thus
$2P' - Q = -4x + 2x = -2x$ and
$P'' - Q' + R = -2 + 2 + \alpha(\alpha+1) = \alpha(\alpha+1)$, and thus, from
Prob. 46, $(1-x^2)\mu'' - 2x\mu' + \alpha(\alpha+1)\mu = 0$ is the adjoint
of the given D.E.

50. Write the adjoint D.E. given in Prob. 46 as
$\hat{P}\mu'' + \hat{Q}\mu' + \hat{R}\mu = 0$ where $\hat{P} = P$, $\hat{Q} = 2P' - Q$, and
$\hat{R} = P'' - Q' + R$. The adjoint of this equation, namely
the adjoint of the adjoint, is

$\hat{P}y" + (2\hat{P}' - \hat{Q})y' + (\hat{P}" - \hat{Q}' + \hat{R})y = 0$. After

substituting for $\hat{P}$, $\hat{Q}$, and $\hat{R}$ and simplifying, we obtain
$Py" + Qy' + Ry = 0$. This is the same as the original
equation.

51. From Prob. 46 the adjoint of $Py" + Qy' + Ry = 0$ is
$P\mu" + (2P' - Q)\mu' + (P" - Q' + R)\mu = 0$. The two equations
are the same if $2P' - Q = Q$ and $P" - Q' + R = R$. This
will be true if $P' = Q$. Hence the original D.E. is self-
adjoint if $P' = Q$. For Prob. 47, $P(x) = x^2$ so
$P'(x) = 2x$ and $Q(x) = x$. Hence the Bessel equation of
order v is not self-adjoint. In a similar manner we find
that Problems 48 and 49 are self-adjoint.

Section 3.3, Page 163

1. $\exp(1+2i) = e^{1+2i} = ee^{2i} = e(\cos 2 + i\sin 2)$.

5.   Recall that
$2^{1-i} = e^{\ln(2^{1-i})} = e^{(1-i)\ln 2} = e^{\ln 2}e^{-i\ln 2} = 2(\cos\ln 2 - i\sin\ln 2)$.

7.   As in Sect. 3.1, we seek solutions of the form $y = e^{rt}$.
Substituting this into the D.E. yields the characteristic
equation $r^2 - 2r + 2 = 0$, which has the roots $r_1 = 1 + i$
and $r_2 = 1 - i$, using the quadratic formula. Thus $\lambda = 1$
and $\mu = 1$ and from Eq.(24) the general solution is
$y = c_1 e^t \cos t + c_2 e^t \sin t$.

11. The characteristic equation is $r^2 + 6r + 13 = 0$, which
has the roots $r = \dfrac{-6 \pm \sqrt{-16}}{2} = -3 \pm 2i$. Thus $\lambda = -3$ and
$\mu = 2$, so Eq.(24) becomes $y = c_1 e^{-3t}\cos 2t + c_2 e^{-3t}\sin 2t$.

14. The characteristic equation is $9r^2 + 9r - 4$, which has
the real roots $-4/3$ and $1/3$. Thus the solution has the
same form as in Section 3.1, $y(t) = c_1 e^{t/3} + c_2 e^{-4t/3}$.

18. The characteristic equation is $r^2 + 4r + 5 = 0$, which has
the roots $r_1, r_2 = -2 \pm i$. Thus $y = c_1 e^{-2t}\cos t + c_2 e^{-2t}\sin t$
and $y' = (-2c_1+c_2)e^{-2t}\cos t + (-c_1-2c_2)e^{-2t}\sin t$, so that

$y(0) = c_1 = 1$ and
$y'(0) = -2c_1 + c_2 = 0$,
or $c_2 = 2$. Hence

$y = e^{-2t}(\cos t + 2\sin t)$.
The oscillation is hard
to see on this graph, but
$y(t)$ does cross the t axis
at $t = \tan^{-1}(-.5) = 2.68$ and
periodically after that.

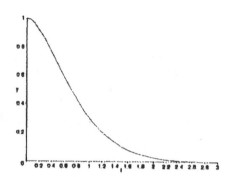

22. The characteristic equation is
    $r^2 + 2r + 2 = 0$, so
    $r_1, r_2 = -1 \pm i$. Since the I.C.
    are given at $\pi/4$ we want to
    alter Eq.(24) by letting
    $c_1 = e^{\pi/4}d_1$ and $c_2 = e^{\pi/4}d_2$.
    Thus, for $\lambda = -1$ and $\mu = 1$ we
    have $y = e^{-(t-\pi/4)}(d_1\cos t + d_2\sin t)$;
    so $y' = -e^{-(t-\pi/4)}(d_1\cos t + d_2\sin t) + e^{-(t-\pi/4)}(-d_1\sin t + d_2\cos t)$.

    Hence $y(\frac{\pi}{4}) = \sqrt{2}\,d_1/2 + \sqrt{2}\,d_2/2 = 2$ and

    $y'(\frac{\pi}{4}) = -\sqrt{2}\,d_1 = -2$ and thus $y = \sqrt{2}\,e^{-(t-\pi/4)}(\cos t + \sin t)$.

23a. The characteristic equation is $3r^2 - r + 2 = 0$, which has

    the roots $r_1, r_2 = \dfrac{1}{6} \pm \dfrac{\sqrt{23}}{6}i$. Thus

    $u(t) = e^{t/6}(c_1\cos\dfrac{\sqrt{23}}{6}t + c_2\sin\dfrac{\sqrt{23}}{6}t)$ and we obtain

    $u(0) = c_1 = 2$ and $u'(0) = \dfrac{1}{6}c_1 + \dfrac{\sqrt{23}}{6}c_2 = 0$. Solving

    for c_2 we find $u(t) = e^{t/6}(2\cos\dfrac{\sqrt{23}}{6}t - \dfrac{2}{\sqrt{23}}\sin\dfrac{\sqrt{23}}{6}t)$.

23b. To estimate the first time that $|u(t)| = 10$ plot the graph
    of $u(t)$ as found in part (a). Use this estimate in an
    appropriate computer software program to find $t = 10.7598$.

25a. The characteristic equation is $r^2 + 2r + 6 = 0$, so
    $r_1, r_2 = -1 \pm \sqrt{5}\,i$ and $y(t) = e^{-t}(c_1\cos\sqrt{5}\,t + c_2\sin\sqrt{5}\,t)$.
    Thus $y(0) = c_1 = 2$ and $y'(0) = -c_1 + \sqrt{5}\,c_2 = \alpha$ and hence

$$y(t) = e^{-t}(2\cos\sqrt{5}\,t + \frac{\alpha+2}{\sqrt{5}}\sin\sqrt{5}\,t).$$

25b. $y(1) = e^{-1}(2\cos\sqrt{5} + \frac{\alpha+2}{\sqrt{5}}\sin\sqrt{5}) = 0$ and hence

$$\alpha = -2 - \frac{2\sqrt{5}}{\tan\sqrt{5}} = 1.50878.$$

25c. For $y(t) = 0$ we must have $2\cos\sqrt{5}\,t + \frac{\alpha+2}{\sqrt{5}}\sin\sqrt{5}\,t = 0$ or

$\tan\sqrt{5}\,t = \frac{-2\sqrt{5}}{\alpha+2}$. For $\alpha \geq 0$ (actually, for $\alpha > -2$) this

yields $\sqrt{5}\,t = \pi - \arctan\frac{2\sqrt{5}}{\alpha+2}$ since arctanx is an odd

function.

25d. From part (c) $\arctan\frac{2\sqrt{5}}{\alpha+2} \to 0$ as $\alpha \to \infty$, so $t \to \pi/\sqrt{5}$.

31. Let $r = \lambda+i\mu$, then $\frac{d}{dt}(e^{rt}) = \frac{d}{dt}[e^{\lambda t}(\cos\mu t + i\sin\mu t)]$

$= \lambda e^{\lambda t}(\cos\mu t + i\sin\mu t) + e^{\lambda t}(-\mu\sin\mu t + i\mu\cos\mu t)$

$= \lambda e^{\lambda t}(\cos\mu t + i\sin\mu t) + i\mu e^{\lambda t}(i\sin\mu t + \cos\mu t)$

$= e^{\lambda t}(\lambda+i\mu)(\cos\mu t + i\sin\mu t) = re^{rt}.$

33. Suppose that $t = a$ and $t = b$ $(b>a)$ are consecutive zeros
of y_1. We must show that y_2 vanishes once and only once
in the interval $a < t < b$. Assume that it does not
vanish. Then we can form the quotient y_1/y_2 on the
interval $a \leq t \leq b$. Note $y_2(a) \neq 0$ and $y_2(b) \neq 0$,
otherwise y_1 and y_2 would not be a fundamental set of
solutions. Next, y_1/y_2 vanishes at $t = a$ and $t = b$ and
has a derivative in $a < t < b$. By Rolles theorem, the
derivative must vanish at an interior point. But

$$(\frac{y_1}{y_2})' = \frac{y_1'y_2 - y_2'y_1}{y_2^2} = \frac{-W(y_1,y_2)}{y_2^2}, \text{ which cannot be zero}$$

since y_1 and y_2 are fundamental solutions. Hence we have
a contradiction and conclude that y_2 must vanish at a
point between a and b. Finally, we show that it can
vanish at only one point between a and b. Suppose that
it vanishes at two points c and d between a and b. By
the argument we have just given we can show that y_1 must
vanish between c and d. But this contradicts the
hypothesis that a and b are consecutive zeros of y_1.

34a. Using the Chain Rule we have $\dfrac{dy}{dt} = \dfrac{dy}{dx}\dfrac{dx}{dt} = \dfrac{1}{t}\dfrac{dx}{dt}$ and

$$\frac{d^2y}{dt^2} = \frac{-1}{t^2}\frac{dy}{dx} + \frac{1}{t}\frac{d^2y}{dx^2}\frac{dx}{dt} = \frac{-1}{t^2}\frac{dy}{dx} + \frac{1}{t^2}\frac{d^2y}{dx^2}.$$

36. Using the result of Prob.34(b) we have

$\dfrac{d^2y}{dx^2} + (4-1)\dfrac{dy}{dx} + 2y = 0$, which has the characteristic

equation $r^2 + 3r + 2 = 0$. Thus $y(x) = c_1e^{-2x} + c_2e^{-x}$ so that

$y(t) = c_1e^{-2\ln t} + c_2e^{-\ln t} = \dfrac{c_1}{t^2} + \dfrac{c_2}{t}$.

40. Again, if $x = \ln t$, the D.E. becomes $\dfrac{d^2y}{dx^2} + (-1-1)\dfrac{dy}{dx} + 5y = 0$,

so the characteristic equation is $r^2 - 2r + 5 = 0$. Finding
the roots, we then have $y(x) = e^x(c_1\cos 2x + c_2\sin 2x)$ or
$y(t) = t[c_1\cos(2\ln t) + c_2\sin(2\ln t)]$.

44. We use the result of Prob.43. Note that $p(t) = t$ and
$q(t) = e^{-t^2} > 0$ for $-\infty < t < \infty$. Thus $(q' + 2pq)/q^{3/2} = 0$
and the D.E. can be transformed into an equation with
constant coefficients by letting $x = u(t) = \int e^{-t^2/2}dt$.
Substituting $x = u(t)$ in the differential equation found in
part (b) of Prob.43 we obtain, after dividing by the
coefficient of d^2y/dx^2, the D.E. $d^2y/dx^2 + y = 0$. Hence the
general solution of the original D.E. is
$y(t) = c_1\cos x(t) + c_2\sin x(t)$, $x(t) = \int e^{-t^2/2}dt$.

Section 3.4, Page 171

1. Substituting $y = e^{rt}$ into the D.E., we find that
$r^2 - 2r + 1 = 0$, which gives $r_1 = 1$ and $r_2 = 1$. Since the
roots are equal, the second fundamental solution is te^t,
see Eq.(26), and thus the general solution is
$y = c_1e^t + c_2te^t$.

5. The characteristic equation is $r^2 - 2r + 10 = 0$, and thus
$r = 1 \pm 3i$. The general solution, from Sect. 3.3, is
$y(t) = e^t(c_1\cos 3t + c_2\sin 3t)$.

9. The characteristic equation is $25r^2 - 20r + 4 = 0$, which
    may be written as $(5r-2)^2 = 0$ and hence the roots are
    $r_1, r_2 = 2/5$. Thus $y = c_1 e^{2t/5} + c_2 t e^{2t/5}$.

12. The characteristic equation is
    $r^2 - 6r + 9 = (r-3)^2$, which has
    the repeated root $r = 3$. Thus
    $y = c_1 e^{3t} + c_2 t e^{3t}$, which gives
    $y(0) = c_1 = 0$, $y'(t) = c_2(e^{3t}+3te^{3t})$
    and $y'(0) = c_2 = 2$. Hence
    $y(t) = 2te^{3t}$, which becomes
    positively unbounded as $t \to \infty$.

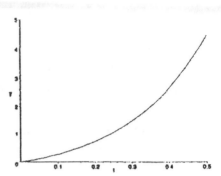

14. The characteristic equation is
    $r^2 + 4r + 4 = (r+2)^2 = 0$, which
    has the repeated root $r = -2$.
    Since the I.C. are given at
    $t = -1$, write the general
    solution as

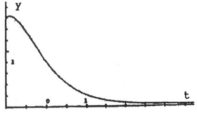

    $y = d_1 e^{-2(t+1)} + d_2 t e^{-2(t+1)}$. Then
    $y' = -2d_1 e^{-2(t+1)} + d_2 e^{-2(t+1)} - 2d_2 t e^{-2(t+1)}$ and hence
    $d_1 - d_2 = 2$ and $-2d_1 + 3d_2 = 1$ which yield $d_1 = 7$ and $d_2 = 5$.
    Thus $y = 7e^{-2(t+1)} + 5te^{-2(t+1)}$, a decaying exponential as
    shown in the graph.

17a. The characteristic equation is $4r^2 + 4r + 1 = (2r+1)^2 = 0$,
     so we have $y(t) = (c_1 + c_2 t)e^{-t/2}$. Thus $y(0) = c_1 = 1$ and
     $y'(0) = -c_1/2 + c_2 = 2$ and hence $c_2 = 5/2$ and
     $y(t) = (1 + 5t/2)e^{-t/2}$.

17b. From part(a), $y'(t) = -\dfrac{1}{2}(1 + 5t/2)e^{-t/2} + \dfrac{5}{2}e^{-t/2} = 0$, when
     $-\dfrac{1}{2} - \dfrac{5t}{4} + \dfrac{5}{2} = 0$, or $t_M = \dfrac{8}{5}$ and $y_M = 5e^{-4/5}$.

17c. From part(a), c_1 is the same and $y'(0) = -\dfrac{1}{2} + c_2 = b$ or
     $c_2 = b + \dfrac{1}{2}$ and $y(t) = [1 + (b + \dfrac{1}{2})t]e^{-t/2}$.

17d. From part(c), $y'(t) = -\dfrac{1}{2}[1 + (b+\dfrac{1}{2})t]e^{-t/2} + (b+\dfrac{1}{2})e^{-t/2} = 0$
     which yields $t_M = \dfrac{4b}{2b+1} \to 2$ as $b \to \infty$ and

$y_M = (1 + \dfrac{2b+1}{2} \cdot \dfrac{4b}{2b+1})e^{-2b/(2b+1)} = (1 + 2b)e^{-2b/(2b+1)}$. Since

$e^{-2b/(2b+1)} = e^{-2/(2+b^{-1})} \rightarrow e^{-1}$ as $b \rightarrow \infty$, $y_M \rightarrow \infty$ as $b \rightarrow \infty$.

19. If $r_1 = r_2$ then $y(t) = (c_1 + c_2 t)e^{r_1 t}$. Since the exponential is never zero, $y(t)$ can be zero only if $c_1 + c_2 t = 0$, which yields at most one positive value of t if c_1 and c_2 differ in sign. If $r_2 > r_1$ then

    $y(t) = c_1 e^{r_1 t} + c_2 e^{r_2 t} = e^{r_1 t}(c_1 + c_2 e^{(r_2 - r_1)t})$. Again, this is zero only if c_1 and c_2 differ in sign, in which case

    $t = \dfrac{\ln(-c_1/c_2)}{(r_2 - r_1)}$.

21. If $r_2 \neq r_1$ then $\phi(t;r_1,r_2) = (e^{r_2 t} - e^{r_1 t})/(r_2 - r_1)$ is defined for all t. Note that ϕ is a linear combination of the fundamental solutions, $e^{r_1 t}$ and $e^{r_2 t}$, of the D.E. and thus ϕ is a solution of the D.E. Think of r_1 as fixed and let $r_2 \rightarrow r_1$. The limit of ϕ as $r_2 \rightarrow r_1$ is indeterminate. If we use L'Hospital's rule (with r_2 as the variable), we

    find $\lim\limits_{r_2 \rightarrow r_1} \dfrac{e^{r_2 t} - e^{r_1 t}}{r_2 - r_1} = \lim\limits_{r_2 \rightarrow r_1} \dfrac{te^{r_2 t}}{1} = te^{r_1 t}$. Hence, the

    solution $\phi(t;r_1,r_2) \rightarrow te^{r_1 t}$ as $r_2 \rightarrow r_1$.

25. Let $y_2 = v/t$. Then $y_2' = v'/t - v/t^2$ and
    $y_2'' = v''/t - 2v'/t^2 + 2v/t^3$. Substituting in the D.E. we obtain
    $t^2(v''/t - 2v'/t^2 + 2v/t^3) + 3t(v'/t - v/t^2) + v/t = 0$. The
    terms involving v add to zero since $\dfrac{1}{t}$ is a solution. The
    left side then reduces to $tv'' + v' = 0$, which is linear
    in v', so $v' = c_1/t$. Thus $v = c_1\ln t + c_2$ so a second
    solution is $y_2(t) = (c_1\ln t + c_2)/t$. However, we may set
    $c_2 = 0$ and $c_1 = 1$ without loss of generality and thus we
    have $y_2(t) = (\ln t)/t$ as a second solution. Note that in
    the form we actually calculated, $y_2(t)$ is a linear
    combination of $1/t$ and $\ln t/t$, and hence is the general
    solution.

27. In this case the calculations are somewhat easier if we
    do not use the explicit form for $y_1(x) = \sin x^2$ at the

beginning but simply set $y_2(x) = y_1 v$. Substituting this form for y_2 in the D.E. gives $x(y_1 v)'' - (y_1 v)' + 4x^3(y_1 v) = 0$. On carrying out the differentiations and making use of the fact that y_1 is a solution, we obtain

$xy_1 v'' + (2xy_1' - y_1)v' = 0$. Let $w = v'$ and separate

variables to find $\dfrac{dw}{w} = (\dfrac{1}{x} - \dfrac{2y_1'}{y_1})dx$. Integration yields

$\ln w = \ln x - 2\ln y_1 + C$, so $w = v' = cx/(\sin x^2)^2$. Setting $u = x^2$ allows integration of this to get $v = c_1 \cot x^2 + c_2$. Setting $c_1 = 1$, $c_2 = 0$ and multiplying by $y_1 = \sin x^2$ we obtain $y_2(x) = \cos x^2$ as the second solution of the D.E.

30. Substituting $y_2(x) = y_1(x)v(x)$ in the D.E. gives

$x^2(y_1 v)'' + x(y_1 v)' + (x^2 - \dfrac{1}{4})y_1 v = 0$. On carrying out the

differentiations and making use of the fact that y_1 is a solution, we obtain $x^2 y_1 v'' + (2x^2 y_1' + xy_1)v' = 0$. This is a first order linear equation for $w = v'$,
$w' + (2y_1'/y_1 + 1/x)w = 0$, with solution (by separating variables)

$w = v'(x) = c\exp[-\int(2\dfrac{y_1'}{y_1} + \dfrac{1}{x})dx] = c\exp[-2\ln y_1 - \ln x]$

$= c\dfrac{1}{xy_1^2} = \dfrac{c}{x(x^{-1}\sin^2 x)} = c\csc^2 x$, where c is an

arbitrary constant, which we will take to be one. Then
$v(x) = \int \csc^2 x\, dx = -\cot x + k$ where again k is an
arbitrary constant which can be taken equal to zero.
Thus $y_2(x) = y_1(x)v(x) = (x^{-1/2}\sin x)(-\cot x) = -x^{-1/2}\cos x$.
The second solution is usually taken to be $x^{-1/2}\cos x$.
Note that $c = -1$ would have given this solution.

31b. Let $y_2(x) = e^x v(x)$, then $y_2' = e^x v' + e^x v$, and
$y_2'' = e^x v'' + 2e^x v' + e^x v$. Substituting in the D.E. we
obtain $xe^x v'' + (xe^x - Ne^x)v' = 0$, or $v'' + (1-N/x)v' = 0$.
This is a first order linear D.E. for v' with integrating
factor $\mu(x) = \exp[\int(1-N/x)dx] = x^{-N}e^x$. Hence
$(x^{-N}e^x v')' = 0$, and $v' = cx^N e^{-x}$ which gives

$v(x) = c\int x^N e^{-x} dx + k$. On taking $k = 0$ we obtain as the second solution $y_2(x) = ce^x \int x^N e^{-x} dx$. The integral can be evaluated by using the method of integration by parts. At each stage let $u = x^N$ or x^{N-1}, or whatever the power of x that remains, and let $dv = e^{-x}$. Note that this dv is not related to the $v(x)$ in $y_2(x)$. For $N = 2$ we have

$$y_2(x) = ce^x \int x^2 e^{-x} dx = ce^x [x^2 \frac{e^{-x}}{-1} - \int 2x \frac{e^{-x}}{-1} dx]$$

$$= -cx^2 + ce^x [2x \frac{e^{-x}}{-1} - \int 2 \frac{e^{-x}}{-1} dx]$$

$$= c(-x^2 - 2x - 2) = -2c(1 + x + x^2/2!).$$

Choosing $c = -1/2!$ gives the desired result. For the general case $c = -1/N!$

33. $(y_2/y_1)' = (y_1 y_2' - y_1' y_2)/y_1^2 = W(y_1, y_2)/y_1^2$. Abel's identity is $W(y_1, y_2) = c \exp[-\int_{t_0}^t p(r) dr]$. Hence $(y_2/y_1)' = cy_1^{-2} \exp[-\int_{t_0}^t p(r) dr]$. Integrating and setting $c = 1$ (since a solution y_2 can be multiplied by any constant) and taking the constant of integration to be zero we obtain

$$y_2(t) = y_1(t) \int_{t_0}^t \frac{\exp[-\int_{s_0}^s p(r) dr]}{[y_1(s)]^2} ds.$$

35. From Prob. 33 and Abel's formula we have

$(\frac{y_2}{y_1})' = \frac{\exp[\int (1/t) dt]}{\sin^2(t^2)} = \frac{e^{\ln t}}{\sin^2(t^2)} = t\csc^2(t^2)$. Thus $y_2/y_1 = -(1/2)\cot(t^2)$ and hence we can choose $y_2 = \cos(t^2)$ since $y_1 = \sin^2(t^2)$.

38. The general solution of the D.E. is $y = c_1 e^{r_1 t} + c_2 e^{r_2 t}$ where $r_1, r_2 = (-b \pm \sqrt{b^2 - 4ac})/2a$ provided $b^2 - 4ac \neq 0$. In this case there are two possibilities. If $b^2 - 4ac > 0$ then $(b^2 - 4ac)^{1/2} < b$ and r_1 and r_2 are real and _negative_. Consequently $e^{r_1 t} \to 0$ and $e^{r_2 t} \to 0$; and hence $y \to 0$, as $t \to \infty$. If $b^2 - 4ac < 0$ then r_1 and r_2 are complex conjugates with _negative_ real part. Again $e^{r_1 t} \to 0$ and $e^{r_2 t} \to 0$; and hence $y \to 0$, as $t \to \infty$.

Finally, if $b^2 - 4ac = 0$, then $y = c_1e^{r_1t} + c_2te^{r_1t}$ where $r_1 = -b/2a < 0$. Hence, again $y \to 0$ as $t \to \infty$. This conclusion does not hold if either $b = 0$ (since, in this case, $y(t) = c_1\cos\omega t + c_2\sin\omega t$, where $\omega^2 = \dfrac{c}{a}$) or $c = 0$ (since one of the solutions would be $y_1(t) = c_1$).

42. Substituting $x = \ln t$ into the D.E. gives
$$\frac{d^2y}{dx^2} + \frac{dy}{dx} + 0.25y = 0, \text{ which has the solution}$$
$y(x) = c_1e^{-x/2} + c_2xe^{-x/2}$ so that $y(t) = c_1t^{-1/2} + c_2t^{-1/2}\ln t$.

46. Again $x = \ln t$, so $\dfrac{d^2y}{dx^2} + 4\dfrac{dy}{dx} + 13y = 0$. The roots of the characteristic are $r_{1,2} = -2 \pm 3i$ and thus
$y(x) = (c_1\cos 3x + c_2\sin 3x)e^{-2x}$ which gives
$y(t) = [c_1\cos(3\ln t) + c_2\sin(3\ln t)]t^{-2}$.

Section 3.5, Page 183

1. First we find the solution of the homogeneous D.E., which has the characteristic equation $r^2-2r-3 = (r-3)(r+1) = 0$. Hence $y_c = c_1e^{3t} + c_2e^{-t}$ and we can assume $Y = Ae^{2t}$ for the particular solution. Thus $Y' = 2Ae^{2t}$ and $Y'' = 4Ae^{2t}$ and substituting into the D.E. yields
$4Ae^{2t} - 2(2Ae^{2t}) - 3(Ae^{2t}) = 3e^{2t}$. Thus $-3A = 3$ and $A = -1$, yielding $y = c_1e^{3t} + c_2e^{-t} - e^{2t}$.

4. Initially we might assume $Y = A + B\sin 2t + C\cos 2t$. However, since a constant is a solution of the related homogeneous D.E. we must modify Y by multiplying the constant A by t and thus the correct form is $Y = At + B\sin 2t + C\cos 2t$.

6. Since $y_c = c_1e^{-t} + c_2te^{-t}$ we must assume $Y = At^2e^{-t}$, so that $Y' = 2Ate^{-t} - At^2e^{-t}$ and $y'' = 2Ae^{-t} - 4Ate^{-t} + At^2e^{-t}$. Substituting in the D.E. gives $(At^2-4At+2A)e^{-t} + 2(-At^2+2At)e^{-t} + At^2e^{-t} = 2e^{-t}$. Notice that all terms on

the left involving t^2 and t add to zero and we are left
with $2A = 2$, or $A = 1$. Hence $y = c_1 e^{-t} + c_2 t e^{-t} + t^2 e^{-t}$.

8. The assumed form is $Y = (At + B)\sin 2t + (Ct + D)\cos 2t$,
    which is appropriate for both terms appearing on the
    right side of the D.E. Since none of the terms appearing
    in Y are solutions of the homogeneous equation, we do not
    need to modify Y. Calculating Y' and Y'' we have
    $Y' = A\sin 2t + C\cos 2t + 2(At+B)\cos 2t - 2(Ct+D)\sin 2t$ and
    $Y'' = 4A\cos 2t - 4C\sin 2t - 4(At+B)\sin 2t - 4(Ct+D)\cos 2t$. Thus
    $Y''+Y = -3At\sin 2t - 3Ct\cos 2t - (3B+4C)\sin 2t - (4A-3D)\cos 2t$.
    Equating like coefficients yields $A = 0, -3B-4C = 3, -3C = 1$,
    and $4A-3D = 0$. Hence $Y(t) = -(5/9)\sin 2t -(1/3)t\cos 2t$.

11. First solve the homogeneous D.E. Substituting $y = e^{rt}$
    gives $r^2 + r + 4 = 0$. Hence $y_c = e^{-t/2}[c_1\cos(\sqrt{15}\,t/2) +$
    $c_2\sin(\sqrt{15}\,t/2)]$. We replace $\sinh t$ by $(e^t - e^{-t})/2$ and
    then assume $Y(t) = Ae^t + Be^{-t}$. Since neither e^t nor e^{-t}
    are solutions of the homogeneous equation, there is no
    need to modify our assumption for Y. Substituting in the
    D.E., we obtain $6Ae^t + 4Be^{-t} = e^t - e^{-t}$. Hence, $A = 1/6$
    and $B = -1/4$. The general solution is
    $y = e^{-t/2}[c_1\cos(\sqrt{15}\,t/2) + c_2\sin(\sqrt{15}\,t/2)] + e^t/6 - e^{-t}/4$.
    [For this problem we could also have found a particular
    solution as a linear combination of $\sinh t$ and $\cosh t$:
    $Y(t) = A\cosh t + B\sinh t$. Substituting this in the D.E.
    gives $(5A + B)\cosh t + (A + 5B)\sinh t = 2\sinh t$. The
    solution is $A = -1/12$ and $B = 5/12$. A simple calculation
    shows that $-(1/12)\cosh t + (5/12)\sinh t = e^t/6 - e^{-t}/4$.]

13. $y_c = c_1 e^{-2t} + c_2 e^t$ so for the particular solution we
    assume $Y = At + B$. Since neither At or B are solutions of
    the homogeneous equation it is not necessary to modify
    the original assumption. Substituting Y in the D.E. we
    obtain $0 + A - 2(At+B) = 2t$ or $-2A = 2$ and $A - 2B = 0$.
    Solving for A and B we obtain $y = c_1 e^{-2t} + c_2 e^t - t - 1/2$
    as the general solution. $y(0) = 0 \Rightarrow c_1 + c_2 - 1/2 = 0$
    and $y'(0) = 1 \Rightarrow -2c_1 + c_2 - 1 = 1$, which yield $c_1 = -1/2$
    and $c_2 = 1$. Thus $y = e^t - (1/2)e^{-2t} - t - 1/2$.

16. Since the characteristic equation is $r^2 - 2r - 3 = 0$,
    $y_c = c_1 e^{3t} + c_2 e^{-t}$. The nonhomogeneous term is the product
    of a linear polynomial and an exponential, so assume Y of
    the same form: $Y = (At+B)e^{2t}$, which we do not need to
    modify since these terms are not in y_c. Thus
    $Y' = Ae^{2t} + 2(At+B)e^{2t}$ and $Y'' = 4Ae^{2t}+4(At+B)e^{2t}$.
    Substituting into the D.E. we find $-3At = 3t$ and
    $2A - 3B = 0$, yielding $A = -1$ and $B = -2/3$. Thus, the
    general solution is $y = c_1 e^{3t} + c_2 e^{-t} - \frac{2}{3}e^{2t} - te^{2t}$.

19a. The solution of the homogeneous D.E. is $y_c = c_1 e^{-3t} + c_2$.

    After inspection of the nonhomogeneous term, for $2t^4$ we
    must assume a fourth order polynominial, for $t^2 e^{-3t}$ we
    must assume a quadratic polynomial times the exponential,
    and for $\sin 3t$ we must assume $C\sin 3t + D\cos 3t$. Thus
    $Y(t) = (A_0 t^4 + A_1 t^3 + A_2 t^2 + A_3 t + A_4) + (B_0 t^2 + B_1 t + B_2)e^{-3t} + C\sin 3t + D\cos 3t$.
    However, since e^{-3t} and a constant are solutions of the
    homogeneous D.E., we must multiply the coefficient of e^{-3t}
    and the polynomial by t. The correct form, then, is
    $Y(t) = t(A_0 t^4 + A_1 t^3 + A_2 t^2 + A_3 t + A_4) +$
    $\qquad\qquad t(B_0 t^2 + B_1 t + B_2)e^{-3t} + C\sin 3t + D\cos 3t$.

22a. The solution of the homgeneous D.E. is
    $y_c = e^{-t}[c_1 \cos t + c_2 \sin t]$. After inspection of the
    nonhomogeneous term, we assume
    $Y(t) = Ae^{-t} + (B_0 t^2 + B_1 t + B_2)e^{-t}\cos t + (C_0 t^2 + C_1 t + C_2)e^{-t}\sin t$.
    Since $e^{-t}\cos t$ and $e^{-t}\sin t$ are solutions of the
    homogeneous D.E., it is necessary to multiply both the
    last two terms by t. Hence the correct form is
    $Y(t) = Ae^{-t} + t(B_0 t^2 + B_1 t + B_2)e^{-t}\cos t + t(C_0 t^2 + C_1 t + C_2)e^{-t}\sin t$.

27a. Calculating Y' and Y'' and substituting into the D.E. we
    get $(v''-2v'+v)e^{-t} - 3(v'-v)e^{-t} - 4ve^{-t} = 2e^{-t}$. This
    reduces to $v'' - 5v' = 2$, which is a 1st order D.E. for v'.

27b. The linear D.E. for w has an integrating factor of e^{-5t}
    and thus $(e^{-5t}w)' = 2e^{-5t}$, which gives $w = v' = -\frac{2}{5} + c_1 e^{5t}$.

27c. Integration then gives $v = -\dfrac{2}{5}t + \dfrac{c_1}{5}e^{5t} + c_2$ so

$Y(t) = ve^{-t} = -\dfrac{2}{5}te^{-t} + \dfrac{c_1}{5}e^{4t} + c_2e^{-t}$. The last two terms can be thought of as y_c and the first term as a particular solution.

29. First solve the I.V.P. $y'' + y = t$, $y(0) = 0$, $y'(0) = 1$ for $0 \le t \le \pi$. The solution of the homogeneous D.E. is $y_c(t) = c_1\cos t + c_2\sin t$. The correct form for $Y(t)$ is $y(t) = A_0t + A_1$. Substituting in the D.E. we find $A_0 = 1$ and $A_1 = 0$. Hence, $y = c_1\cos t + c_2\sin t + t$. Applying the I.C., we obtain $y = t$, for $0 \le t \le \pi$.

For $t > \pi$ we have $y'' + y = \pi e^{\pi-t}$ so the form, now, for $Y(t)$ is $Y(t) = Ee^{\pi-t}$. Substituting $Y(t)$ in the D.E., we obtain $Ee^{\pi-t} + Ee^{\pi-t} = \pi e^{\pi-t}$ so $E = \pi/2$. Hence the general solution for $t > \pi$ is $Y = D_1\cos t + D_2\sin t + (\pi/2)e^{\pi-t}$. If y and y' are to be continuous at $t = \pi$, then the solutions and their derivatives for $t \le \pi$ and $t > \pi$ must have the same value at $t = \pi$. These conditions require $\pi = -D_1 + \pi/2$ and $1 = -D_2 - \pi/2$. Hence $D_1 = -\pi/2$, $D_2 = -(1 + \pi/2)$, and

$$y = \phi(t) = \begin{cases} t, & 0 \le t \le \pi \\ -(\pi/2)\cos t - (1 + \pi/2)\sin t + (\pi/2)e^{\pi-t}, & t > \pi. \end{cases}$$

The graphs of the nonhomogeneous term and ϕ follow.

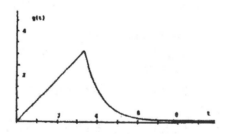
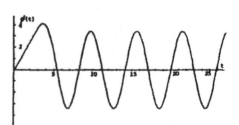

31. According to Theorem 3.5.1, the difference of any two solutions of the linear second order nonhomogeneous D.E. is a solution of the corresponding homogeneous D.E. Hence $Y_1 - Y_2$ is a solution of $ay'' + by' + cy = 0$. In Prob. 38 of Section 3.4 we showed that if $a > 0$, $b > 0$, and $c > 0$ then every solution of this D.E. goes to zero

as $t \to \infty$. If $b = 0$, then y_c involves only sines and cosines, so $Y_1 - Y_2$ does not approach zero as $t \to \infty$.

34. From Prob. 33 we write the D.E. as $(D-4)(D+1)y = 3e^{2t}$. Thus let $(D+1)y = u$ and then $(D-4)u = 3e^{2t}$. This last equation is the same as $du/dt - 4u = 3e^{2t}$, which may be solved by multiplying both sides by e^{-4t} and integrating (see Sect. 2.1). This yields $u = (-3/2)e^{2t} + Ce^{4t}$. Substituting this form of u into $(D+1)y = u$ we obtain $dy/dt + y = (-3/2)e^{2t} + Ce^{4t}$. Again, multiplying by e^{t} and integrating gives $y = (-1/2)e^{2t} + C_1e^{4t} + C_2e^{-t}$, where $C_1 = C/5$.

Section 3.6, Page 189

2. Two linearly independent solutions of the homogeneous D.E. are $y_1(t) = e^{2t}$ and $y_2(t) = e^{-t}$. Assume $Y = u_1(t)e^{2t} + u_2(t)e^{-t}$, then $Y'(t) = [2u_1(t)e^{2t} - u_2(t)e^{-t}] + [u_1'(t)e^{2t} + u_2'(t)e^{-t}]$. We set $u_1'(t)e^{2t} + u_2'(t)e^{-t} = 0$. Then $Y'' = 4u_1e^{2t} + u_2e^{-t} + 2u_1'e^{2t} - u_2'e^{-t}$ and substituting in the D.E. gives $2u_1'(t)e^{2t} - u_2'(t)e^{-t} = 2e^{-t}$ (the terms involving u_1 and u_2 add to zero since e^{-t} and e^{2t} are solutions ofthe homogneous equation). Thus we have two algebraic equations for $u_1'(t)$ and $u_2'(t)$ with the solution $u_1'(t) = 2e^{-3t}/3$ and $u_2'(t) = -2/3$. Hence $u_1(t) = -2e^{-3t}/9$ and $u_2(t) = -2t/3$. Substituting in the expression for $Y(t)$ we obtain $Y(t) = (-2e^{-3t}/9)e^{2t} + (-2t/3)e^{-t} = (-2e^{-t}/9) - (2te^{-t}/3)$. Since e^{-t} is a solution of the homogeneous D.E., we can choose $Y(t) = -2te^{-t}/3$.

5. Since cost and sint are solutions of the homogeneous D.E., we assume $Y = u_1(t)\cos t + u_2(t)\sin t$. Thus $Y' = -u_1(t)\sin t + u_2(t)\cos t$, after setting $u_1'(t)\cos t + u_2'(t)\sin t = 0$. Finding Y'' and substituting

into the D.E. then yields $-u_1'(t)\sin t + u_2'(t)\cos t = \tan t$.
The two equations for $u_1'(t)$ and $u_2'(t)$ have the solution:
$u_1'(t) = -\sin^2 t/\cos t = -\sec t + \cos t$ and
$u_2'(t) = \sin t$. Thus $u_1(t) = \sin t - \ln(\tan t + \sec t)$ and
$u_2(t) = -\cos t$, which when substituted into the assumed
form for Y, simplified, and added to the homogeneous
solution yields
$y = c_1\cos t + c_2\sin t - (\cos t)\ln(\tan t + \sec t)$.

11. Two linearly independent solutions of the homogeneous
    D.E. are $y_1(t) = e^{3t}$ and $y_2(t) = e^{2t}$. Applying Theorem
    3.6.1 with $W(y_1,y_2)(t) = -e^{5t}$, we obtain
    $$Y(t) = -e^{3t}\int \frac{e^{2s}g(s)}{-e^{5s}}\, ds + e^{2t}\int \frac{e^{3s}g(s)}{-e^{5s}}ds$$
    $$= \int [e^{3(t-s)} - e^{2(t-s)}]g(s)ds.$$
    The complete solution is then obtained by adding
    $c_1e^{3t} + c_2e^{2t}$ to $Y(t)$. Note: since we are taking e^{3t} and e^{2t}
    under the integral the integration variable can't be t.

14. That t and te^t are solutions of the homogeneous D.E. can
    be verified by direction substitution. Thus we assume
    $Y = tu_1(t) + te^t u_2(t)$. Following the pattern of earlier
    problems we find $tu_1'(t) + te^t u_2'(t) = 0$, Eq.(21), and
    $u_1'(t) + (t+1)e^t u_2' = 2t$, Eq.(25). [Note that $g(t) = 2t$,
    since the D.E. must be put into the form of Eq.(16)].
    The solution of these equations gives $u_1'(t) = -2$ and
    $u_2'(t) = 2e^{-t}$. Hence, $u_1(t) = -2t$ and $u_2(t) = -2e^{-t}$, and
    $Y(t) = t(-2t) + te^t(-2e^{-t}) = -2t^2 - 2t$. However, since t
    is a solution of the homogeneous D.E. we can choose as
    our particular solution $Y(t) = -2t^2$.

18. For this problem, and for many others, it is probably
    easier to rederive Eqs.(26) without using the explicit
    form for $y_1(x)$ and $y_2(x)$ and then to substitute for $y_1(x)$
    and $y_2(x)$ in Eqs.(26). In this case if we take
    $y_1 = x^{-1/2}\sin x$ and $y_2 = x^{-1/2}\cos x$, then $W(y_1,y_2) = -1/x$.
    If the D.E. is put in the form of Eq.(16), then
    $g(x) = 3x^{-1/2}\sin x$ and thus $u_1'(x) = 3\sin x\cos x$ and

$u_2'(x) = -3\sin^2 x = 3(-1 + \cos 2x)/2.$ Hence

$u_1(x) = (3\sin^2 x)/2$ and $u_2(x) = -3x/2 + 3(\sin 2x)/4$, and

$$Y(x) = \frac{3\sin^2 x}{2}\frac{\sin x}{\sqrt{x}} + (-\frac{3x}{2} + \frac{3\sin 2x}{4})\frac{\cos x}{\sqrt{x}}$$

$$= \frac{3\sin^2 x}{2}\frac{\sin x}{\sqrt{x}} + (-\frac{3x}{2} + \frac{3\sin x\cos x}{2})\frac{\cos x}{\sqrt{x}}$$

$$= \frac{3\sin x}{2\sqrt{x}} - \frac{3\sqrt{x}\cos x}{2}.$$

The first term is a multiple of $y_1(x)$ and thus can be neglected for $Y(x)$.

22. Putting limits on the integrals of Eq.(28)

$$Y(t) = -y_1(t)\int_{t_0}^{t}\frac{y_2(s)g(s)ds}{W(y_1,y_2)(s)} + y_2(t)\int_{t_0}^{t}\frac{y_1(s)g(s)ds}{W(y_1,y_2)(s)}$$

$$= \int_{t_0}^{t}\frac{-y_1(t)y_2(s)g(s)ds}{W(y_1,y_2)(s)} + \int_{t_0}^{t}\frac{y_2(t)y_1(s)g(s)ds}{W(y_1,y_2)(s)}$$

$$= \int_{t_0}^{t}\frac{[y_1(s)y_2(t) - y_1(t)y_2(s)]g(s)}{y_1(s)y_2'(s) - y_1'(s)y_2(s)} ds.$$ To show

that $Y(t)$ satisfies $L[y] = g(t)$ we must take the derivative of Y using Leibnitz's rule, which says that if

$$Y(t) = \int_{t_0}^{t}G(t,s)ds, \text{ then } Y'(t) = G(t,t) + \int_{t_0}^{t}\frac{\partial G}{\partial t}(t,s)ds.$$

Letting $G(t,s)$ be the above integrand, then $G(t,t) = 0$

and $\dfrac{\partial G}{\partial t} = \dfrac{y_1(s)y_2'(t) - y_1'(t)y_2(s)}{W(y_1,y_2)(s)} g(s).$ Likewise

$$Y'' = \frac{\partial G(t,t)}{\partial t} + \int_{t_0}^{t}\frac{\partial^2 G}{\partial t^2}(t,s)ds$$

$$= g(t) + \int_{t_0}^{t}\frac{y_1(s)y_2''(t) - y_1''(t)y_2(s)}{W(y_1,y_2)(s)}g(s)ds.$$

Since y_1 amd y_2 are solutions of $L[y] = 0$, we have $L[Y] = g(t)$ since all the terms involving the integral will add to zero. Clearly $Y(t_0) = 0$ and $Y'(t_0) = 0$.

25. Note that $y_1 = e^{\lambda t}\cos\mu t$ and $y_2 = e^{\lambda t}\sin\mu t$ and thus $W(y_1,y_2) = \mu e^{2\lambda t}$. From Prob. 22 we then have:

$$Y(t) = \int_{t_0}^{t}\frac{e^{\lambda s}\cos\mu s\, e^{\lambda t}\sin\mu t - e^{\lambda t}\cos\mu t\, e^{\lambda s}\sin\mu s}{\mu e^{2\lambda s}} g(s)ds$$

$$= \mu^{-1} \int_{t_0}^{t} e^{\lambda(t-s)} [\cos\mu s \ \sin\mu t - \cos\mu t \ \sin\mu s] g(s) ds$$

$$= \mu^{-1} \int_{t_0}^{t} e^{\lambda(t-s)} [\sin\mu(t-s)] g(s) ds.$$

29. First, we put the D.E. in standard form by dividing by t^2: $y'' - 2y'/t + 2y/t^2 = 4$. Assuming that $y = tv(t)$ and substituting in the D.E. we obtain $tv'' = 4$. Hence $v'(t) = 4\ln t + c_2$ and $v(t) = 4\int \ln t \ dt + c_2 t + c_1 = 4(t\ln t - t) + c_2 t + c_1$, using integration by parts. Thus $y = 4t^2\ln t + c_3 t^2 + c_1 t$, where $c_3 = c_2 - 4$. Since $y_1 = c_1 t$, we can take $y_2 = 4t^2\ln t + c_3 t^2$, where $c_3 t^2$ represents the second fundamental solution of the related homogeneous equation and $4t^2\ln t$ is the particular solution.

Section 3.7, Page 202

2. From Eq.(15) we have $R\cos\delta = -1$, and $R\sin\delta = \sqrt{3}$. Thus $R = \sqrt{1+3} = 2$ and $\delta = \tan^{-1}(-\sqrt{3}) + \pi = 2\pi/3 \approx 2.09440$. Note that we have to "add" π to the inverse tangent value since δ must be a second quadrant angle. Thus $u = 2\cos(t-2\pi/3)$.

6. The motion is an undamped free vibration. The units are in the CGS system. The spring constant [see Eq.(2)] is $k = (100 \ \text{gm})(980 \text{cm/sec}^2)/5\text{cm}$. Hence the D.E. for the motion is $100u'' + [(100 \cdot 980)/5]u = 0$ where u is measured in cm and time in sec. We obtain $u'' + 196u = 0$ so $u = A\cos 14t + B\sin 14t$. The I.C. are $u(0) = 0 \Rightarrow A = 0$ and $u'(0) = 10 \text{ cm/sec} \Rightarrow B = 10/14 = 5/7$. Hence $u(t) = (5/7)\sin 14t$, which first reaches equilibrium when $14t = \pi$, or $t = \pi/14$.

8. We use Eq.(33) without R and E(t) (there is no resistor or impressed voltage) and with $L = 1$ henry and $1/C = 4\times10^6$ since $C = .25\times10^{-6}$ farads. Thus the I.V.P. is $Q'' + 4\times10^6 Q = 0$, $Q(0) = 10^{-6}$ coulombs and $Q'(0) = 0$.

9. The spring constant is $k = (20)(980)/5 = 3920$ dyne/cm. The I.V.P. for the motion is $20u'' + 400u' + 3920u = 0$ or $u'' + 20u' + 196u = 0$ and $u(0) = 2$, $u'(0) = 0$. Here u is measured in cm and t in sec. The general solution of the D.E. is $u = Ae^{-10t}\cos 4\sqrt{6}\,t + Be^{-10t}\sin 4\sqrt{6}\,t$. The I.C. $u(0) = 2 \Rightarrow A = 2$ and $u'(0) = 0 \Rightarrow -10A + 4\sqrt{6}B = 0$. The solution is $u = e^{-10t}[2\cos 4\sqrt{6}\,t + 5(\sin 4\sqrt{6}\,t)/\sqrt{6}\,]$cm.

The quasi frequency is $\mu = 4\sqrt{6}$, the quasi period is
$T_d = 2\pi\mu = \pi/2\sqrt{6}$ and $T_d/T = 7/2\sqrt{6}$ since
$T = 2\pi/14 = \pi/7$. To find an upper bound for τ, write u
in the form of Eq. (26): $u(t) = \sqrt{4+25/6}\, e^{-10t}\cos(4\sqrt{6}\, t-\delta)$.
Now, since $|\cos(4\sqrt{6}\, t-\delta)| \le 1$, we have $|u(t)| < .05 \Rightarrow$
$\sqrt{4+25/6}\, e^{-10t} < .05$, which

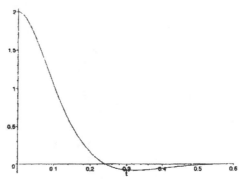

yields $\tau = .4046$. A more
precise answer can be
obtained with a computer
algebra system, which in
this case yields $\tau = .4045$.
The original estimate was
unusually close for this
problem since
$\cos(4\sqrt{6}\, t-\delta) = -0.9996$
for $t = .4046$.

12. Substituting the given values for L, C and R in Eq.(33),
we obtain the D.E. $.2Q'' + 3\text{x}10^2\, Q' + 10^5\, Q = 0$. The I.C.
are $Q(0) = 10^{-6}$ and $Q'(0) = I(0) = 0$. The roots of the
characteristic equation are $r_1 = -500$ and $r_2 = -1000$.
Thus $Q = c_1 e^{-500t} + c_2 e^{-1000t}$ and hence $Q(0) = 10^{-6} \Rightarrow$
$c_1 + c_2 = 10^{-6}$ and $Q'(0) = 0 \Rightarrow -500c_1 - 1000c_2 = 0$.
Solving for c_1 and c_2 yields the solution.

17. The mass is $8/32$ lb-sec^2/ft, and the spring constant is
$8/(1/8) = 64$ lb/ft. Hence $(1/4)u'' + \gamma u' + 64u = 0$ or
$u'' + 4\gamma u' + 256u = 0$, where u is measured in ft, t in sec
and the units of γ are lb-sec/ft. The characteristic
equation is $r^2 + 4\gamma r + 256 = 0$, so
$r_1, r_2 = [-4\gamma \pm \sqrt{16\gamma^2 - 1024}\,]/2$. The system will be
overdamped, critically damped or underdamped as
$(16\gamma^2 - 1024)$ is > 0, $= 0$, or < 0, respectively. Thus
the system is critically damped when $\gamma = 8$ lb-sec/ft.

19. The general solution of the D.E. is $u = Ae^{r_1 t} + Be^{r_2 t}$
where $r_1, r_2 = [-\gamma \pm (\gamma^2 - 4km)^{1/2}]/2m$ provided
$\gamma^2 - 4km \ne 0$, and where A and B are determined by the
I.C. When the motion is overdamped, $\gamma^2 - 4km > 0$ and
$r_1 > r_2$. Setting $u = 0$, we obtain $Ae^{r_1 t} = -Be^{r_2 t}$ or
$e^{(r_1-r_2)t} = -B/A$. Since the exponential function is a
monotone function, there is at most one value of t
(when $B/A < 0$) for which this equation can be satisfied.
Hence u can vanish at most once. If the system is
critically damped, the general solution is

$u(t) = (A + Bt)e^{-\gamma t/2m}$. The exponential function is never zero; hence u can vanish only if $A + Bt = 0$. If $B = 0$ then u never vanishes; if $B \neq 0$ then u vanishes once at $t = -A/B$ provided $A/B < 0$.

20. The general solution of Eq.(21) for the case of critical damping is $u = (A + Bt)e^{-\gamma t/2m}$. The I.C. $u(0) = u_0 \Rightarrow$

    $A = u_0$ and $u'(0) = v_0 \Rightarrow A(-\gamma/2m) + B = v_0$. Hence $u = [u_0 + (v_o + \gamma u_0/2m)t]e^{-\gamma t/2m}$. If $v_0 = 0$, then

    $u = u_0(1 + \gamma t/2m)e^{-\gamma t/2m}$, which is never zero since γ and m are postive. By L'Hospital's Rule $u \rightarrow 0$ as $t \rightarrow \infty$. Finally for $u_0 > 0$, we want the condition which will insure that $v = 0$ at least once. Since the exponential function is never zero we require $u_0 + (v_0 + \gamma u_0/2m)t = 0$ at a positive value of t. This requires that $v_0 + \gamma u_0/2m \neq 0$ and that $t = -u_0(v_0 + u_0\gamma/2m)^{-1} > 0$. We know that $u_0 > 0$ so we must have $v_0 + \gamma u_0/2m < 0$ or $v_0 < -\gamma u_0/2m$.

23. From Prob. 21: $\Delta = \dfrac{2\pi\gamma}{\mu(2m)} = T_d\gamma/2m$. Substituting the given values (and $m = 1/4$ from Prob.17) we find

    $\gamma = \dfrac{(1/2)(3)}{.3} = 5$ lb sec/ft.

24. From Eq.(13) $\omega_0^2 = \dfrac{2k}{3}$ so $P = 2\pi/\sqrt{2k/3} = \pi \Rightarrow k = 6$.

    Thus $u(t) = c_1\cos 2t + c_2\sin 2t$ and $u(0) = 2 \Rightarrow c_1 = 2$ and $u'(0) = v \Rightarrow c_2 = v/2$. Hence

    $u(t) = 2\cos 2t + \dfrac{v}{2}\sin 2t = \sqrt{4 + \dfrac{v^2}{4}}\cos(2t - \gamma)$.

    Thus $\sqrt{4 + \dfrac{v^2}{4}} = 3$ and $v = \pm 2\sqrt{5}$.

27. First, consider the static case (which is the equilibrium position). Let Δl denote the length of the block below the surface of the water. The weight of the block, which is a downward force, is $w = \rho l^3 g$. This is balanced by an equal and opposite buoyancy force B, which is equal to the weight of the

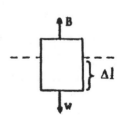

displaced water. Thus $B = (\rho_0 l^2 \Delta l)g = \rho l^3 g$. Now let u(t) be the displacement of the block from its equilibrium position. We take downward as the positive direction. In a displaced position the forces acting on the block are its weight, which acts downward and is unchanged, and the buoyancy force which is now $\rho_0 l^2(\Delta l + u)g$ and acts upward. The resultant force must be equal to the mass of the block times the acceleration, namely $\rho l^3 u''$. Hence $\rho l^3 g - \rho_0 l^2(\Delta l + u)g = \rho l^3 u''$. Hence the D.E. for the motion of the block is $\rho l^3 u'' + \rho_0 l^2 gu = 0$ or $u'' + \dfrac{\rho_0 g}{\rho l}u = 0$. This gives a simple harmonic motion with frequency $(\rho_0 g/\rho l)^{1/2}$ and natural period $T = 2\pi(\rho l/\rho_0 g)^{1/2}$.

29a. The characteristic equation is $4r^2 + r + 8 = 0$, so $r = (-1 \pm \sqrt{127})/8$ and hence
$u(t) = e^{-t/8}(c_1 \cos \dfrac{\sqrt{127}}{8}t + c_2 \sin \dfrac{\sqrt{127}}{8}t)$. $u(0) = 0 \Rightarrow c_1 = 0$ and $u'(0) = 2 \Rightarrow \dfrac{\sqrt{127}}{8}c_2 = 2$. Thus
$u(t) = \dfrac{16}{\sqrt{127}}e^{-t/8}\sin \dfrac{\sqrt{127}}{8}t$.

29c. The phase plot is the spiral shown and the direction of motion is clockwise since the graph starts at (0,2) and u increases initially.

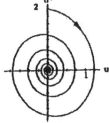

30a. The kinetic energy of a mass is given by $\dfrac{1}{2}mv^2$, so at t = 0 we have $v = u'(0) = b$ and thus $\dfrac{1}{2}mb^2$ is the initial kenetic energy. The work done deforming a spring an amount y from its undeformed state is stored in the spring and is known as the elastic potential energy. For our example, then, the potential energy is given by $\displaystyle\int_0^x F dy = \int_0^x ky dy = \dfrac{1}{2}kx^2$. For x = u(0) = a, this becomes $\dfrac{1}{2}ka^2$, as the initial potential energy.

30c. From part (a), the total energy in the system is
$ku^2/2 + m(u')^2/2$. Using $u(t)$ as found in part(b), calculate
u' and show that $ku^2/2 + m(u')^2/2 = (ka^2 + mb^2)/2$ for all t.
This confirms the principle of conservation of energy when
there is no damping.

Section 3.8, Page 215

1. We use the trigonometric identities
   $$\cos(A + B) = \cos A \cos B - \sin A \sin B$$
   $$\cos(A - B) = \cos A \cos B + \sin A \sin B$$
   to obtain $\cos(A + B) - \cos(A - B) = -2\sin A \sin B$. If we
   choose $A + B = 9t$ and $A - B = 7t$, then $A = 8t$ and $B = t$.
   Substituting in the formula just derived, we obtain
   $\cos 9t - \cos 7t = -2\sin 8t \sin t$.

5. The mass $m = 4/32 = 1/8$ lb-sec^2/ft and the spring
   constant $k = 4/(1/8) = 32$ lb/ft. Since there is no
   damping, the I.V.P. is $(1/8)u'' + 32u = 2\cos 3t$,
   $u(0) = 1/6$, $u'(0) = 0$ where u is measured in ft and t in
   sec.

7a. From Prob. 5, we have $m = 1/8$, $F_0 = 2$, $\omega_0^2 = 256$, and
   $\omega^2 = 9$, so Eq.(18) becomes
   $u = c_1\cos 16t + c_2\sin 16t + \dfrac{16}{247} \cos 3t$. The I.C.
   $u(0) = 1/6 \Rightarrow c_1 + 16/247 = 1/6$ and $u'(0) = 0 \Rightarrow 16c_2 = 0$,
   and thus $u = (151/1482)\cos 16t + (16/247)\cos 3t$ ft.

7b.

7c. Resonance occurs when the frequency ω of the forcing
   function $4\sin\omega t$ is the same as the natural frequency ω_0
   of the system. Since $\omega_0 = 16$, the system will resonate
   when $\omega = 16$ rad/sec.

10. The I.V.P. is .25u" + 16u = 8sin8t, or u" + 64u = 32sin8t,

u(0) = $\dfrac{1}{4}$ and u'(0) = 0. Since u_c = $c_1$cos8t + $c_2$sin8t, we

assume the form U(t) = t(Acos8t + Bsin8t) and resonance
occurs. Substituting U into the D.E. we find
U" + 64U = -16Asin8t + 16Bcos8t and thus A = -2, B = 0.
Therefore u(t) = $c_1$cos8t + $c_2$sin8t - 2tcos8t. The I.C.
yield c_1 = 1/4 and 8c_2 - 2 = 0 and hence
u(t) = (cos8t + sin8t - 8tcos8t)/4. The velocity will be
zero when u'(t) = 8sin8t(8t-1) = 0. Note that 8t-1 = 0
gives the first zero and sin8t gives all the others.

11a. For this problem the mass m = 8/32 lb-sec^2/ft and the
spring constant k = 8/(1/2) = 16 lb/ft, so the D.E. is
0.25u" + 0.25u' + 16u = 4cos2t where u is measured in ft
and t in sec. To determine the steady state response we
need only compute a particular solution of the
nonhomogeneous D.E. since the solutions of the
homogeneous D.E. decay to zero as t → ∞. We assume
u(t) = Acos2t + Bsin2t, and substitute in the D.E.:
- Acos2t - Bsin2t + (1/2)(-Asin2t + Bcos2t) + 16(Acos2t +
Bsin2t) = 4cos2t. Hence 15A + (1/2)B = 4 and
-(1/2)A + 15B = 0, from which we obtain A = 240/901 and
B = 8/901. Thus U(t) = (240cos2t + 8sin2t)/901.

11b. In order to determine the value of m that maximizes the
steady state response, we note that the present problem
has exactly the form of Eq.(8) considered in the text.
Referring to Eqs.(11) and (12), the magnitude of the
response, R, is a maximum when Δ is a minimum since F_0 is
constant. Δ, as given in Eq.(12), will be a minimum when
f(m) = $m^2(\omega_0{}^2 - \omega^2)^2 + \gamma^2\omega^2$, where $\omega_0{}^2$ = k/m, is a
minimum. We calculate df/dm and set this quantity equal
to zero to obtain m = k/ω^2. We verify that this value of
m gives a minimum of f(m) by the second derivative test.
For this problem k = 16 lb/ft and ω = 2 rad/sec so the
value of m that maximizes the response of the system is
m = 4 slugs.

14. Since U(t) = Rcos(ωt-δ) we have U'(t) = $\dfrac{-F_0\omega}{\Delta}$sin(ωt-δ),

where Δ is given by Eq.(12). Since F_0 is a constant,

differentiate $\dfrac{\omega}{\Delta}$ with respect to ω and set it equal to

zero. Alternatively, you can minimize $(\Delta/\omega)^2$, which
simplifies the differentiation.

15. We must solve the three I.V.P.: $(1) u_1'' + u_1 = F_0 t$,

$0 < t < \pi$, $u_1(0) = u_1'(0) = 0$; (2) $u_2'' + u_2 = F_0(2\pi-t)$,

$\pi < t < 2\pi$, $u_2(\pi) = u_1(\pi)$, $u_2'(\pi) = u_1'(\pi)$; and

(3) $u_3'' + u_3 = 0$, $2\pi < t$, $u_3(2\pi) = u_2(2\pi)$, $u_3'(2\pi) = u_2'(2\pi)$.
The conditions at π and 2π insure the continuity of u and
u' at those points. The general solutions of the D.E.
are $u_1 = b_1 \cos t + b_2 \sin t + F_0 t$, $u_2 = c_1 \cos t + c_2 \sin t + F_0(2\pi-t)$, and $u_3 = d_1 \cos t + d_2 \sin t$. The I.C. and matching
conditions, in order, give $b_1 = 0$, $b_2 + F_0 = 0$,
$-b_1 + \pi F_0 = -c_1 + \pi F_0$, $-b_2 + F_0 = -c_2 - F_0$, $c_1 = d_1$, and
$c_2 - F_0 = d_2$. [Note that the form of the particular
solution for the first two cases is not the most general,
but they do yield a solution]. Solving these equations we
obtain

$$u = F_0 \begin{cases} t - \sin t & , \ 0 \leq t \leq \pi \\ (2\pi - t) - 3\sin t & , \ \pi < t \leq 2\pi \\ - 4\sin t & , \ 2\pi < t. \end{cases}$$

16. From Eq.(33) of Sect. 3.7, the I.V.P. is
$Q'' + 5 \times 10^3 Q' + 4 \times 10^6 Q = 12$, $Q(0) = 0$, and $Q'(0) = 0$. The
particular solution is of the form $Q = A$, so that upon
substitution into the D.E. we obtain $4 \times 10^6 A = 12$ or
$A = 3 \times 10^{-6}$. The general solution of the D.E. is
$Q = c_1 e^{r_1 t} + c_2 e^{r_2 t} + 3 \times 10^{-6}$, where r_1 and r_2 satisfy
$r^2 + 5 \times 10^3 r + 4 \times 10^6 = 0$ and thus $r_1 = -1000$ and
$r_2 = -4000$. The I.C. yield $c_1 = -4 \times 10^{-6}$ and $c_2 = 10^{-6}$ and
thus $Q = 10^{-6}(e^{-4000t} - 4e^{-1000t} + 3)$ coulombs. Substituting
t = .001 sec we obtain
$Q(.001) = 10^{-6}(e^{-4} - 4e^{-1} + 3) = 1.5468 \times 10^{-6}$ coulombs.
Since the exponentials are to a negative power
$Q(t) \to 3 \times 10^{-6}$ coulombs as $t \to \infty$, which is the steady
state charge.

22. The steady-state response is 12sin2t and thus the
amplitude of the steady state response is four times the
amplitude of the forcing term. This large an increase is
due to the fact that the forcing function has the same
frequency as the natural frequency, $\omega_0 = 2$, of the
system. The graph also shows a phase lag of approximately
1/4 of a period. That is, the maximum of the response
occurs 1/4 of a period after the maximum of the forcing

function. Both these results are substantially different
than those of either Probs. 21 or 23.

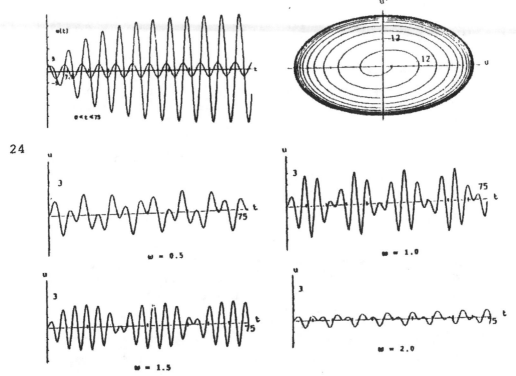

24

From viewing the above graphs, it appears that the system
exhibits a beat near ω = 1.5, while the pattern for
ω = 1.0 is more irregular. However, the system exhibits
the resonance characteristic of the linear system for ω
near 1, as the amplitude of the response is the largest
here.

CHAPTER 4

Section 4.1, Page 224

2. Writing the equation in standard form of Eq.(2), we
    obtain $y''' + [(\sin t)/t]y'' + (3/t)y = \cos t/t$. The
    functions $p_1(t) = \sin t/t$, $p_3(t) = 3/t$ and $g(t) = \cos t/t$
    each have a discontinuity at $t = 0$. Hence Theorem 4.1.1
    guarantees that a solution exists for $t < 0$ or for $t > 0$.

4. The equation is in standard form with $p_1(t)$, $p_2(t)$ and
    $p_3(t)$ being continuous for all t. However, $g(t) = \ln t$ is
    defined and continuous only for $t > 0$.

8. We have $W(f_1,f_2,f_3) = \begin{vmatrix} 2t-3 & 2t^2+1 & 3t^2+t \\ 2 & 4t & 6t+1 \\ 0 & 4 & 6 \end{vmatrix} = 0$ for all t.

    Thus by the extension of Theorem 3.3.1 (or by the
    discussion following Eq.(8), the given functions are
    linearly dependent. To find a linear relation we have
    $c_1(2t-3) + c_2(2t^2+1) + c_3(3t^2+t) =$
    $(2c_2+3c_3)t^2 + (2c_1+c_3)t + (-3c_1+c_2)$ which is zero when
    $(2c_2+3c_3) = 0$, $2c_1+c_3 = 0$ and $-3c_1+c_2 = 0$. Solving, we find
    $c_1 = 1$, $c_2 = 3$ and $c_3 = -2$ and hence $(2t-3)+3(2t^2+1)-2(3t^2+t) = 0$.

13. That e^t, e^{-t}, and e^{-2t} are solutions can be verified by
    direct substitution. For $y = e^{-2t}$, $y' = -2e^{-2t}$,
    $y'' = 4e^{-2t}$, and $y''' = -8e^{-2t}$ and thus
    $(-8e^{-2t}) + 2(4e^{-2t}) - (-2e^{-2t}) - 2(e^{-2t}) = 0$, verifying
    that e^{-2t} is a solution. Computing the Wronskian we
    obtain,

    $$W(e^t,e^{-t},e^{-2t}) = \begin{vmatrix} e^t & e^{-t} & e^{-2t} \\ e^t & -e^{-t} & -2e^{-2t} \\ e^t & e^{-t} & 4e^{-2t} \end{vmatrix} = e^{-2t}\begin{vmatrix} 1 & 1 & 1 \\ 1 & -1 & -2 \\ 1 & 1 & 4 \end{vmatrix} = -6e^{-2t}$$

17. To show that the given Wronskian is zero, it is helpful,
    in evaluating the Wronskian, to note that
    $(\sin^2 t)' = 2\sin t\cos t = \sin 2t$.
    The result can also be obtained directly since
    $\sin^2 t = (1 - \cos 2t)/2 = \frac{1}{10}(5) + (-1/2)\cos 2t$ and hence

$\sin^2 t$ is a linear combination of 5 and $\cos 2t$. Thus the functions are linearly dependent and their Wronskian is zero.

19a. If $y = t^n$ then $y' = nt^{n-1}$, $y'' = n(n-1)t^{n-2}$, $\cdots$
$y^{n-1} = [n(n-1)(n-2)\cdots 2]t$, so $y^{(n)} = [n(n-1)(n-2)\cdots(2)(1)]$.

19b. The nth derivative of e^{rt} is $y^{(n)} = r^n e^{rt}$.

19c. If we let $L[y] = y^{(4)} - 5y'' + 4y$ and if we use the result of Prob. 19b, we have $L[e^{rt}] = (r^4 - 5r^2 + 4)e^{rt}$. Thus e^{rt} will be a solution of the D.E. provided $(r^2-4)(r^2-1) = 0$. Solving for r, we obtain the four solutions e^t, e^{-t}, e^{2t} and e^{-2t}. Since $W(e^t, e^{-t}, e^{2t}, e^{-2t}) \neq 0$, the four functions form a fundamental set of solutions.

23. Writing the equation in the form of Eq.(2), we have
$$p_1(t) = \frac{2}{t} \text{ and from Prob. 20, } W = ce^{-\int \frac{2dt}{t}} = \frac{c}{t^2}.$$

25a. On $0 < t < 1$, $f(t) = t^3$ and $g(t) = t^3$. Hence there are nonzero constants, $c_1 = 1$ and $c_2 = -1$, such that $c_1 f(t) + c_2 g(t) = 0$ for each t in $(0,1)$. On $-1 < t < 0$, $f(t) = -t^3$ and $g(t) = t^3$; thus $c_1 = c_2 = 1$ defines constants such that $c_1 f(t) + c_2 g(t) = 0$ for each t in $(-1,0)$. Thus f and g are linearly dependent on $0 < t < 1$ and on $-1 < t < 0$.

25b. We will show that $f(t)$ and $g(t)$ are linearly independent on $-1 < t < 1$ by demonstrating that it is impossible to find constants c_1 and c_2, not both zero, such that $c_1 f(t) + c_2 g(t) = 0$ for all t in $(-1,1)$. Assume that there are two such nonzero constants and choose two points t_0 and t_1 in $-1 < t < 1$ such that $t_0 < 0$ and $t_1 > 0$. Then $-c_1 t_0^3 + c_2 t_0^3 = 0$ and $c_1 t_1^3 + c_2 t_1^3 = 0$. These equations have a nontrivial solution for c_1 and c_2 only if the determinant of coefficients is zero. But the determinant of coefficients is $-2t_0^3 t_1^3 \neq 0$ for t_0 and t_1 as specified. Hence $f(t)$ and $g(t)$ are linearly independent on $-1 < t < 1$.

25c. For $-1 < t < 0$, $W(f,g) = \begin{vmatrix} -t^3 & t^3 \\ -3t^2 & 3t^2 \end{vmatrix} = 0$ and for $0 \leq t < 1$,

$W(f,g) = \begin{vmatrix} t^3 & t^3 \\ 3t^2 & 3t^2 \end{vmatrix} = 0$. This shows that f and g cannot

be solutions of an equation $y'' + p(t)y' + q(t)y = 0$ with p and q continuous on $-1 < t < 1$.

27. Differentiating e^t and substituting in the D.E. we verify that $y = e^t$ is a solution: $(2-t)e^t + (2t-3)e^t - te^t + e^t = 0$. Now, as in Prob. 26, we let $y = v(t)e^t$. Differentiating three times and substituting into the D.E. yields $(2-t)e^t v''' + (3-t)e^t v'' = 0$. Dividing by $(2-t)e^t$ and letting $w = v''$ we obtain the first order separable

equation $w' = -\dfrac{t-3}{t-2}w = (-1 + \dfrac{1}{t-2})w$. Separating t and w,

integrating, and then solving for w yields $w = v'' = c_1(t-2)e^{-t}$. Integrating this twice then gives $v = c_1 t e^{-t} + c_2 t + c_3$ so that $y = ve^t = c_1 t + c_2 t e^t + c_3 e^t$, which is the complete solution, since it contains the given $y_1(t)$ and three constants.

Section 4.2, Page 231

2. If $-1 + i\sqrt{3} = R(\cos\theta + i\sin\theta) = Re^{i\theta}$, then $R\cos\theta = -1$ and $R\sin\theta = \sqrt{3}$. Thus $R^2 = (-1)^2 + (\sqrt{3})^2 = 4$ and the angle θ is given by $R\cos\theta = 2\cos\theta = -1$ and $R\sin\theta = 2\sin\theta = \sqrt{3}$. Hence $\cos\theta = -1/2$ and $\sin\theta = \sqrt{3}/2$ which has the solution $\theta = 2\pi/3$. The angle θ is only determined up to an additive integer multiple of $\pm 2\pi$.

8. Writing $(1-i)$ in the form $Re^{i\theta}$, we have $R\cos\theta = 1$ and $R\sin\theta = -1$, which yield $R = \sqrt{2}$ and $\theta = -\pi/4$. Thus $(1-i) = \sqrt{2}\,e^{i(-\pi/4+2m\pi)}$ (where m is any integer) and hence $(1-i)^{1/2} = [2^{1/2}e^{i(-\pi/4+2m\pi)}]^{1/2} = 2^{1/4}e^{i(-\pi/8+m\pi)}$. We obtain the two square roots by setting $m = 0,1$. They are $2^{1/4}e^{-i\pi/8}$ and $2^{1/4}e^{i7\pi/8}$. Note that any other integer value of m gives one of these two values. Also note that $1-i$ could be written as $1-i = \sqrt{2}\,e^{i(7\pi/4 + 2m\pi)}$.

12. We look for solutions of the form $y = e^{rt}$. Substituting
    in the D.E., we obtain the characteristic equation
    $r^3 - 3r^2 + 3r - 1 = 0$ which has roots $r = 1,1,1$. Since
    the roots are repeated, the general solution is
    $y = c_1 e^t + c_2 t e^t + c_3 t^2 e^t$.

15. We look for solutions of the form $y = e^{rt}$. Substituting
    in the D.E. we obtain the characteristic equation
    $r^6 + 1 = 0$. The six roots of $-1 = e^{i\pi}$ are obtained by
    setting $m = 0,1,2,3,4,5$ in $(-1)^{1/6} = e^{i(\pi+2m\pi)/6}$. They are
    $e^{i\pi/6} = (\sqrt{3} + i)/2$, $e^{i\pi/2} = i$, $e^{i5\pi/6} = (-\sqrt{3} + i)/2$,
    $e^{i7\pi/6} = (-\sqrt{3} - i)/2$, $e^{i3\pi/2} = -i$, and
    $e^{i11\pi/6} = (\sqrt{3} - i)/2$. Note that there are three pairs of
    conjugate roots. The general solution is
    $$y = e^{\sqrt{3}t/2}[c_1\cos(t/2) + c_2\sin(t/2)]$$
    $$+ e^{-\sqrt{3}t/2}[c_3\cos(t/2) + c_4\sin(t/2)] + c_5\cos t + c_6\sin t.$$

23. The characteristic equation is $r^3 - 5r^2 + 3r + 1 = 0$.
    Using the procedure suggested following Eq. (12) we try,
    since $a_0 = a_n = 1$, $r = 1$ as a root and find that indeed
    it is. Factoring out $(r-1)$ we are then left with
    $r^2 - 4r - 1 = 0$, which has the roots $2 \pm \sqrt{5}$.

27. The characteristic equation in this case is
    $12r^4 + 31r^3 + 75r^2 + 37r + 5 = 0$. Using an equation
    solver we find $r = -\dfrac{1}{4}$, $-\dfrac{1}{3}$, $-1 \pm 2i$. Thus
    $y = c_1 e^{-t/4} + c_2 e^{-t/3} + e^{-t}(c_3\cos 2t + c_4\sin 2t)$. As in
    Prob. 23, it is possible to find the first two of these
    roots without using an equation solver. Factoring then
    reduces the characteristic equation to a quadratic, which
    can be solved for the other two roots.

29. The characteristic equation is $r^3 + r = 0$ and hence
    $r = 0, +i, -i$ are the roots and the general solution is
    $y(t) = c_1 + c_2\cos t + c_3\sin t$. $y(0) = 0$ implies
    $c_1 + c_2 = 0$, $y'(0) = 1$ implies $c_3 = 1$ and $y''(0) = 2$
    implies $-c_2 = 2$. Use this last equation in the first to
    find $c_1 = 2$ and thus $y(t) = 2 - 2\cos t + \sin t$, which
    continues to oscillate about $y = 2$ as $t \to \infty$.

29.   30.

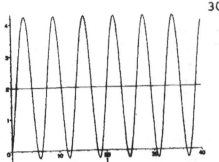
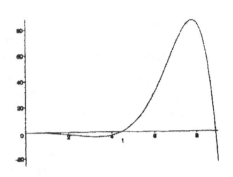

30. The general solution is given by Eq. (21).

31. The characteristic equation is $r^4 - 4r^3 + 4r^2 = 0$, which
    has the roots $r = 0,0,2,2$. Thus the general solution
    would normally be written $y(t) = c_1 + c_2t + c_3e^{2t} + c_4te^{2t}$.
    However, in order to evaluate the c's when the initial
    conditions are given at $t = 1$, it is advantageous to
    rewrite this as $y(t) = c_1 + c_2t + c_5e^{2(t-1)} + c_6(t-1)e^{2(t-1)}$,
    which also satisfies the given D.E.

31.   34.

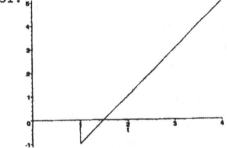
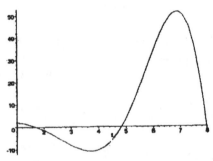

34. The characteristic equation is $4r^3 + r + 5 = 0$, which has
    roots -1, $\frac{1}{2} \pm i$, where $r_1 = -1$ can be found as in
    Prob.23. Thus $y(t) = c_1e^{-t} + e^{t/2}(c_2\cos t + c_3\sin t)$,
    $y'(t) = -c_1e^{-t} + e^{t/2}[(c_2/2 + c_3)\cos t + (-c_2 + c_3/2)\sin t]$
    and
    $y''(t) = c_1e^{-t} + e^{t/2}[(-3c_2/4 + c_3)\cos t + (-c_2 - 3c_3/4)\sin t]$.
    The I.C. then yield $c_1 + c_2 = 2$, $-c_1 + c_2/2 + c_3 = 1$ and
    $c_1 - 3c_2/4 + c_3 = -1$. Solving these last three equations
    give $c_1 = 2/13$, $c_2 = 24/13$ and $c_3 = 3/13$.

37. The approach for solving the D.E. would normally yield
    $y(t) = c_1\cos t + c_2\sin t + c_5e^t + c_6e^{-t}$ as the solution.

Since $\cosh t = (e^t + e^{-t})/2$ and $\sinh t = (e^t - e^{-t})/2$, $y(t)$ can be written as $y(t) = c_1 \cos t + c_2 \sin t + c_3 \cosh t + c_4 \sinh t$, where c_3 and c_4 can be written in terms of c_5 and c_6. It is convenient to use $\cosh t$ and $\sinh t$ rather than e^t and e^{-t} because the I.C. are given at $t = 0$. Since $\cosh t$ and $\sinh t$ and all of their derivatives are either 0 or 1 at $t = 0$, the algebra in satisfying the I.C. is greatly simplified.

38a. Since $p_1(t) = 0$, $W = ce^{-\int 0 dt} = c$.

39a. As in Sect.3.7, the force that the spring designated by k_1 exerts on mass m_1 is $-3u_1$. By an analysis similar to that shown in Sect.3.7, the middle spring exerts a force of $-2(u_1 - u_2)$ on mass m_1 and a force of $-2(u_2 - u_1)$ on mass m_2. Thus Newton's Law gives $m_1 u_1'' = -3u_1 - 2(u_1 - u_2)$ and $m_2 u_2'' = -2(u_2 - u_1)$, where u_1 and u_2 are measured from their equilibrium positions. Setting the masses equal to 1 and rewriting each equation yields Eqs.(i). In all cases the positive direction is taken in the direction shown in Figure 4.2.4.

39b. The characteristic equation for (ii) is $r^4 + 7r^2 + 6 = 0$, or $(r^2 + 1)(r^2 + 6) = 0$. Thus the general solution of Eq.(ii) is $u_1(t) = c_1 \cos t + c_2 \sin t + c_3 \cos \sqrt{6}\,t + c_4 \sin \sqrt{6}\,t$.

39c. From Eq.(iii) and u_1 from 39b we have $u_1(0) = c_1 + c_3 = 1$ and $u_1'(0) = c_2 + \sqrt{6}\,c_4 = 0$. From Eq.(i) we have $u_1''(0) = 2u_2(0) - 5u_1(0) = -1$ and $u_1'''(0) = 2u_2'(0) - 5u_1'(0) = 0$, hence $-c_1 - 6c_3 = -1$ and $-c_2 - 6\sqrt{6}\,c_4 = 0$. Solving the four equations for the c_i we find $c_1 = 1$ and $c_2 = c_3 = c_4 = 0$, so that $u_1 = \cos t$. The first of Eqs.(i) then gives $2u_2 = u_1'' + 5u_1 = 4\cos t$ and thus $u_2 = 2\cos t$.

Section 4.3, Page 237

1. First solve the homogeneous D.E. The characteristic equation is $r^3 - r^2 - r + 1 = 0$, and the roots are $r = -1$, 1, 1; hence $y_c(t) = c_1 e^{-t} + c_2 e^t + c_3 t e^t$. Using the

superposition principle, we can write a particular solution as the sum of particular solutions corresponding to the D.E. $y''' - y'' - y' + y = 2e^{-t}$ and $y''' - y'' - y' + y = 3$. Our initial choice for a particular solution, Y_1, of the first equation is Ae^{-t}; but e^{-t} is a solution of the homogeneous equation so we multiply by t. Thus, $Y_1(t) = Ate^{-t}$. For the second equation we choose $Y_2(t) = B$, and there is no need to modify this choice. The constants are determined by substituting into the individual equations. We obtain $A = 1/2$, $B = 3$. Thus, the general solution is

$$y = c_1 e^{-t} + c_2 e^t + c_3 te^t + (te^{-t})/2 + 3.$$

5.   The characteristic equation is $r^4 - 4r^2 = r^2(r^2 - 4) = 0$, so $y_c(t) = c_1 + c_2 t + c_3 e^{-2t} + c_4 e^{2t}$. For the particular solution correspnding to t^2 we assume $Y_1 = t^2(At^2 + Bt + C)$ and for the particular solution corresponding to e^t we assume $Y_2 = De^t$. Substituting Y_1, in the D.E. yields $-48A = 1$, $B = 0$ and $24A - 8C = 0$ and substituting Y_2 yields $-3D = 1$. Solving for A, B, C and D gives the desired solution.

9.   The characteristic equation for the related homogeneous D.E. is $r^3 + 4r = 0$ with roots $r = 0, +2i, -2i$. Hence $y_c(t) = c_1 + c_2\cos 2t + c_3\sin 2t$. The initial choice for $Y(t)$ is $At + B$, but since B is a solution of the homogeneous equation we must multiply by t and assume $Y(t) = t(At+B)$. A and B are found by substituting in the D.E., which gives $A = 1/8$, $B = 0$, and thus the general solution is $y(t) = c_1 + c_2\cos 2t + c_3\sin 2t + (1/8)t^2$. Applying the I.C. we have $y(0) = 0 \Rightarrow c_1 + c_2 = 0$, $y'(0) = 0 \Rightarrow 2c_3 = 0$, and $y''(0) = 1 \Rightarrow -4c_2 + 1/4 = 1$,

which have the solution $c_1 = 3/16$, $c_2 = -3/16$, $c_3 = 0$. For $t = \pi$, 2π... the graph will be tangent to $t^2/8$ and for large t the graph will be approximated by $t^2/8$.

13. The characteristic equation for the homogeneous D.E. is $r^3 - 2r^2 + r = 0$ with roots $r = 0,1,1$. Hence the complementary solution is $y_c(t) = c_1 + c_2e^t + c_3te^t$. We consider the differential equations $y''' - 2y'' + y' = t^3$ and $y''' - 2y'' + y' = 2e^t$ separately. Our initial choice for a particular solution, Y_1, of the first equation is $A_0t^3 + A_1t^2 + A_2t + A_3$; but since a constant is a solution of the homogeneous equation we must multiply by t. Thus $Y_1(t) = t(A_0t^3 + A_1t^2 + A_2t + A_3)$. For the second equation we first choose $Y_2(t) = Be^t$, but since both e^t and te^t are solutions of the homogeneous equation, we multiply by t^2 to obtain $Y_2(t) = Bt^2e^t$. Then $Y(t) = Y_1(t) + Y_2(t)$ by the superposition principle and $y(t) = y_c(t) + Y(t)$.

17. The characteristic equation is $r^4 - r^3 - r^2 + r = r(r-1)(r^2-1) = 0$, so the complementary solution is $y_c(t) = c_1 + c_2e^{-t} + c_3e^t + c_4te^t$. The superposition principle allows us to consider separately the D.E. $y^{(4)} - y''' - y'' + y' = t^2 + 4$ and $y^{(4)}s - y''' - y'' + y' = t\sin t$. For the first equation our initial choice is $Y_1(t) = A_0t^2 + A_1t + A_2$; but this must be multiplied by t since a constant is a solution of the homogeneous D.E. Hence $Y_1(t) = t(A_0t^2 + A_1t + A_2)$. For the second equation our initial choice that $Y_2 = (B_0t + B_1)\cos t + + (C_0t + C_1)\sin t$ does not need to be modified. Hence
$$Y(t) = t(A_0t^2 + A_1t + A_2) + (B_0t + B_1)\cos t + (C_0t + C_1)\sin t.$$

20. $(D-a)(D-b)f = (D-a)(Df-bf) = D^2f - (a+b)Df + abf$ and $(D-b)(D-a)f = (D-b)(Df-af) = D^2f - (b+a)Df + baf$. Since $a+b = b+a$ and $ab = ba$, we find the given equation holds for any function f.

22a. The D.E. of Prob.13 can be written as $D(D-1)^2y = t^3 + 2e^t$. Since D^4 annihilates t^3 and $(D-1)$ annihilates $2e^t$, we have $D^5(D-1)^3y = 0$, which corresponds to Eq.(ii) of Prob. 21. The solution of this equation is $y(t) = A_1t^4 + A_2t^3 + A_3t^2 + A_4t + A_5 + (B_1t^2 + B_2t + B_3)e^{-t}$. Since A_5 and $(B_2t + B_3)e^{-t}$ are solutions of the homogeneous equation related to the original D.E., they

may be deleted and thus
$$Y(t) = A_1t^4 + A_2t^3 + A_3t^2 + A_4t + B_1t^2e^{-t}.$$

22b. If $y = te^{-t}$ then $Dy = -te^{-t} + e^{-t}$ and $D^2y = te^{-t} - 2e^{-t}$,
so $(D+1)^2y = (D^2+2D+1)y = 0$ and thus $(D+1)^2$ annihilates
te^{-t}. Likewise D^2-1 annihilates $2\cos t$. Thus $(D+1)^2(D^2+1)$
annihilates the right side of the D.E. of Prob.14.

22e. $D^3(D^2+1)^2$ annihilates the right side of the D.E.of Prob.17.

Section 4.4, Page 242

1. The complementary solution is $y_c = c_1 + c_2\cos t + c_3\sin t$
    and thus we assume a particular solution of the form
    $Y = u_1(t) + u_2(t)\cos t + u_3(t)\sin t$. Differentiating and
    assuming Eq.(5), we obtain $Y' = -u_2\sin t + u_3\cos t$ and
    $$u_1' + u_2'\cos t + u_3'\sin t = 0 \qquad (a).$$
    Continuing this process we obtain $Y'' = -u_2\cos t - u_3\sin t$,
    and
    $$-u_2'\sin t + u_3'\cos t = 0 \qquad\qquad (b)$$
    and $Y''' = u_2\sin t - u_3\cos t - u_2'\cos t - u_3'\sin t$.
    Substituting Y and its derivatives, as given above, into
    the D.E. we obtain the third equation:
    $$-u_2'\cos t - u_3'\sin t = \tan t \qquad (c).$$
    Equations (a), (b) and (c) constitute Eqs.(10) of the
    text for this problem and may be solved to give
    $u_1' = \tan t$, $u_2' = -\sin t$, and $u_3' = -\sin^2t/\cos t$. Thus
    $u_1 = -\ln\cos t$, $u_2 = \cos t$ and $u_3 = \sin t - \ln(\sec t + \tan t)$
    and substitution into Y above gives
    $Y = -\ln\cos t + 1 - (\sin t)\ln(\sec t + \tan t)$, since
    $\sin^2t + \cos^2t = 1$ Note that the constant 1 can be
    absorbed in c_1 in y_c above.

4. Replace $\tan t$ in Eq.(c) of Prob. 1 by $\sec t$ and use
    Eqs.(a) and (b) as in Prob.1 to obtain $u_1' = \sec t$,
    $u_2' = -1$ and $u_3' = -\sin t/\cos t$.

5. Replace $\sec t$ in Prob. 7 with $e^{-t}\sin t$.

7. Since e^t, cost and sint are solutions of the related homogenous equation we have

$Y(t) = u_1 e^t + u_2 \cos t + u_3 \sin t$. Eqs. (10) then are

$u_1' e^t + u_2' \cos t + u_3' \sin t = 0$

$u_1' e^t - u_2' \sin t + u_3' \cos t = 0$

$u_1' e^t - u_2' \cos t - u_3' \sin t = \sec t$.

Using Abel's identity, $W(t) = c\exp(-\int p_1(t)dt) = ce^t$.

Using the above equations, $W(0) = \begin{vmatrix} 1 & 1 & 0 \\ 1 & 0 & 1 \\ 1 & -1 & 0 \end{vmatrix} = 2$, so

$c = 2$ and $W(t) = 2e^t$. From Eq.(11), we have

$u'_1(t) = \dfrac{\sec t\, W_1(t)}{2e^t}$, where $W_1 = \begin{vmatrix} 0 & \cos t & \sin t \\ 0 & -\sin t & \cos t \\ 1 & -\cos t & -\sin t \end{vmatrix} = 1$

and thus

$u_1'(t) = \dfrac{1}{2}e^{-t}/\cos t$. Likewise

$u_2' = \dfrac{\sec t\, W_2(t)}{2e^t} = -\dfrac{1}{2}\sec t(\cos t - \sin t)$ and

$u_3' = \dfrac{\sec t\, W_3(t)}{2e^t} = -\dfrac{1}{2}\sec t(\sin t + \cos t)$. Thus

$u_1 = \dfrac{1}{2}\int_{t_0}^{t} \dfrac{e^{-s}ds}{\cos(s)}$, $u_2 = -\dfrac{1}{2}t - \dfrac{1}{2}\ln(\cos t)$ and $u_3 = -\dfrac{1}{2}t + \dfrac{1}{2}\ln(\cos t)$

which, when substituted into the assumed form for Y, yields the desired solution.

11. Since the D.E. is the same as in Prob. 7, we may use the complete solution from that, with $t_0 = 0$. Thus

$y(0) = c_1 + c_2 = 2$, $y'(0) = c_1 + c_3 - \dfrac{1}{2} + \dfrac{1}{2} = -1$ and

$y''(0) = c_1 - c_2 + \dfrac{1}{2} - 1 + \dfrac{1}{2} = 1$.

A computer algebra system may be used to find the respective derivatives. Note that the solution is valid only for

$0 \le t < \dfrac{\pi}{2}$, where we see

the vertical asymptote.

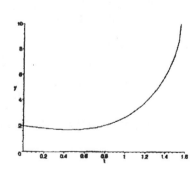

14. Since a fundamental set of solutions of the homogeneous
    D.E. is $y_1 = e^t$, $y_2 = \cos t$, $y_3 = \sin t$, a particular
    solution is of the form

    $Y(t) = e^t u_1(t) + (\cos t)u_2(t) + (\sin t)u_3(t)$. Differentiating
    and making the same assumptions that lead to Eqs.(10), we
    obtain

    $$u_1'e^t + u_2'\cos t + u_3'\sin t = 0$$
    $$u_1'e^t - u_2'\sin t + u_3'\cos t = 0$$
    $$u_1'e^t - u_2'\cos t - u_3'\sin t = g(t)$$

    Solving these equations using either determinants or by
    elimination, we obtain $u_1' = (1/2)e^{-t}g(t)$,

    $u_2' = (1/2)(\sin t - \cos t)g(t), u_3' = -(1/2)(\sin t + \cos t)g(t)$.
    Integrating these and substituting into Y yields

    $$Y(t) = \frac{1}{2}\{e^t\int_{t_0}^{t} e^{-s}g(s)ds + \cos t \int_{t_0}^{t}[\sin(s) - \cos(s)]g(s)ds$$
    $$-\sin t\int_{t_0}^{t}[\sin(s) + \cos(s)]g(s)ds\}.$$

    Putting e^t, $\cos t$ and $\sin t$ inside the respective integrals
    yields

    $$Y(t) = (1/2)\int_{t_0}^{t}[e^{t-s} + \cos t\sin(s) - \cos t\cos(s) -$$
    $$\sin t\sin(s) - \sin t\cos(s)]g(s)ds.$$

    If we use the trigonometric identities
    $\sin(A-B) = \sin A\cos B - \cos A\sin B$ and
    $\cos(A-B) = \cos A\cos B + \sin A\sin B$, we obtain the desired
    result. Note: Eqs.(11) and (12) of this section give the
    same result, but it is not recommended to memorize these
    equations.

16. The characteristic equation has the repeated roots
    $r = 1,1,1$ and thus the particular solution has the form
    $Y = e^t u_1(t) + te^t u_2(t) + t^2 e^t u_3(t)$. Differentiating,
    making the same assumptions as in the earlier problems,
    and solving the three linear equations for u_1', u_2', and u_3'
    yields
    $u_1' = (1/2)t^2 e^{-t}g(t)$, $u_2' = -te^{-t}g(t)$ and $u_3' = (1/2)e^{-t}g(t)$.
    Integrating and substituting into Y yields the desired
    solution. For instance

    $$te^t u_2 = -te^t\int_{t_0}^{t} se^{-s}g(s)ds = -\frac{1}{2}\int_{t_0}^{t} 2tse^{(t-s)}g(s)ds, \text{ and}$$

likewise for u_1 and u_3. If $g(t) = t^{-2}e^t$ then $g(s) = e^s/s^2$,

and thus $e^{(t-s)}g(s) = \dfrac{e^t}{s^2}$ and the integration with repect

to s is accomplished using the power rule. Note that
terms involving t_0 become part of the complimentary
solution.

CHAPTER 5

Section 5.1, Page 249

2. Use the ratio test:

$$\lim_{n \to \infty} \frac{\left|(n+1)x^{n+1}/2^{n+1}\right|}{\left|nx^n/2^n\right|} = \lim_{n \to \infty} \frac{n+1}{n} \frac{1}{2} |x| = \frac{|x|}{2}.$$

Therefore the series converges absolutely for $|x| < 2$. For $x = 2$ and $x = -2$ the nth term does not approach zero as $n \to \infty$ so the series diverge. Hence the radius of convergence is $\rho = 2$.

5. Use the ratio test:

$$\lim_{n \to \infty} \frac{\left|(2x+1)^{n+1}/(n+1)^2\right|}{\left|(2x+1)^n/n^2\right|} = \lim_{n \to \infty} \frac{n^2}{(n+1)^2}|2x+1| = |2x+1|.$$

Therefore the series converges absolutely for $|2x+1| < 1$, or $|x+1/2| < 1/2$. At $x = 0$ and $x = -1$ the series also converge absolutely. However, for $|x+1/2| > 1/2$ the series diverges by the ratio test. The radius of convergence is $\rho = 1/2$.

9. For this problem $f(x) = \sin x$, so $f'(x) = \cos x$, $f''(x) = -\sin x$, $f'''(x) = -\cos x$, $f^{iv}(x) = \sin x...$, and thus $f(0) = 0$, $f'(0) = 1$, $f''(0) = 0$, $f'''(0) = -1,...$. The even terms in the series will vanish and the odd terms will alternate in sign. We obtain $\sin x = \sum_{n=0}^{\infty} (-1)^n x^{2n+1}/(2n+1)!$.

Now, $\lim_{n \to \infty} \frac{\left|(-1)^{n+1}x^{2n+3}/(2n+3)!\right|}{\left|(-1)^n x^{2n+1}/(2n+1)!\right|} = \lim_{n \to \infty} x^2 \frac{1}{(2n+3)(2n+2)} = 0,$

so the series converges for all x and hence $\rho = \infty$.

12. For this problem $f(x) = x^2$. Hence $f'(x) = 2x$, $f''(x) = 2$, and $f^{(n)}(x) = 0$ for $n > 2$. Then $f(-1) = 1$, $f'(-1) = -2$, $f''(-1) = 2$ and $x^2 = 1 - 2(x+1) + 2(x+1)^2/2! = 1 - 2(x+1) + (x+1)^2$. Since the series terminates after a finite number of terms, it converges for all x. Thus $\rho = \infty$.

13. For this problem $f(x) = \ln x$. Hence $f'(x) = 1/x$, $f''(x) = -1/x^2$, $f'''(x) = 1 \cdot 2/x^3$, ... , and $f^{(n)}(x) = (-1)^{n+1}(n-1)!/x^n$. Then $f(1) = 0$, $f'(1) = 1$, $f''(1) = -1$, $f'''(1) = 1 \cdot 2$, ... , $f^{(n)}(1) = (-1)^{n+1}(n-1)!$ The Taylor series

is $\ln x = (x-1) - (x-1)^2/2 + (x-1)^3/3 - \ldots =$

$\displaystyle\sum_{n=1}^{\infty} (-1)^{n+1}(x-1)^n/n$. It follows from the ratio test that

the series converges absolutely for $|x-1| < 1$. However, the series diverges at $x = 0$ so $\rho = 1$.

18. Writing the individual terms of y, we have

$y = a_0 + a_1 x + a_2 x^2 + \ldots + a_n x^n + \ldots$, so

$y' = a_1 + 2a_2 x + 3a_3 x^2 + \ldots + (n+1)a_{n+1}x^n + \ldots$, and

$y'' = 2a_2 + 3\cdot2a_3 x + 4\cdot3a_4 x^2 + \ldots + (n+2)(n+1)a_{n+2}x^n + \ldots$.

If $y'' = y$, we then equate coefficients of like powers of x to obtain $2a_2 = a_0$, $3\cdot2a_3 = a_1$, $4\cdot3a_4 = a_2$, $\ldots (n+2)(n+1)a_{n+2} = a_n$.

Thus $a_2 = \dfrac{a_0}{2}$, $a_3 = \dfrac{a_1}{6}$, $a_4 = \dfrac{a_2}{4\cdot3} = \dfrac{a_0}{4!}$, $\cdots a_{n+2} = \dfrac{a_n}{(n+2)(n*1)}$.

These yield the desired results for $n = 0,1,2,3, \ldots$.

19. Set $m = n-1$ on the right hand side of the equation. Then $n = m+1$ and when $n = 1$, $m = 0$. Thus the right hand side

becomes $\displaystyle\sum_{m=0}^{\infty} a_m(x-1)^{m+1}$, which is the same as the left hand

side when m is replaced by n.

23. Multiplying each term of the first series by x yields

$x\displaystyle\sum_{n=1}^{\infty} na_n x^{n-1} = \sum_{n=1}^{\infty} na_n x^n = \sum_{n=0}^{\infty} na_n x^n$, where the last equality

can be verified by writing out the first few terms (or noting that $na_n = 0$ for $n = 0$). Changing the index from k to n ($n=k$) in the second series then yields

$\displaystyle\sum_{n\neq0}^{\infty} na_n x^n + \sum_{n=0}^{\infty} a_n x^n = \sum_{n=0}^{\infty} (n+1)a_n x^n$.

25. $\displaystyle\sum_{m=2}^{\infty} m(m-1)a_m x^{m-2} + x\sum_{k=1}^{\infty} ka_k x^{k-1} = \sum_{n=0}^{\infty} (n+2)(n+1)a_{n+2}x^n + \sum_{k=1}^{\infty} ka_k x^k$

$= \displaystyle\sum_{n=0}^{\infty} [(n+2)(n+1)a_{n+2} + na_n]x^n$. In the first case we have

let $n = m - 2$ in the first summation and multiplied each term of the second summation by x. In the second case we have let $n = k$ and noted that for $n = 0$, $na_n = 0$.

28. If we shift the index of summation in the first sum by letting m = n-1, we have

$$\sum_{n=1}^{\infty} na_n x^{n-1} = \sum_{m=0}^{\infty} (m+1)a_{m+1}x^m.$$ Substituting this into the given equation and letting m = n again, we obtain:

$$\sum_{n=0}^{\infty} (n+1)a_{n+1} x^n + 2\sum_{n=0}^{\infty} a_n x^n = 0, \text{ or}$$

$$\sum_{n=0}^{\infty} [(n+1)a_{n+1} + 2a_n]x^n = 0.$$

Hence $a_{n+1} = -2a_n/(n+1)$ for n = 0,1,2,3,... . Thus $a_1 = -2a_0$, $a_2 = -2a_1/2 = 2^2 a_0/2$, $a_3 = -2a_2/3 = -2^3 a_0/2\cdot 3 = -2^3 a_0/3!$... and $a_n = (-1)^n 2^n a_0/n!$. Notice that for n = 0 this formula reduces to a_0 so we can write

$$\sum_{n=0}^{\infty} a_n x^n = \sum_{n=0}^{\infty} (-1)^n 2^n a_0 x^n/n! = a_0 \sum_{n=0}^{\infty} (-2x)^n/n! = a_0 e^{-2x}.$$

Section 5.2, Page 259

2a. $y = \sum_{n=0}^{\infty} a_n x^n$; $y' = \sum_{n=1}^{\infty} na_n x^{n-1}$ and since we must multiply y' by x in the D.E. we do not shift the index; and

$$y'' = \sum_{n=2}^{\infty} n(n-1)a_n x^{n-2} = \sum_{n=0}^{\infty} (n+2)(n+1)a_{n+2}x^n.$$ Substituting in the D.E., we obtain

$$\sum_{n=0}^{\infty} (n+2)(n+1)a_{n+2}x^n - \sum_{n=1}^{\infty} na_n x^n - \sum_{n=0}^{\infty} a_n x^n = 0.$$ In order to have the starting point the same in all three summations, we let n = 0 in the first and third terms to obtain the following

$$(2\cdot 1\, a_2 - a_0)x^0 + \sum_{n=1}^{\infty} [(n+2)(n+1)a_{n+2} - (n+1)a_n]x^n = 0.$$

Thus $a_{n+2} = a_n/(n+2)$ for n = 1,2,3,... . Note that the recurrence relation is also correct for n = 0.

2b. From the recurrence relation we have $a_2 = a_0/2$,

$a_4 = a_2/4 = a_0/2 \cdot 4$, $a_6 = a_4/6 = a_0/2 \cdot 4 \cdot 6$, so

$y_1 = 1 + x^2/2 + x^4/2 \cdot 4 + x^6/2 \cdot 4 \cdot 6 + \cdots$, and

$a_3 = a_1/3$, $a_5 = a_3/5 = a_1/3 \cdot 5$, $a_7 = a_5/7 = a_1/3 \cdot 5 \cdot 7$, so

$y_2 = (x + x^3/3 + x^5/3 \cdot 5 + x^7/3 \cdot 5 \cdot 7 + \cdots$.

2c. $W(y_1, y_2)(0) = \begin{vmatrix} 1 & 0 \\ 0 & 1 \end{vmatrix} = 1$ and thus y_1, y_2 form a fundamental set of solutions.

2d. From Part(b) we see the even coefficients can be written as $a_{2m} = a_0/2^m\, m!$. For the odd coefficients notice that $a_3 = 2a_1/(2 \cdot 3) = 2a_1/3!$, that $a_5 = 2 \cdot 4a_1/(2 \cdot 3 \cdot 4 \cdot 5) = 2^2 \cdot 2a_1/5!$, and that $a_7 = 2 \cdot 4 \cdot 6a_1/(2 \cdot 3 \cdot 4 \cdot 5 \cdot 6 \cdot 7) = 2^3 \cdot 3!\, a_1/7!$. Likewise $a_9 = a_7/9 = 2^3 \cdot 3!\, a_1/(7!)9 = 2^3 \cdot 3!\, 8a_1/9! = 2^4 \cdot 4!\, a_1/9!$. Continuing we have $a_{2m+1} = 2^m m!\, a_1/(2m+1)!$.

Thus $y = a_0 \sum_{m=0}^{\infty} \frac{x^{2m}}{2^m m!} + a_1 \sum_{m=0}^{\infty} \frac{2^m m!\, x^{2m+1}}{(2m+1)!}$.

3a. $y = \sum_{n=0}^{\infty} a_n(x-1)^n$; $y' = \sum_{n=1}^{\infty} na_n(x-1)^{n-1} = \sum_{n=0}^{\infty} (n+1)a_{n+1}(x-1)^n$,

and

$y'' = \sum_{n=2}^{\infty} n(n-1)a_n(x-1)^{n-2} = \sum_{n=0}^{\infty} (n+2)(n+1)a_{n+2}(x-1)^n$.

Substituting in the D.E. and setting $x = 1 + (x-1)$ we obtain

$$\sum_{n=0}^{\infty} (n+2)(n+1)a_{n+2}(x-1)^n - \sum_{n=0}^{\infty} (n+1)a_{n+1}(x-1)^n - \sum_{n=1}^{\infty} na_n(x-1)^n$$

$$- \sum_{n=0}^{\infty} a_n(x-1)^n = 0,$$

where the third term comes from:

$-(x-1)y' = -\sum_{n=0}^{\infty} (n+1)a_{n+1}(x-1)^{n+1} = -\sum_{n=1}^{\infty} na_n(x-1)^n$.

Letting $n = 0$ in the first, second, and the fourth sums, we obtain

$(2 \cdot 1 \cdot a_2 - 1 \cdot a_1 - a_0)(x-1)^0 +$

$$\sum_{n=1}^{\infty} \left[(n+2)(n+1)a_{n+2} - (n+1)a_{n+1} - (n+1)a_n \right](x-1)^n = 0.$$

Setting the terms in the square brackets equal to zero and dividing by $(n-1)$ yields $(n+2)a_{n+2} - a_{n+1} - a_n = 0$ for $n = 1,2,3,\ldots,$ (which also holds for $n = 0$). This recurrance relation can be used to solve for a_2 in terms of a_0 and a_1, then for a_3 in terms of a_0 and a_1, etc.

3b. In many cases it is easier to first take $a_0 = 0$ and generate one solution and then take $a_1 = 0$ and generate a second solution. Thus, choosing $a_0 = 0$ we find that $a_2 = a_1/2$, $a_3 = (a_2+a_1)/3 = a_1/2$, $a_4 = (a_3+a_2)/4 = a_1/4$, $a_5 = (a_4+a_3)/5 = 3a_1/20,\ldots$. This yields the solution $y_2(x) = (x-1) + (x-1)^2/2 + (x-1)^3/2 + (x-1)^4/4 + \ldots$. The second solution may be obtained by choosing $a_1 = 0$. Then $a_2 = a_0/2$, $a_3 = (a_2+a_1)/3 = a_0/6$, $a_4 = (a_3+a_2)/4 = a_0/6$, $a_5 = (a_4+a_3)/5 = a_0/15,\ldots$. This yields the solution $y_1(x) = 1+(x-1)^2/2+(x-1)^3/6+(x-1)^4/6+(x-1)^5/15+\ldots$.

3c. $W(y_1,y_2)(1) = \begin{vmatrix} 1 & 0 \\ 0 & 1 \end{vmatrix} = 1$ and thus y_1, y_2 form a fundamental set of solutions.

3d. A general term is not easily found in this case.

5. $y = \sum_{n=0}^{\infty} a_n x^n$; $y' = \sum_{n=1}^{\infty} na_n x^{n-1}$; and $y'' = \sum_{n=2}^{\infty} n(n-1)a_n x^{n-2}$.

Substituting in the D.E. and shifting the index in both summations for y'' gives

$$\sum_{n=0}^{\infty} (n+2)(n+1)a_{n+2}x^n - \sum_{n=1}^{\infty} (n+1)n\, a_{n+1}x^n + \sum_{n=0}^{\infty} a_n x^n =$$

$$(2\cdot 1\cdot a_2 + a_0)x^0 + \sum_{n=1}^{\infty} \left[(n+2)(n+1)a_{n+2} - (n+1)na_{n+1} +a_n \right]x^n = 0.$$

Thus $a_2 = -a_0/2$ and $a_{n+2} = na_{n+1}/(n+2) - a_n/(n+2)(n+1)$, $n = 1,2,\ldots$. Choosing $a_0 = 0$ yields $a_2 = 0$, $a_3 = -a_1/6$, $a_4 = 2a_3/4 = -a_1/12$, $a_5 = 3a_4/5 - a_3/20 = -a_1/24 \ldots$, and hence $y_2(x) = a_1(x - x^3/6 - x^4/12 - x^5/24 + \ldots)$. A second linearly independent solution is obtained by choosing $a_1 = 0$. Then

$a_2 = -a_0/2$, $a_3 = a_2/3 = -a_0/6$, $a_4 = 2a_3/4 - a_2/12 = -a_0/24, \ldots$
which gives $y_1(x) = a_0(1 - x^2/2 - x^3/6 - x^4/24 + \ldots)$.

8.   If $y = \displaystyle\sum_{n=0}^{\infty} a_n(x-1)^n$ then

$$xy = [1+(x-1)]y = \sum_{n=0}^{\infty} a_n(x-1)^n + \sum_{n=0}^{\infty} a_n(x-1)^{n+1},$$

$$y' = \sum_{n=1}^{\infty} na_n(x-1)^{n-1}, \text{ and}$$

$$xy'' = [1+(x-1)]y''$$

$$= \sum_{n=2}^{\infty} n(n-1)a_n(x-1)^{n-2} + \sum_{n=2}^{\infty} n(n-1)a_n(x-1)^{n-1}.$$

14.   You will need to rewrite $x+1$ as $3 + (x-2)$ in order to
       multiply $x+1$ times y' as a power series about $x_0 = 2$.

16a. From Prob. 6 we have

$$y(x) = c_1(1 - x^2 + \frac{1}{6}x^4 + \ldots) + c_2(x - \frac{1}{4}x^3 + \frac{7}{160}x^5 + \ldots).$$

Now $y(0) = c_1 = -1$ and $y'(0) = c_2 = 3$ and thus

$$y(x) = -1 + x^2 - \frac{1}{6}x^4 + \ldots + 3x - \frac{3}{4}x^3 + \cdots$$

$$= -1 + 3x + x^2 - \frac{3}{4}x^3 - \frac{1}{6}x^4 + \ldots .$$

16b.

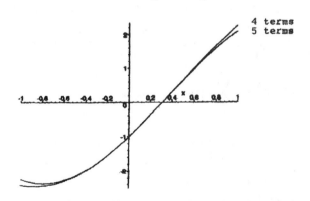

16c. It appears that f is a reasonable approximation for
      $|x| < 0.7$. In fact, the magnitude of the difference in
      the two graphs is .02 for $|x| = .6$ and .04 for $|x| = .7$

19. Letting $t = x-1$ yields $(x-1)^2 = t^2$ and $(x^2-1) = t^2+2t$. Now let $u(t) = y(t+1)$ and hence $u' = y'$ and $u'' = y''$. Thus the D.E. transforms into $u''(t) + t^2u'(t) + (t^2+2t)u(t) = 0$.

Assuming that $u(t) = \sum_{n=0}^{\infty} a_nt^n$, we have $u'(t) = \sum_{n=1}^{\infty} na_nt^{n-1}$ and

$u''(t) = \sum_{n=2}^{\infty} n(n-1)a_nt^{n-2}$. Substituting in the D.E. and

shifting indices yields

$$\sum_{n=0}^{\infty} (n+2)(n+1)a_{n+2}t^n + \sum_{n=2}^{\infty} (n-1)a_{n-1}t^n + \sum_{n=2}^{\infty} a_{n-2}t^n$$

$$+ \sum_{n=1}^{\infty} 2a_{n-1}t^n = 0,$$

$$2 \cdot 1 \cdot a_2t^0 + (3 \cdot 2 \cdot a_3 + 2 \cdot a_0)t^1 + \sum_{n=2}^{\infty} [(n+2)(n+1)a_{n+2}$$

$$+ (n+1)a_{n-1} + a_{n-2}]t^n = 0.$$

It follows that $a_2 = 0$, $a_3 = -a_0/3$ and $a_{n+2} = -a_{n-1}/(n+2) - a_{n-2}/[(n+2)(n+1)]$, $n = 2,3,4...$. We obtain one solution by choosing $a_1 = 0$. Then $a_4 = -a_0/12$, $a_5 = -a_2/5 - a_1/20 = 0$, $a_6 = -a_3/6 - a_2/30 = a_0/18,...$. Thus one solution is $u_1(t) = a_0(1 - t^3/3 - t^4/12 + t^6/18 + ...)$ so $y_1(x) = u_1(x-1) = 1 - (x-1)^3/3 - (x-1)^4/12 + (x-1)^6/18 + ...$. We obtain a second solution by choosing $a_0 = 0$. Then $a_4 = -a_1/4$, $a_5 = -a_2/5 - a_1/20 = -a_1/20$, $a_6 = -a_3/6 - a_2/30 = 0$, $a_7 = -a_4/7 - a_3/42 = a_1/28,...$. Thus $u_2(t) = t - t^4/4 - t^5/20 + t^7/28 + ...$ or $y_2(x) = u_2(x-1)$

$$= (x-1) - (x-1)^4/4 - (x-1)^5/20 + (x-1)^7/28 + ...$$

The Taylor series for $x^2 - 1$ about $x = 1$ may be obtained by writing $x = 1 + (x-1)$ so $x^2 = 1 + 2(x-1) + (x-1)^2$ and $x^2 - 1 = 2(x-1) + (x-1)^2$. The D.E. now appears as $y'' + (x-1)^2y' + [(x-1)^2 + 2(x-1)]y = 0$ which is identical to the transformed equation with $t = x - 1$.

22b. $y = a_0 + a_1x + a_2x^2 + \ldots$, $y^2 = a_0^2 + 2a_0a_1x + (2a_0a_2 + a_1^2)x^2$

$+ \ldots$, $y' = a_1 + 2a_2x + 3a_3x^2 + \ldots$, and

$(y')^2 = a_1^2 + 4a_1a_2x + (6a_1a_3 + 4a_2^2)x^2 + \ldots$. Substituting

these into $(y')^2 = 1 - y^2$ and collecting coefficients of

like powers of x yields $(a_1^2 + a_0^2 - 1) + (4a_1a_2 + 2a_0a_1)x +$

$(6a_1a_3 + 4a_2^2 + 2a_0a_2 + a_1^2)x^2 + \ldots = 0$. As in the earlier

problems, each coefficient must be zero. The I.C. $y(0) = 0$

requires that $a_0 = 0$, and thus $a_1^2 + a_0^2 - 1 = 0$ gives $a_1^2 = 1$.

However, the D.E. indicates that y' is always positive, so

$y'(0) = a_1 > 0$ implies $a_1 = 1$. Then $4a_1a_2 + 2a_0a_1 = 0$ implies

that $a_2 = 0$; and $6a_1a_3 + 4a_2^2 + 2a_0a_2 + a_1^2 = 6a_1a_3 + a_1^2 = 0$

implies that $a_3 = -1/6$. Thus $y = x - x^3/3! + \ldots$, which are

the first two terms of the Taylor series for sinx.

23. We have $y(x) = a_0Y_1 + a_1Y_2$, where y_1 and y_2 are found in

Prob.2. Now, $y(0) = a_0 = 1$ and $y'(0) = a_1 = 0$ and thus

$y(x) = 1 + x^2/2! + x^4/4! + x^6/6! + \cdots$.

23.   26.

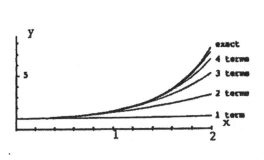
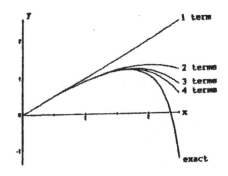

26. Again, $y(x) = a_0Y_1 + a_1Y_2$, where y_1 and y_2 are found in

Prob.10, and $y(0) = a_0 = 0$ and $y'(0) = a_1 = 1$.

Thus $y(x) = x - \dfrac{x^3}{12} - \dfrac{x^5}{240} - \dfrac{x^7}{2240}$.

Section 5.3, Page 265

1. The D.E. can be solved for y'' to yield $y'' = -xy' - y$. If

$y = \phi(x)$ is a solution, then $\phi''(x) = -x\phi'(x) - \phi(x)$ and

thus setting $x = 0$ we obtain $\phi''(0) = -0 - 1 = -1$.

Differentiating the equation for y'' yields
$y''' = -xy'' - 2y'$ and hence setting $y = \phi(x)$ again yields
$\phi'''(0) = -0 - 0 = 0$. In a similar fashion
$y^{(4)} = -xy''' - 3y''$ and thus $\phi^{(4)}(0) = -0 - 3(-1) = 3$.
The process can be continued to calculate higher
derivatives of $\phi(x)$.

3.   We have $y'' = -\dfrac{1+x}{x}y' - \dfrac{3\ln x}{x^2}y$, so $\phi''(1) = 2\phi'(1) - 0\phi(1) = 0$,

and $y''' = -\dfrac{1+x}{x}y'' + (\dfrac{1-3\ln x}{x^2})y' + (\dfrac{6\ln x-3}{x^3})y$ so

$\phi'''(1) = -2\phi''(1) + \phi'(1) - 3\phi(1) = -6$.

6.   The zeros of $P(x) = x^2 - 2x - 3$ are $x = -1$ and $x = 3$. For
$x_0 = 4$, $x_0 = -4$, and $x_0 = 0$ the distance to the nearest
zero of $P(x)$ is $1,3$, and 1, respectively. Thus a lower
bound for the radius of convergence for series solutions
in powers of $(x-4)$, $(x+4)$, and x is $\rho = 1$, $\rho = 3$, and
$\rho = 1$, respectively.

7.   If $x^3 = -1$, then $x = e^{i(\pi+2k\pi)/3}$, $k = 0,1,2$. Thus
$x_1 = e^{i\pi/3} = (1+i\sqrt{3})/2$, $x_2 = -1$ and $x_3 = (1-i\sqrt{3})/2$. Thus
for $x_0 = 0$ we have $\rho = 1$ and for $x_0 = 2$ we have

$\rho = \sqrt{(3/2)^2 + 3/4} = \sqrt{3}$.

9a. Since $P(x) = 1$ has no zeros, the radius of convergence
about $x_0 = 0$ is $\rho = \infty$.

9f. Since $P(x) = 2 + x^2$ has zeros at $x = \pm\sqrt{2}\,i$, the lower
bound for the radius of convergence of the series
solution about $x_0 = 0$ is $\rho = \sqrt{2}$.

9h. $P(x) = x$ has a zero at $x = 0$ and since $x_0 = 1$, $\rho = 1$.

10a. If we assume that $y = \displaystyle\sum_{n=0}^{\infty} a_n x^n$, then $y' = \displaystyle\sum_{n=1}^{\infty} n a_n x^{n-1}$ and

$y'' = \displaystyle\sum_{n=2}^{\infty} n(n-1)a_n x^{n-2}$. Substituting in the D.E. gives:

$$\sum_{n=2}^{\infty} n(n-1)a_n x^{n-2} - \sum_{n=2}^{\infty} n(n-1)a_n x^n - \sum_{n=1}^{\infty} n a_n x^n + \alpha^2 \sum_{n=0}^{\infty} a_n x^n = 0.$$

Shifting indices of summation and collecting coefficients of like powers of x yields the equation:

$$(2 \cdot 1 \cdot a_2 + \alpha^2 a_0)x^0 + [3 \cdot 2 \cdot a_3 + (\alpha^2 - 1)a_1]x^1$$

$$+ \sum_{n=2}^{\infty} [(n+2)(n+1)a_{n+2} + (\alpha^2 - n^2)a_n]x^n = 0.$$

Hence the recurrence relation is
$a_{n+2} = (n^2 - \alpha^2)a_n/(n+2)(n+1)$, $n = 0, 1, 2, \ldots$. For the first solution we choose $a_1 = 0$. We find that
$a_2 = -\alpha^2 a_0/2 \cdot 1$, $a_3 = 0$, $a_4 = (2^2 - \alpha^2)a_2/4 \cdot 3 = -(2^2 - \alpha^2)\alpha^2 a_0/4!$
$\ldots$, $a_{2m} = -[(2m-2)^2 - \alpha^2] \ldots (2^2 - \alpha^2)\alpha^2 a_0/(2m)!$,

and $a_{2m+1} = 0$, so $y_1(x) = 1 - \dfrac{\alpha^2}{2!}x^2 - \dfrac{(2^2 - \alpha^2)\alpha^2}{4!}x^4 - \ldots$

$$-\frac{[(2m-2)^2 - \alpha^2] \ldots (2^2 - \alpha^2)\alpha^2}{(2m)!}x^{2m} - \ldots,$$

where we have set $a_0 = 1$. For the second solution we take $a_0 = 0$ and $a_1 = 1$ in the recurrence relation to obtain the desired solution.

10b. If α is an even integer $2k$ then $(2m-2)^2 - \alpha^2 = 4(m-1)^2 - 4k^2$. Thus when $m = k+1$ all terms in the series for $y_1(x)$ are zero after the x^{2k} term. A similar argument shows that if $\alpha = 2k+1$ then all terms in $y_2(x)$ are zero after the x^{2k+1} term.

11. The Taylor series about $x = 0$ for $\sin x$ is
$\sin x = x - x^3/3! + x^5/5! - \ldots$. Assuming that
$y = \displaystyle\sum_{n=2}^{\infty} a_n x^n$ we find $y'' + (\sin x)y = 2a_2 + 6a_3 x + 12a_4 x^2$
$+ 20a_5 x^3 + 30a_6 x^4 + 42a_7 x^5 + \ldots$
$+ (x - x^3/3! + x^5/5! - \ldots)(a_0 + a_1 x + a_2 x^2 + a_3 x^3 + a_4 x^4 + \ldots)$
$= 2a_2 + (6a_3 + a_0)x + (12a_4 + a_1)x^2 + (20a_5 + a_2 - a_0/6)x^3 +$
$(30a_6 + a_3 - a_1/6)x^4 + (42a_7 + a_4 - a_2/3! + a_0/5!)x^5 + \ldots = 0.$
Hence $a_2 = 0$, $a_3 = -a_0/6$, $a_4 = -a_1/12$, $a_5 = a_0/120$,
$a_6 = (a_1 + a_0)/180$, $a_7 = -a_0/7! + a_1/504$, $\ldots$. We set
$a_0 = 1$ and $a_1 = 0$ and obtain
$y_1(x) = (1 - x^3/6 + x^5/120 + x^6/180 + \ldots)$. Next we set

$a_0 = 0$ and $a_1 = 1$ and obtain

$y_2(x) = (x - x^4/12 + x^6/180 + x^7/504 + ...)$. Since $p(x) = 1$ and $q(x) = \sin x$ both have $\rho = \infty$, the solution in this case converges for all x, that is, $\rho = \infty$

18. We know that $e^x = 1 + x + x^2/2! + x^3/3! + ...$, and therefore $e^{x^2} = 1 + x^2 + x^4/2! + x^6/3! + ...$. Hence, if $y = \sum_{n=0}^{\infty} a_n x^n$, we have $y' = \sum_{n=1}^{\infty} n a_n x^{n-1}$, so

$a_1 + 2a_2 x + 3a_3 x^2 + ... = (1 + x^2 + x^4/2 + ...)(a_0 + a_1 x + a_2 x^2 + ...)$

$$= a_0 + a_1 x + (a_0 + a_2) x^2 + ...$$

Thus, $a_1 = a_0$ $2a_2 = a_1$ and $3a_3 = a_0 + a_2$, which yield the desired solution.

20. Substituting $y = \sum_{n=0}^{\infty} a_n x^n$ into the D.E. we obtain

$\sum_{n=1}^{\infty} n a_n x^{n-1} - \sum_{n=0}^{\infty} a_n x^n = x^2$. Shifting indices in the summation

yields $\sum_{n=0}^{\infty} [(n+1)a_{n+1} - a_n] x^n = x^2$. Equating coefficients of

both sides then gives: $a_1 - a_0 = 0$, $2a_2 - a_1 = 0$, $3a_3 - a_2 = 1$ and $(n+1)a_{n+1} = a_n$ for $n = 3, 4, ...$. Thus $a_1 = a_0$, $a_2 = a_1/2 = a_0/2$, $a_3 = 1/3 + a_2/3 = 1/3 + a_0/2 \cdot 3$, $a_4 = a_3/4 = 1/3 \cdot 4 + a_0/2 \cdot 3 \cdot 4 = 2/4! + a_0/4!$, and in general $a_n = a_{n-1}/n = 2/n! + a_0/n!$. Hence

$$y(x) = a_0 (1 + x + \frac{x^2}{2!} + ... + \frac{x^n}{n!} ...) + 2(\frac{x^3}{3!} + \frac{x^4}{4!} + ... + \frac{x^n}{n!} + ...).$$

Using the power series for e^x, the first and second sums can be rewritten as $a_0 e^x + 2(e^x - 1 - x - x^2/2)$, which is the same solution as found using methods of Chapt.2.

22. Substituting $y = \sum_{n=0}^{\infty} a_n x^n$ into the Legendre equation,

shifting indices, and collecting coefficients of like powers of x yields

$$[2 \cdot 1 \cdot a_2 + \alpha(\alpha+1)a_0]x^0 + \{3 \cdot 2 \cdot a_3 - [2 \cdot 1 - \alpha(\alpha+1)]a_1\}x^1 +$$

$$\sum_{n=2}^{\infty} \{(n+2)(n+1)a_{n+2} - [n(n+1) - \alpha(\alpha+1)]a_n\}x^n = 0. \quad \text{Thus}$$

$a_2 = -\alpha(\alpha+1)a_0/2!$, $a_3 = [2 \cdot 1 - \alpha(\alpha+1)]a_1/3! =$
$-(\alpha-1)(\alpha+2)a_1/3!$ and the recurrence relation is
$(n+2)(n+1)a_{n+2} = -[\alpha(\alpha+1) - n(n+1)]a_n = -(\alpha-n)(\alpha+n+1)a_n$,
$n = 2,3,\ldots$. Setting $a_1 = 0$, $a_0 = 1$ yields a solution
with $a_3 = a_5 = a_7 = \ldots = 0$ and
$a_4 = \alpha(\alpha-2)(\alpha+1)(\alpha+3)/4!,\ldots$, and

$a_{2m} = (-1)^m[\alpha(\alpha-2)\ldots(\alpha-2m+2)][(\alpha+1)\ldots(\alpha+2m-1)]/(2m)!$.
The second linearly independent solution is obtained by
setting $a_0 = 0$ and $a_1 = 1$. The coefficients are
$a_2 = a_4 = a_6 = \ldots = 0$ and $a_3 = -(\alpha-1)(\alpha+2)/3!$, and
$a_5 = -(\alpha-3)(\alpha+4)a_3/5 \cdot 4 = (\alpha-1)(\alpha-3)(\alpha+2)(\alpha+4)/5!$.

26. Using the chain rule we have:
$$\frac{dF(\phi)}{d\phi} = \frac{dF[\phi(x)]}{dx}\frac{dx}{d\phi} = -f'(x)\sin\phi(x) = -f'(x)\sqrt{1-x^2},$$

$$\frac{d^2F(\phi)}{d\phi^2} = \frac{d}{dx}[-f'(x)\sqrt{1-x^2}]\frac{dx}{d\phi} = (1-x^2)f''(x) - xf'(x),$$

which when substituted into the D.E. yields the desired
result.

28. Since $[(1-x^2)y']' = (1-x^2)y'' - 2xy'$, the Legendre
Equation, from Prob. 22, can be written as shown. Thus,
carrying out the multiplications indicated yields the two
equations:

$P_m[(1-x^2)P_n']' = -n(n+1)P_nP_m$

$P_n[(1-x^2)P_m']' = -m(m+1)P_nP_m$.

As long as $n \neq m$ the second equation can be subtracted
from the first and the result integrated from -1 to 1 to
obtain

$$\int_{-1}^{1} \{P_m[(1-x^2)P_n']'-P_n[(1-x^2)P_m']'\}dx = [m(m+1)-n(n+1)]\int_{-1}^{1} P_nP_mdx$$

The left side may be integrated by parts to yield

$$[P_m(1-x^2)P_n' - P_n(1-x^2)P_m']_{-1}^{1} + \int_{-1}^{1}[P_m'(1-x^2)P_n' - P_n'(1-x^2)P_m']dx,$$

which is zero. Thus $\int_{-1}^{1} P_n(x)P_m(x)dx = 0$ for $n \neq m$.

Section 5.4, Page 276

2. This equation is of the form of an Euler equation with x
    replaced by x + 1, so we seek solutions of the form
    $y = (x+1)^r$ for x + 1 > 0. Substitution of y into the D.E.
    yields $F(r) = [r(r-1) + 3r + 3/4](x+1)^r = 0$. Thus
    $r^2 + 2r + 3/4 = 0$, which gives r = -3/2, -1/2. The general
    solution of the D.E. is then
    $y = c_1|x+1|^{-1/2} + c_2|x+1|^{-3/2}$, x ≠ -1.

4. If $y = x^r$ then $F(r) = r(r-1) + 3r + 5 = 0$.
    So $r^2 + 2r + 5 = 0$ and $r = (-2 \pm \sqrt{4-20})/2 = -1 \pm 2i$.
    Thus the general solution of the D.E. is
    $y = c_1x^{-1}\cos(2\ln|x|) + c_2x^{-1}\sin(2\ln|x|)$, x ≠ 0.

9. Again let $y = x^r$ to obtain $F(r) = r(r-1) - 5r + 9 = 0$, or
    $(r-3)^2 = 0$. Thus the roots are x = 3,3 and
    $y = c_1x^3 + c_2x^3\ln|x|$, x ≠ 0, is the solution of the D.E.

13. In this csae $F(r) = 2r(r-1) + r - 3 = 2r^2 - r - 3 =$
    $(2r-3)(r+1) = 0$, so $y = c_1x^{3/2} + c_2x^{-1}$ (since $x_0 = 1$, we
    don't need $|x|$) and $y' = \frac{3}{2}c_1x^{1/2} - c_2x^{-2}$. Setting x = 1
    in y and y' we obtain $c_1 + c_2 = 1$ and $\frac{3}{2}c_1 - c_2 = 4$, which
    yield $c_1 = 2$ and $c_2 = -1$. Hence $y = 2x^{3/2} - x^{-1}$. As x → 0^+
    we have y → -∞ due to the second term.

16. We have $F(r) = r(r-1) + 3r + 5 = r^2 + 2r + 5 = 0$. Thus
    $r_1, r_2 = -1 \pm 2i$ and $y = x^{-1}[c_1\cos(2\ln x) + c_2\sin(2\ln x)]$.
    Then $y(1) = c_1 = 1$ and $y' = -x^{-2}[\cos(2\ln x) + c_2\sin(2\ln x)]$
    $+ x^{-1}[-\sin(2\ln x)2/x + c_2\cos(2\ln x)2/x]$ so that
    $y'(1) = -1+2c_2 = -1$, or $c_2 = 0$. Hence $y = x^{-1}\cos(2\ln x)$
    for x > 0. As x → 0^+ this will oscillate rapidly, with
    large amplitudes.

17. Since the coefficients of y, y' and y″ have no common
    factors and since P(x) vanishes only at x = 0 we conclude
    that x = 0 is a singular point. Writing the D.E. in the
    form y″ + p(x)y' + q(x)y = 0, we obtain p(x) = (1-x)/x
    and q(x) = 1. Thus for the singular point we have
    $\lim_{x \to 0} x\, p(x) = \lim_{x \to 0} 1-x = 1$, $\lim_{x \to 0} x^2 q(x) = 0$ and thus x = 0

is a regular singular point.

21. Writing the D.E. in the form $y'' + p(x)y' + q(x)y = 0$, we
find $p(x) = x/(1-x)(1+x)^2$ and $q(x) = (1+x)/(1-x^2)$.
Therefore $x = \pm 1$ are singular points. Since
$\lim\limits_{x \to 1} (x-1)p(x)$ and $\lim\limits_{x \to 1} (x-1)^2 q(x)$ both exist, we conclude
$x = 1$ is a regular singular point. Finally, since
$\lim\limits_{x \to -1} (x+1)p(x)$ does not exist, we conclude that $x = -1$ is
an irregular singular point.

28. Writing the D.E. in the form $y'' + p(x)y' + q(x)y = 0$, we
see that $p(x) = e^x/x$ and $q(x) = (3\cos x)/x$. Thus $x = 0$ is
a singular point. Since $xp(x) = e^x$ is analytic at $x = 0$
and $x^2 q(x) = 3x\cos x$ is analytic at $x = 0$ the point $x = 0$
is a regular singular point.

33. Writing the D.E. in the form $y'' + p(x)y' + q(x)y = 0$, we
see that $p(x) = \dfrac{x}{\sin x}$ and $q(x) = \dfrac{4}{\sin x}$. Since $\lim\limits_{x \to 0} q(x)$
does not exist, the point $x_0 = 0$ is a singular point and
since neither $\lim\limits_{x \to \pm n\pi} p(x)$ nor $\lim\limits_{x \to \pm n\pi} q(x)$ exist, either, the
points $x_0 = \pm n\pi$ are also singular points. To determine
whether the singular points are regular or irregular we
must use Eq.(31) and the result #7 of multiplication and
division of power series from Section 5.1. For $x_0 = 0$,
we have

$$xp(x) = \frac{x^2}{\sin x} = \frac{x^2}{x - \dfrac{x^3}{6} + \dots} = x[1 + \frac{x^2}{6} + \dots]$$

$$= x + \frac{x^3}{6} + \dots,$$

which converges about $x_0 = 0$ and thus $xp(x)$ is analytic
at $x_0 = 0$. $x^2 q(x)$, by similar steps, is also analytic at
$x_0 = 0$ and thus $x_0 = 0$ is a regular singular point. For
$x_0 = n\pi$, we have

$$(x-n\pi)p(x) = \frac{(x-n\pi)x}{\sin x} = \frac{(x-n\pi)[(x-n\pi) + n\pi]}{\pm(x-n\pi) + \dfrac{-(x-n\pi)^3}{6} \pm \dots}$$

$$= \pm[(x-n\pi) + n\pi][1 + \frac{(x-n\pi)^2}{6} + \dots], \text{ which}$$

converges about $x_0 = n\pi$ and thus $(x-n\pi)p(x)$ is analytic at $x = n\pi$. Similarly $(x+n\pi)p(x)$ and $(x\pm n\pi)^2 q(x)$ are analytic and thus $x_0 = \pm n\pi$ are regular singular points.

35. Substituting $y = x^r$, we find that $r(r-1) + \alpha r + 5/2 = 0$ or $r^2 + (\alpha-1)r + 5/2 = 0$. Thus $r_1, r_2 = [-(\alpha-1) \pm \sqrt{(\alpha-1)^2 - 10}]/2$. In order for solutions to approach zero as $x \to 0$ it is necessary that the real parts of r_1 and r_2 be positive. Suppose that $\alpha > 1$, then $\sqrt{(\alpha-1)^2 - 10}$ is either imaginary or real and less than $\alpha - 1$; hence the real parts of r_1 and r_2 will be negative. Suppose that $\alpha = 1$, then $r_1, r_2 = \pm i\sqrt{10}$ and the solutions are oscillatory. Suppose that $\alpha < 1$, then $\sqrt{(\alpha-1)^2 - 10}$ is either imaginary or real and less than $|\alpha-1| = 1 - \alpha$; hence the real parts of r_1 and r_2 will be positive. Thus if $\alpha < 1$ the solutions of the D.E. will approach zero as $x \to 0$.

39. In all cases the roots of $F(r) = 0$ are given by Eq.(6) and the forms of the solution are given in Eqs.(25),(26) and (27).

39a. The real part of the root must be positive so, from Eq.(6), $\alpha < 1$. Also $\beta > 0$, since the $\sqrt{(\alpha-1)^2 - 4\beta}$ term must be less than $|\alpha-1|$.

39d. The real part of the root must be negative, so $\alpha > 1$, with $\beta \geq 0$ (for $\beta = 0$ one root is zero, which is bounded as $x \to \infty$). If $\alpha = 1$, then the roots are $\pm\sqrt{-4\beta}$, so $\beta > 0$ will yield oscillatory solutions as $x \to \infty$, which are bounded.

40. Assume that $y = v(x)x^{r_1}$. Then $y' = v(x)r_1 x^{r_1-1} + v'(x)x^{r_1}$ and $y'' = v(x)r_1(r_1-1)x^{r_1-2} + 2v'(x)r_1 x^{r_1-1} + v''(x)x^{r_1}$. Substituting in the D.E. and collecting terms yields $x^{r_1+2} v'' + (\alpha + 2r_1)x^{r_1+1} v' + [r_1(r_1-1) + \alpha r_1 + \beta]x^{r_1} v = 0$. Now we make use of the fact that r_1 is a double root of $f(r) = r(r-1) + \alpha r + \beta$. This means that $f(r_1) = 0$ and $f'(r_1) = 2r_1 - 1 + \alpha = 0$. Hence the D.E. for v reduces to $x^{r_1+2} v'' + x^{r_1+1} v'$. Since $x > 0$ we may divide by x^{r_1+1}

to obtain $xv'' + v' = 0$.   Thus $v(x) = \ln x$ and a second
solution is $y = x^{r_1}\ln x$.

41. Substituting $y = \sum\limits_{n=0}^{\infty} a_n x^n$ into the D.E. yields

$2\sum\limits_{n=2}^{\infty} n(n-1)a_n x^{n-1} + 3\sum\limits_{n=1}^{\infty} na_n x^{n-1} + \sum\limits_{n=0}^{\infty} a_n x^{n+1} = 0$.   The last sum

becomes $\sum\limits_{n=2}^{\infty} a_{n-2} x^{n-1}$ (let $m = n+2$ and then replace m by n),

the first term of the middle sum is $3a_1$, and thus we have

$3a_1 + \sum\limits_{n=2}^{\infty} \{[2n(n-1)+3n]a_n + a_{n-2}\}x^{n-1} = 0$.   Hence $a_1 = 0$ and

$a_n = \dfrac{-a_{n-2}}{n(2n+1)}$, which is the desired recurrance relation.

Thus all even coefficients are found in terms of a_0 and
all odd coefficients are zero, thereby yielding only one
solution of the desired form.

43. If $\xi = 1/x$ then
$\dfrac{dy}{dx} = \dfrac{dy}{d\xi}\dfrac{d\xi}{dx} = -\dfrac{1}{x^2}\dfrac{dy}{d\xi} = -\xi^2\dfrac{dy}{d\xi}$,

$\dfrac{d^2y}{dx^2} = \dfrac{d}{d\xi}(-\xi^2\dfrac{dy}{d\xi})\dfrac{d\xi}{dx} = (-2\xi\dfrac{dy}{d\xi} - \xi^2\dfrac{d^2y}{d\xi^2})(-\dfrac{1}{x^2})$

$\qquad = \xi^4\dfrac{d^2y}{d\xi^2} + 2\xi^3\dfrac{dy}{d\xi}$.

Substituting in the D.E. we have

$P(1/\xi)[\xi^4\dfrac{d^2y}{d\xi^2} + 2\xi^3\dfrac{dy}{d\xi}] + Q(1/\xi)[-\xi^2\dfrac{dy}{d\xi}] + R(1/\xi)y = 0$, or

$\xi^4 P(1/\xi)\dfrac{d^2y}{d\xi^2} + [2\xi^3 P(1/\xi) - \xi^2 Q(1/\xi)]\dfrac{dy}{d\xi} + R(1/\xi)y = 0$.

The result then follows from the theory of singular
points at $\xi = 0$.

45. Since $P(x) = x^2$, $Q(x) = x$ and $R(x) = -4$ we have
$f(\xi) = [2P(1/\xi)/\xi - Q(1/\xi)/\xi^2]/P(1/\xi) = 2/\xi - 1/\xi = 1/\xi$
and $g(\xi) = R(1/\xi)/\xi^4 P(1/\xi) = -4/\xi^2$.   Thus the point at
infinity is a singular point.   Since both $\xi f(\xi)$ and
$\xi^2 g(\xi)$ are analytic at $\xi = 0$, the point at infinity is a

regular singular point.

47. Since $P(x) = x^2$, $Q(x) = x$, and $R(x) = x^2 - v^2$,
    $f(\xi) = [2P(1/\xi)/\xi - Q(1/\xi)/\xi^2]/P(1/\xi) = 2/\xi - 1/\xi = 1/\xi$
    and $g(\xi) = R(1/\xi)/\xi^4 P(1/\xi) = (1/\xi^2 - v^2)/\xi^2 = 1/\xi^4 - v^2/\xi^2$.
    Thus the point at infinity is a singular point. Although
    $\xi f(\xi) = 1$ is analytic at $\xi = 0$, $\xi^2 g(\xi) = 1/\xi^2 - v^2$ is not,
    so the point at infinity is an irregular singular point.

Section 5.5, Page 282

2a. If the D.E. is put in the standard form
    $y'' + p(x)y + q(x)y = 0$, then $p(x) = x^{-1}$ and
    $q(x) = 1 - 1/9x^2$. Thus $x = 0$ is a singular point. Since
    $xp(x) \to 1$ and $x^2 q(x) \to -1/9$ as $x \to 0$ it follows that
    $x = 0$ is a regular singular point.

2b. In determining a series solution of the D.E. it is more
    convenient to leave the equation in the form given rather
    than divide by x^2, the coefficient of y''. If we
    substitute

$$y = \sum_{n=0}^{\infty} a_n x^{n+r}, \text{ we have}$$

$$\sum_{n=0}^{\infty} (n+r)(n+r-1)a_n x^{n+r} + \sum_{n=0}^{\infty} (n+r)a_n x^{n+r} + (x^2 - \frac{1}{9})\sum_{n=0}^{\infty} a_n x^{n+r} = 0.$$

Note that $x^2 \sum_{n=0}^{\infty} a_n x^{n+r} = \sum_{n=0}^{\infty} a_n x^{n+r+2} = \sum_{n=2}^{\infty} a_{n-2}x^{n+r}$. Thus we

have $[r(r-1) + r - \frac{1}{9}]a_0 x^r + [(r+1)r + (r+1) - \frac{1}{9}]a_1 x^{r+1} +$

$$\sum_{n=2}^{\infty} \{[(n+r)(n+r-1) + (n+r) - \frac{1}{9}]a_n + a_{n-2}\} x^{n+r} = 0. \text{ From}$$

the first term, the indicial equation is $r^2 - 1/9 = 0$
with roots $r_1 = 1/3$ and $r_2 = -1/3$. For either value of
r it is necessary to take $a_1 = 0$ in order that the
coefficient of x^{r+1} be zero. The recurrence relation is
$a_n = -a_{n-2}./[(n+r)^2 - 1/9]$.

2c. For $r = 1/3$ we have

$$a_n = \frac{-a_{n-2}}{(n + \frac{1}{3})^2 - (\frac{1}{3})^2} = -\frac{a_{n-2}}{(n + \frac{2}{3})n}, \quad n = 2,3,4,\ldots .$$

Since $a_1 = 0$ it follows from the recurrence relation that $a_3 = a_5 = a_7 = \ldots = 0$. For the even coefficients it is convenient to let $n = 2m$, $m = 1,2,3,\ldots$. Then $a_{2m} = -a_{2m-2}/2^2 m(m + \frac{1}{3})$. The first few coefficients are given by

$$a_2 = \frac{(-1)a_0}{2^2(1 + \frac{1}{3})1}, \quad a_4 = \frac{(-1)a_2}{2^2(2 + \frac{1}{3})2} = \frac{a_0}{2^4(1 + \frac{1}{3})(2 + \frac{1}{3})2!}$$

$$a_6 = \frac{(-1)a_4}{2^2(3 + \frac{1}{3})3} = \frac{(-1)a_0}{2^6(1 + \frac{1}{3})(2 + \frac{1}{3})(3 + \frac{1}{3})3!}, \quad \text{and the}$$

coefficent of x^{2m} for $m = 1, 2, \ldots$ is

$$a_{2m} = \frac{(-1)^m a_0}{2^{2m} m!(1 + \frac{1}{3})(2 + \frac{1}{3}) \ldots (m + \frac{1}{3})}. \quad \text{Thus one}$$

solution (on setting $a_0 = 1$) is

$$y_1(x) = x^{1/3}[1 + \sum_{m=1}^{\infty} \frac{(-1)^m}{m!(1 + \frac{1}{3})(2 + \frac{1}{3})\ldots(m + \frac{1}{3})}(\frac{x}{2})^{2m}].$$

2d. Since $r_2 = -1/3 \neq r_1$ and $r_1 - r_2 = 2/3$ is not an integer, we can calculate a second series solution corresponding to $r = -1/3$. The recurrence relation is $n(n-2/3)a_n = -a_{n-2}$, which yields the desired solution following the steps in Part c. Note that $a_1 = 0$, as in the first solution, and thus all the odd coefficients are zero.

4a. Putting the D.E. in the form $y'' + p(x)y' + q(x)y = 0$, we see that $p(x) = 1/x$ and $q(x) = -1/x$. Thus $x = 0$ is a singular point, and since $xp(x) \to 1$ and $x^2 q(x) \to 0$, as $x \to 0$, $x = 0$ is a regular singular point.

4b. Substituting $y = \sum_{n=0}^{\infty} a_n x^{n+r}$ in $xy'' + y' - y = 0$ and shifting indices we obtain

$$\sum_{n=-1}^{\infty} a_{n+1}(r+n+1)(r+n)x^{n+r} + \sum_{n=-1}^{\infty} a_{n+1}(r+n+1)x^{n+r} - \sum_{n=0}^{\infty} a_n x^{n+r} = 0,$$

or $[r(r-1) + r]a_0 x^{-1+r} + \sum_{n=0}^{\infty} [(r+n+1)^2 a_{n+1} - a_n]x^{n+r} = 0.$

From the first coefficient we find $r^2 = 0$ is the indicial equation and from the coefficient of x^{n+r} we find the recurrance relation is $a_{n+1} = a_n/(n+1+r)^2$.

4c. Setting $r = 0$ in the recurrance relation we find $(n+1)^2 a_{n+1} = a_n$, $n = 0,1,2,\ldots$. The coefficients are $a_1 = a_0$, $a_2 = a_1/2^2 = a_0/2^2$, $a_3 = a_2/3^2 = a_0/3^2 \cdot 2^2$, $a_4 = a_3/4^2 = a_0/4^2 \cdot 3^2 \cdot 2^2, \ldots$ and $a_n = a_0/(n!)^2$. Thus one

solution (on setting $a_0 = 1$) is $y = \displaystyle\sum_{n=0}^{\infty} x^n/(n!)^2$.

4d. Since the the indicial equation has only one root, we

only have one solution of the form $y = x^r \displaystyle\sum_{n=0}^{\infty} a_n x^{n+r}$.

11a. If we make the change of variable $t = x-1$ and let $y = u(t)$, then the Legendre equation transforms to $(t^2 + 2t)u''(t) + 2(t+1)u'(t) - \alpha(\alpha+1)u(t) = 0$. Since $x = 1$ is a regular singular point of the original equation, we know that $t = 0$ is a regular singular point

of the transformed equation. Substituting $u = \displaystyle\sum_{n=0}^{\infty} a_n t^{n+r}$

in the transformed equation and shifting indices, we obtain

$$\sum_{n=0}^{\infty} (n+r)(n+r-1)a_n t^{n+r} + 2\sum_{n=-1}^{\infty} (n+r+1)(n+r)a_{n+1}t^{n+r}$$

$$+ 2\sum_{n=0}^{\infty} (n+r)a_n t^{n+r} + 2\sum_{n=-1}^{\infty} (n+r+1)a_{n+1}t^{n+r} - \alpha(\alpha+1)\sum_{n=0}^{\infty} a_n t^{n+r} = 0,$$

or $[2r(r-1) + 2r]a_0 \, t^{r-1} + \sum_{n=0}^{\infty} \{2(n+r+1)^2 a_{n+1}$

$$+ [(n+r)(n+r+1) - \alpha(\alpha+1)]a_n\}t^{n+r} = 0.$$

The indicial equation is $2r^2 = 0$ so $r = 0$ is a double root. Thus there will be only one series solution of the

form $y = \sum_{n=0}^{\infty} a_n t^{n+r}.$

11b. The recurrence relation is

$2(n+1)^2 a_{n+1} = [\alpha(\alpha+1) - n(n+1)]a_n, n = 0,1,2,\ldots$. We have

$a_1 = [\alpha(\alpha+1)]a_0/2 \cdot 1^2$, $a_2 = [\alpha(\alpha+1)][\alpha(\alpha+1) - 1 \cdot 2]a_0/2^2 \cdot 2^2 \cdot 1^2$,

$a_3 = [\alpha(\alpha+1)][\alpha(\alpha+1) - 1 \cdot 2][\alpha(\alpha+1) - 2 \cdot 3]a_0/2^3 \cdot 3^2 \cdot 2^2 \cdot 1^2, \ldots,$

and $a_n = [\alpha(\alpha+1)][\alpha(\alpha+1)-1 \cdot 2] \cdots [\alpha(\alpha+1)-(n-1)n]a_0/2^n(n!)^2.$

Reverting to the variable x it follows that one solution of the Legendre equation in powers of x-1 is

$$y_1(x) = \sum_{n=0}^{\infty} [\alpha(\alpha+1)][\alpha(\alpha+1) - 1 \cdot 2] \cdots$$

$[\alpha(\alpha+1) - (n-1)n](x-1)^n/2^n(n!)^2$ where we have set $a_0 = 1$, which is equivalent to the answer in the text if a (-1) is taken out of each square bracket.

14a. The standard form is $y'' + p(x)y' + q(x)y = 0$, with $p(x) = 1/x$ and $q(x) = 1$. Thus $x = 0$ is a singular point; and since $xp(x) \to 1$ and $x^2q(x) \to 0$ as $x \to 0$, $x = 0$ is a regular singular point.

14b. Substituting $y = \sum_{n=0}^{\infty} a_n x^{n+r}$ into $x^2y'' + xy' + x^2y = 0$ and shifting indices appropriately, we obtain

$$\sum_{n=0}^{\infty} (n+r)(n+r-1)a_n x^{n+r} + \sum_{n=0}^{\infty} (n+r)a_n x^{n+r} + \sum_{n=2}^{\infty} a_{n-2}x^{n+r} = 0, \text{ or}$$

$$[r(r-1)+r]a_0 x^r + [(1+r)r+1+r]a_1 x^{r+1} + \sum_{n=2}^{\infty} [(n+r)^2 a_n + a_{n-2}]x^{n+r} = 0.$$

The indicial equation is $r^2 = 0$ so $r = 0$ is a double root. It is necessary to take $a_1 = 0$ in order that the coefficient of x^{r+1} be zero.

14c. The recurrence relation is $n^2 a_n = -a_{n-2}$, $n = 2, 3, \ldots$.
Since $a_1 = 0$ it follows that $a_3 = a_5 = a_7 = \ldots = 0$. For
the even coefficients we let $n = 2m$, $m = 1, 2, \ldots$. Then
$a_{2m} = -a_{2m-2}/2^2 m^2$ so $a_2 = -a_0/2^2 \cdot 1^2$, $a_4 = a_0/2^2 \cdot 2^2 \cdot 1^2 \cdot 2^2, \ldots$,
and $a_{2m} = (-1)^m a_0/2^{2m}(m!)^2$. Thus one solution of the Bessel

equation of order zero is $J_0(x) = 1 + \sum\limits_{m=1}^{\infty} (-1)^m x^{2m}/2^{2m}(m!)^2$

where we have set $a_0 = 1$.

14d. Using the ratio test it can be shown that the series
converges for all x. Also note that $J_0(x) \to 1$ as $x \to 0$.

15. In order to determine the form of the integral for x near
zero we must study the integrand for x small. Using the
above series for J_0, we have

$$\frac{1}{x[J_0(x)]^2} = \frac{1}{x[1 - x^2/2 + \ldots]^2} = \frac{1}{x[1 - x^2 + \ldots]}$$

$$= \frac{1}{x}[1 + x^2 + \ldots] \text{ for x small. Thus}$$

$$y_2(x) = J_0(x)\int \frac{dx}{x[J_0(x)]^2} = J_0(x)\int [\frac{1}{x} + x + \ldots]dx$$

$$= J_0(x)[\ln x + \frac{x^2}{x} + \ldots], \text{ and it is clear that } y_2(x)$$

will contain a logarithmic term.

16a. Putting the D.E. in the standard form $y'' + p(x)y' + q(x)y = 0$
we see that $p(x) = 1/x$ and $q(x) = (x^2-1)/x^2$. Thus $x = 0$ is a
singular point and since $xp(x) \to 1$ and $x^2 q(x) \to -1$ as $x \to$
0, $x = 0$ is a regular singular point.

16b. Substituting $y = \sum\limits_{n=0}^{\infty} a_n x^{n+r}$ into $x^2 y'' + xy' + (x^2-1)y = 0$,

shifting indices appropriately, and collecting
coefficients of common powers of x we obtain
$[r(r-1) + r - 1]a_0 x^r + [(1+r)r + 1 + r - 1]a_1 x^{r+1}$

$$+ \sum\limits_{n=2}^{\infty} \{[(n+r)^2 - 1]a_n + a_{n-2}\}x^{n+r} = 0.$$

The indicial equation is $r^2-1 = 0$ so the roots are $r_1 = 1$
and $r_2 = -1$.

16c. For either value of r it is necessary to take $a_1 = 0$ in
order that the coefficient of x^{r+1} be zero. The recurrence
relation is $[(n+r)^2 - 1]a_n = -a_{n-2}$, $n = 2,3,4 \ldots$.
For $r = 1$ we have $a_n = -a_{n-2}/[n(n+2)]$, $n = 2,3,4,\ldots$.
Since $a_1 = 0$ it follows that $a_3 = a_5 = a_7 = \ldots = 0$. Let
$n = 2m$. Then $a_{2m} = -a_{2m-2}/2^2 m(m+1)$, $m = 1,2,\ldots$, so
$a_2 = -a_0/2^2 \cdot 1 \cdot 2$, $a_4 = -a_2/2^2 \cdot 1 \cdot 2 \cdot 3 = a_0/2^2 \cdot 2^2 \cdot 1 \cdot 2 \cdot 2 \cdot 3, \ldots$, and
$a_{2m} = (-1)^m a_0/2^{2m} m!(m+1)!$. Thus one solution (set $a_0 = 1/2$)
of the Bessel equation of order one is

$$J_1(x) = (x/2) \sum_{n=0}^{\infty} (-1)^n x^{2n}/(n+1)!n!2^{2n}.$$

16d. The ratio test shows that the series converges for all x.
Also note that $J_1(x) \to 0$ as $x \to 0$.

16e. For $r = -1$ the recurrence relation is
$[(n-1)^2 - 1]a_n = -a_{n-2}$, $n = 2,3,\ldots$, so for $n = 2$ the
coefficient of a_2 is zero and we cannot calculate a_2.
Consequently it is not possible to find a series solution
of the form $x^{-1} \sum_{n=0}^{\infty} b_n x^n$.

Secton 5.6, Page 290

1a. The D.E. has the form $P(x)y'' + Q(x)y' + R(x)y = 0$ with
$P(x) = x$, $Q(x) = 2x$, and $R(x) = 6e^x$. From this we find
$p(x) = Q(x)/P(x) = 2$ and $q(x) = R(x)/P(x) = 6e^x/x$ and
thus $x = 0$ is a singular point. Since $xp(x) = 2x$ and
$x^2q(x) = 6xe^x$ are analytic at $x = 0$ we conclude that
$x = 0$ is a regular singular point.

1b. We have $xp(x) \to 0 = p_0$ and $x^2q(x) \to 0 = q_0$ as $x \to 0$ and
thus Eq.(7), the indicial equation, is $F(r) = r(r-1) = 0$,
which has the roots $r_1 = 1$ and $r_2 = 0$. These are the
exponents of the singularity at $x = 0$.

3a. The equation has the form $P(x)y'' + Q(x)y' + R(x)y = 0$
with $P(x) = x(x-1)$, $Q(x) = 6x^2$ and $R(x) = 3$. Since $P(x)$,
$Q(x)$, and $R(x)$ are polynomials with no common factors and
$P(0) = 0$ and $P(1) = 0$, we conclude that $x = 0$ and $x = 1$

are singular points. The first point, $x = 0$, can be shown
to be a regular singular point using steps similar to
those shown in Prob. 1. For $x = 1$, we must put the D.E.
in the form seen in Ex.(1). To do this, divide the D.E.
by x and multiply by $(x-1)$ to obtain

$(x-1)^2 y'' + 6x(x-1)y + \dfrac{3}{x}(x-1)y = 0$. Comparing this to

Ex.(1) we find that $(x-1)p(x) = 6x$ and
$(x-1)^2 q(x) = 3(x-1)/x$ which are both analytic at
$x = 1$ and hence $x = 1$ is a regular singular point.

3b. These last two expressions approach $p_0 = 6$ and $q_0 = 0$
respectively as $x \to 1$, and thus the indicial equation is
$F(r) = r(r-1) + 6r + 0 = r(r+5) = 0$.

9a. For this D.E., $p(x) = \dfrac{-(1+x)}{x^2(1-x)}$ and $q(x) = \dfrac{2}{x(1-x)}$ and thus
$x = 0, 1$ are singular points. Since $xp(x)$ is not analytic
at $x = 0$, $x = 0$ is not a regular singular point. Looking
at $(x-1)p(x) = \dfrac{1+x}{x^2}$ and $(x-1)^2 q(x) = \dfrac{2(1-x)}{x}$ we see that
$x = 1$ is a regular singular point.

9b. As in Ex.(1) $p_0 = \lim\limits_{x \to 1}(x-1)p(x) = 2$ and
$q_0 = \lim\limits_{x \to 1}(x-1)^2 q(x) = 0$. Thus the indicial equation is
$F(r) = r^2 + r$ and $r_1 = 0$ and $r_2 = -1$.

13a. We have $p(x) = Q/P = 1/x$ and $q(x) = R/P = 1/x$ and thus
$x = 0$ is a singular point. Now $xp(x) = 1$ and $x^2 q(x) = x$
are analytic at $x = 0$, so $x = 0$ is a regular singular
point.

13b. From Part(a) we have $p_0 = 1$ and $q_0 = 0$ so the indicial
equation is $F(r) = r(r-1) + r = r^2$ and thus the exponents
of the singularity are $r_1 = r_2 = 0$.

13c. From Eq.(4) we have $y(x) = \sum\limits_{n=0}^{\infty} a_n x^n$. Differentiating and
substituting into the D.E. yields

$$\sum_{n=2}^{\infty} n(n-1)a_n x^{n-1} + \sum_{n=1}^{\infty} na_n x^{n-1} - \sum_{n=0}^{\infty} a_n x^n = 0.$$ Adjusting the

exponents and the index appropriately then gives

$$(a_1 - a_0) + \sum_{n=1}^{\infty} [(n+1)^2 a_{n+1} - a_n] x^n = 0.$$ Thus $a_1 = a_0$ and

$a_{n+1} = a_n/(n+1)^2$. Setting $n = 1, 2, 3$ we obtain

$$y_1(x) = 1 + x + x^2/4 + x^3/36 + \cdots .$$

For the second solution we use Eq.(18):

$$y = y_1 \ln x + \sum_{n=1}^{\infty} b_n x^n.$$ When this is substituted in the D.E.

we obtain $$2y_1' + \sum_{n=2}^{\infty} n(n-1) b_n x^{n-1} + \sum_{n=1}^{\infty} n b_n x^{n-1} - \sum_{n=0}^{\infty} b_n x^n = 0,$$

where we have used the fact that y_1 is a solution so that
all terms involving $\ln x$ add to zero. Condensing the
terms, as above, we get

$$2(1 + x/2 + x^2/12 + \cdots) + b_1 + \sum_{n=1}^{\infty} [(n+1)^2 b_{n+1} - b_n] x^n = 0.$$ The

coefficient of each power of x must be zero so $b_1 + 2 = 0$,
$1 + 4b_2 - b_1 = 0$ and $(1/6) + 9b_3 - b_2 = 0$. Solving we get $b_1 = -2$,
$b_2 = (b_1 - 1)/4 = -3/4$ and $b_3 = (b_2 - 1/6)/9 = -11/108$, which
give the desired solution.

17a. We have $p(x) = \dfrac{\sin x}{x^2}$ and $q(x) = -\dfrac{\cos x}{x^2}$, so that $x = 0$ is

a singular point. Note that $xp(x) = (\sin x)/x \to 1 = p_0$
as $x \to 0$ and $x^2 q(x) = -\cos x \to -1 = q_0$ as $x \to 0$. In
order to assert that $x = 0$ is a regular singular point we
must demonstrate that $xp(x)$ and $x^2 q(x)$, with $xp(x) = 1$ at
$x = 0$ and $x^2 q(x) = -1$ at $x = 0$, have convergent power
series (are analytic) about $x = 0$. We know that $\cos x$ is
analytic so we need only consider $(\sin x)/x$. Now

$$\sin x = \sum_{n=0}^{\infty} (-1)^n x^{2n+1}/(2n+1)! \text{ for } -\infty < x < \infty \quad \text{so}$$

$$(\sin x)/x = \sum_{n=0}^{\infty} (-1)^n x^{2n}/(2n+1)! \text{ and hence is analytic.}$$

Thus we conclude that $x = 0$ is a regular singular point.

17b. From part a) it follows that the indicial equation is $r(r-1)$
$+ r - 1 = r^2 - 1 = 0$ and the roots are $r_1 = 1$, $r_2 = -1$.

17c. To find the first few terms of the solution corresponding
to $r_1 = 1$, assume that

$$y(x) = x\sum_{n=0}^{\infty} a_n x^n$$

$$= x(a_0 + a_1 x + a_2 x^2 + \ldots) = a_0 x + a_1 x^2 + a_2 x^3 + \ldots .$$

Substituting this series for y in the D.E. and expanding
sinx and cosx about x = 0 yields

$x^2(2a_1 + 6a_2 x + 12a_3 x^2 + 20a_4 x^3 + \ldots) +$

$(x - x^3/3! + x^5/5! - \ldots)(a_0 + 2a_1 x + 3a_2 x^2 + 4a_3 x^3 + 5a_4 x^4 +$

$\ldots) - (1 - x^2/2! + x^4/4! - \ldots)(a_0 x + a_1 x^2 + a_2 x^3 + a_3 x^4 +$

$a_4 x^5 + \ldots) = 0$. Collecting terms we have $(a_0 - a_0)x +$

$(2a_1 + 2a_1 - a_1)x^2 + (6a_2 + 3a_2 - a_0/6 - a_2 + a_0/2)x^3 +$

$(12a_3 + 4a_3 - 2a_1/6 - a_3 + a_1/2)x^4 +$

$(20a_4 + 5a_4 - 3a_2/6 + a_0/120 - a_4 + a_2/2 - a_0/24)x^5 + \ldots = 0$.

Simplifying yields, $3a_1 x^2 + (8a_2 + a_0/3)x^3 + (15a_3 + a_1/6)x^4$

$+ (24a_4 - a_0/30)x^5 + \ldots = 0$. Thus, $a_1 = 0$, $a_2 = -a_0/4!$,

$a_3 = 0$, $a_4 = a_0/6!$, $\ldots$. Hence

$y_1(x) = x - x^3/4! + x^5/6! + \ldots$ where we have set $a_0 = 1$.

For the second solution we use a variation of Eq.(24)
similar to Eq.(18):

$$y_2(x) = ay_1(x)\ln x + x^{-1}(1 + \sum_{n=1}^{\infty} c_n x^n)$$

$$= ay_1(x)\ln x + \frac{1}{x} + c_1 + c_2 x + c_3 x^2 + c_4 x^3 + \ldots, \text{ so}$$

$y_2' = ay_1'\ln x + ay_1 x^{-1} - x^{-2} + c_2 + 2c_3 x + 3c_4 x^2 + \ldots$, and

$y_2'' = ay_1''\ln x + 2ay_1' x^{-1} - ay_1 x^{-2} + 2x^{-3} + 2c_3 + 3c_4 x + \ldots$.

When these are substituted in the given D.E. the terms
including lnx will appear as

$a[x^2 y_1'' + (\sin x)y_1' - (\cos x)y_1]$, which is zero since y_1 is
a solution. For the remainder of the terms, use

$y_1 = x - x^3/24 + x^5/720$ and the cosx and sinx series as
shown earlier to obtain

$-c_1 + (2/3+2a)x + (3c_3+c_1/2)x^2 + (4/45+c_2/3+8c_4)x^3 +... = 0.$
These yield $c_1 = 0$, $a = -1/3$, $c_3 = 0$, and
$c_4 = -c_2/24 - 1/90$. We may take $c_2 = 0$, since this term
will simply generate $y_1(x)$ over again. Thus

$y_2(x) = -\dfrac{1}{3}y_1(x)\ln x + x^{-1} - \dfrac{1}{90}x^3$. If a computer algebra

system is used, then additional terms in each series may
be obtained without much additional effort. The next
terms, in each case, are shown here:

$y_1(x) = x - \dfrac{x^3}{24} + \dfrac{x^5}{720} - \dfrac{43x^7}{1451520} + ...$ and

$y_2(x) = -\dfrac{1}{3}y_1(x)\ln x + \dfrac{1}{x}[1- \dfrac{x^4}{90} + \dfrac{41x^6}{120960} - ...].$

18a. We first write the D.E. in the standard form as given for
Theorem 5.6.1 except that we are expanding in powers of
$(x-1)$ rather than powers of x:

$(x-1)^2y'' + (x-1)[(x-1)/2\ln x]y' + [(x-1)^2/\ln x]y = 0$. Since
$\ln 1 = 0$, $x = 1$ is a singular point. To show it is a
regular singular point of this D.E. we must show that
$(x-1)/\ln x$ is analytic at $x = 1$; it will then follow that
$(x-1)^2/\ln x = (x-1)[(x-1)/\ln x]$ is also analytic at
$x = 1$. If we expand $\ln x$ in a Taylor series about $x = 1$
we find that $\ln x = (x-1) - \dfrac{1}{2}(x-1)^2 + \dfrac{1}{3}(x-1)^3 - ...$.

Thus

$(x-1)/\ln x = [1 - \dfrac{1}{2}(x-1) + \dfrac{1}{3}(x-1)^2 -...]^{-1} = 1 + \dfrac{1}{2}(x-1)+...$

has a power series expansion about $x = 1$, and hence is
analytic.

18b. We can use the above result to obtain the indicial
equation at $x = 1$. We have

$(x-1)^2y'' + (x-1)[\dfrac{1}{2} + \dfrac{1}{4}(x-1) + ...]y' +$

$\qquad\qquad\qquad\qquad [(x-1) + \dfrac{1}{2}(x-1)^2 + ...]y = 0.$

Thus $p_0 = 1/2$, $q_0 = 0$ and the indicial equation is
$r(r-1) + r/2 = 0$. Hence $r = 1/2$ and $r = 0$.

18c. In order to find the first three non-zero terms in a
series solution corresponding to $r = 1/2$, it is better to
keep the differential equation in its original form and
to substitute the above power series for $\ln x$:

$$[(x-1) - \frac{1}{2}(x-1)^2 + \frac{1}{3}(x-1)^3 - \frac{1}{4}(x-1)^4 + \ldots]y'' + \frac{1}{2}y' + y = 0.$$

Next we substitute $y = a_0(x-1)^{1/2} + a_1(x-1)^{3/2} + a_2(x-1)^{5/2} + \ldots$ and collect coefficients of like powers of $(x-1)$ which are then set equal to zero. This requires some algebra before we find that $6a_1/4 + 9a_0/8 = 0$ and $5a_2 + 5a_1/8 - a_0/12 = 0$. These equations yield $a_1 = -3a_0/4$ and $a_2 = 53a_0/480$. With $a_0 = 1$ we obtain the solution

$$y_1(x) = (x-1)^{1/2} - \frac{3}{4}(x-1)^{3/2} + \frac{53}{480}(x-1)^{5/2} + \ldots .$$

18d. Since the radius of convergence of the Taylor Series for $(x-1)/\ln x$ is 1, we would expect $\rho = 1$.

20a. If we write the D.E. in the standard form as given in Theorem 5.6.1 we obtain $x^2y'' + x[\alpha/x]y' + [\beta/x]y = 0$ where $xp(x) = \alpha/x$ and $x^2q(x) = \beta/x$. Neither of these terms are analytic at $x = 0$ so $x = 0$ is an irregular singular point.

20b. Substituting $y = x^r \sum\limits_{n=0}^{\infty} a_n x^n$ in $x^3y'' + \alpha xy' + \beta y = 0$ gives

$$\sum_{n=0}^{\infty}(n+r)(n+r-1)a_n x^{n+r+1} + \alpha\sum_{n=0}^{\infty}(n+r)a_n x^{n+r} + \beta\sum_{n=0}^{\infty}a_n x^{n+r} = 0.$$

Shifting the index in the first series and collecting coefficients of common powers of x we obtain

$$(\alpha r + \beta)a_0 x^r + \sum_{n=1}^{\infty}\{(n+r-1)(n+r-2)a_{n-1} + [\alpha(n+r) + \beta]a_n\}x^{n+r} = 0.$$

Thus the indicial equation is $\alpha r + \beta = 0$ with the single root $r = -\beta/\alpha$.

20c. From part b, the recurrence relation is

$$a_n = -\frac{(n+r-1)(n+r-2)a_{n-1}}{\alpha(n+r) + \beta}, \quad n = 1,2,\ldots$$

$$= -\frac{(n -\frac{\beta}{\alpha} -1)(n -\frac{\beta}{\alpha} -2)a_{n-1}}{\alpha n}, \quad \text{for } r = -\beta/\alpha.$$

For $\dfrac{\beta}{\alpha} = -1$, $a_n = -\dfrac{n(n-1)a_{n-1}}{\alpha n}$, so that $a_1 = 0 \cdot a_0 = 0$.
Since all other a_n are multiples of a_1, and hence are
zero, $y(x) = x$ is the solution. Similarly for $\dfrac{\beta}{\alpha} = 0$,

$a_n = -\dfrac{(n-1)(n-2)}{\alpha n}a_{n-1}$ and again for $n = 1$ $a_1 = 0$ and

$y(x) = 1$ is the solution. Continuing in this fashion, we
see that the series solution will terminate for β/α any
positive integer as well as 0 and -1. For other values

of β/α, we have $\left| \dfrac{a_n}{a_{n-1}} \right| = \dfrac{(n-\dfrac{\beta}{\alpha}-1)(n-\dfrac{\beta}{\alpha}-2)}{\alpha n}$, which

approaches ∞ as $n \to \infty$ and thus the ratio test yields a
zero radius of convergence.

21b. Substituting $y = \displaystyle\sum_{n=0}^{\infty} a_n x^{n+r}$ in the D.E. gives

$$\sum_{n=0}^{\infty}(n+r)(n+r-1)a_n x^{n+r} + \alpha\sum_{n=0}^{\infty}(n+r)a_n x^{n+r+1-s} + \beta\sum_{n=0}^{\infty}a_n x^{n+r+2-t} = 0.$$

If $s = 2$ and $t = 2$ the first term in each of the three
series is $r(r-1)a_0 x^r$, $\alpha r a_0 x^{r-1}$, and $\beta a_0 x^r$, respectively.
Thus the indicial equation is $F(r) = \alpha r a_0 = 0$, which
requires $r = 0$. Hence there is at most one solution of
the assumed form.

21d. In order for the indicial equation to be quadratic in r
it is necessary that the first term in the first series
contribute to the indicial equation. This means that the
first term in the second and the third series cannot have
powers less than x^r. The first terms are $r(r-1)a_0 x^r$,
$\alpha r a_0 x^{r+1-s}$, and $\beta a_0 x^{r+2-t}$, respectively. Thus if $s \le 1$ and
$t \le 2$ the quadratic term will appear in the indicial
equation.

Section 5.7, Page 301

1. It is clear that $x = 0$ is a singular point. The D.E. is
    in the standard form given in Theorem 5.6.1 with
    $xp(x) = 2$ and $x^2 q(x) = x$. Both are analytic at $x = 0$, so
    $x = 0$ is a regular singular point. Substituting

$y = \displaystyle\sum_{n=0}^{\infty} a_n x^{n+r}$ in the D.E., shifting indices appropriately,

and collecting coefficients of like powers of x yields

$$[r(r-1) + 2r]a_0 x^r + \sum_{n=1}^{\infty} [(r+n)(r+n+1)a_n + a_{n-1}]x^{r+n} = 0.$$

The indicial equation is $F(r) = r(r+1) = 0$ with roots $r_1 = 0$, $r_2 = -1$. Treating a_n as a function of r, we see that $a_n(r) = -a_{n-1}(r)/F(r+n)$, $n = 1,2,\ldots$ if $F(r+n) \neq 0$. Thus $a_1(r) = -a_0/F(r+1)$, $a_2(r) = a_0/F(r+1)F(r+2),\ldots$, and $a_n(r) = (-1)^n a_0/F(r+1)F(r+2)\ldots F(r+n)$, provided $F(r+n) \neq 0$ for $n = 1,2,\ldots$. For the case $r_1 = 0$, we have $a_n(0) = (-1)^n a_0/F(1)F(2) \ldots F(n) = (-1)^n a_0/n!(n+1)!$ so

one solution is $y_1(x) = \displaystyle\sum_{n=0}^{\infty} (-1)^n x^n/n!(n+1)!$ where we have

set $a_0 = 1$.

   If we try to use the above recurrence relation for the case $r_2 = -1$ we find that $a_n(-1) = -a_{n-1}/n(n-1)$, which is undefined for $n = 1$. Thus we must follow the procedure described at the end of Sect. 5.6 to calculate a second solution of the form given in Eq.(24). Specifically, we use Eqs.(19) and (20) of Sect.5.6 to calculate a and $c_n(r_2)$, where $r_2 = -1$. Since $r_1 - r_2 = 1 = N$, we have $a_N(r) = a_1(r) = -1/F(r+1)$, with $a_0 = 1$. Hence

$a = \displaystyle\lim_{r \to -1} [(r+1)(-1)/F(r+1)] = \lim_{r \to -1} [-(r+1)/(r+1)(r+2)] = -1.$

Next

$$c_n(-1) = \frac{d}{dr}[(r+1)a_n(r)]\Big|_{r=-1} = (-1)^n \frac{d}{dr}[\frac{(r+1)}{F(r+1) \ldots F(r+n)}]\Big|_{r=-1},$$

where we again have set $a_0 = 1$. Observe that $(r+1)/F(r+1)\ldots F(r+n)=1/[(r+2)^2(r+3)^2\ldots(r+n)^2(r+n+1)]=1/G_n(r)$. Hence $c_n(-1) = (-1)^{n+1}G_n'(-1)/G_n^2(-1)$. Notice that $G_n(-1) = 1^2 \cdot 2^2 \cdot 3^2 \ldots (n-1)^2 n = (n-1)!n!$ and $G_n'(-1)/G_n(-1) = 2[1/1 + 1/2 + 1/3 +\ldots+ 1/(n-1)] + 1/n = H_n + H_{n-1}$. Thus $c_n(-1) = (-1)^{n+1}(H_n + H_{n-1})/(n-1)!n!$. From Eq.(24) of Sect.5.6 we obtain the second solution

$$y_2(x) = -y_1(x)\ln x + x^{-1}[1 - \sum_{n=1}^{\infty} (-1)^n (H_n + H_{n-1})x^n/n!(n-1)!].$$

2. It is clear that $x = 0$ is a singular point. The D.E. is in the standard form given in Theorem 5.6.1 with $xp(x) = 3$ and $x^2q(x) = 1+x$. Both are analytic at $x = 0$, so $x = 0$ is a regular singular point. Substituting

$$y = \sum_{n=0}^{\infty} a_n x^{n+r} \text{ in the D.E., shifting indices}$$

appropriately, and collecting coefficients of like powers of x yields

$$[r(r-1) + 3r + 1]a_0 x^r + \sum_{n=1}^{\infty} \{[(r+n)(r+n+2) + 1]a_n + a_{n-1}\}x^{n+r} = 0.$$

The indicial equation is $F(r) = r^2 + 2r + 1 = (r+1)^2 = 0$ with the double root $r_1 = r_2 = -1$. Treating a_n as a function of r, we see that $a_n(r) = -a_{n-1}(r)/F(r+n)$, $n = 1,2,\ldots$. Thus $a_1(r) = -a_0/F(r+1)$, $a_2(r) = a_0/F(r+1)F(r+2),\ldots$, and $a_n(r) = (-1)^n a_0/F(r+1)F(r+2)\ldots F(r+n)$. Setting $r = -1$ we find that $a_n(-1)=(-1)^n a_0/(n!)^2$, $n = 1,2,\ldots$. Hence one

solution is $y_1(x) = x^{-1}\sum_{n=0}^{\infty} (-1)^n x^n/(n!)^2$ where we have set

$a_0 = 1$. To find a second solution we follow the procedure described in Sect.5.6 for the case when the roots of the indicial equation are equal. Specifically, the second solution will have the form given in Eq.(17) of that section. We must calculate $a_n'(-1)$. If we let $G_n(r) = F(r+1)\ldots F(r+n) = (r+2)^2(r+3)^2\ldots (r+n+1)^2$ and take $a_0 = 1$, then $a_n'(-1) = (-1)^n[1/G_n(r)]'$ evaluated $r = -1$. Hence $a_n'(-1) = (-1)^{n+1}G_n'(-1)/G_n^2(-1)$. But $G_n(-1) = (n!)^2$ and $G_n'(-1)/G_n(-1) = 2[1/1 + 1/2 + 1/3 + \ldots 1/n]$
$= 2H_n$. Thus a second

solution is $y_2(x) = y_1(x)\ln x - 2x^{-1}\sum_{n=1}^{\infty} (-1)^n H_n x^n/(n!)^2.$

3.   The roots of the indicial equation are r_1 and $r_2 = 0$ and
      thus the analysis is similar to that for Prob. 2.

4.   The roots of the indicial equation are $r_1 = -1$ and
      $r_2 = -2$ and thus the analysis is similar to that for
      Prob.1.

5.   Since $x = 0$ is a regular singular point, substitute

      $$y = \sum_{n=0}^{\infty} a_n x^{n+r}$$ in the D.E., shift indices appropriately,

      and collect coefficients of like powers of x to obtain
      $$[r^2 - 9/4]a_0 x^r + [(r+1)^2 - 9/4]a_1 x^{r+1}$$

      $$+ \sum_{n=2}^{\infty} \{[(r+n)^2 - 9/4]a_n + a_{n-2}\} x^{n+r} = 0.$$

      The indicial equation is $F(r) = r^2 - 9/4 = 0$ with roots
      $r_1 = 3/2$, $r_2 = -3/2$. Treating a_n as a function of r we
      see that $a_n(r) = -a_{n-2}(r)/F(r+n)$, $n = 2,3,..$ if
      $F(r+n) \neq 0$. For the case $r_1 = 3/2$, $F(r_1+1)$, which is the
      coefficient of x^{r_1+1} is $\neq 0$ so we must set $a_1 = 0$. It
      follows that $a_3 = a_5 = ... = 0$. For the even
      coefficients, set $n = 2m$ so
      $a_{2m}(3/2) = -a_{2m-2}(3/2)/F(3/2 + 2m) = -a_{2m-2}/2^2 m(m+3/2)$,
      $m = 1,2...$. Thus $a_2(3/2) = - a_0/2^2 \cdot 1(1 + 3/2)$,
      $a_4(3/2) = a_0/2^4 \cdot 2!(1 + 3/2)(2 + 3/2),...,$ and
      $a_{2m}(3/2) = (-1)^m/2^{2m} m! \cdot (1 + 3/2)...(m + 3/2)$. Hence one
      solution is

      $$y_1(x) = x^{3/2}[1 + \sum_{m=1}^{\infty} \frac{(-1)^m}{m!(1 + 3/2)(2 + 3/2)...(m + 3/2)} (\frac{x}{2})^{2m}],$$

      where we have set $a_0 = 1$. For this problem, the roots r_1
      and r_2 of the indicial equation differ by an integer: $r_1 - r_2$
      $= 3/2 - (-3/2) = 3$. Hence we can anticipate that there may
      be difficulty in calculating a second solution corresponding
      to $r = r_2$. This difficulty will occur in calculating
      $a_3(r) = - a_1(r)/F(r+3)$ since when $r = r_2 = -3/2$ we have
      $F(r_2+3) = F(r_1) = 0$. However, in this problem we are
      fortunate because $a_1 = 0$ and it will not be necessary to use
      the theory described at the end of Sect.5.6. Notice for

$r = r_2 = -3/2$ that the coefficient of x^{r_2+1} is

$[(r_2+1)^2 - 9/4]a_1$, which does not vanish unless $a_1 = 0$. Thus
the recurrence relation for the odd coefficients yields
$a_5 = -a_3/F(7/2)$, $a_7 = -a_5/F(11/2) = a_3/F(11/2)F(7/2)$ and so
forth. Substituting these terms into the assumed form we
see that a multiple of $y_1(x)$ has been obtained and thus we
may take $a_3 = 0$ without loss of generality. Hence
$a_3 = a_5 = a_7 = \ldots = 0$. The even coefficients are given
by $a_{2m}(-3/2) = -a_{2m-2}(-3/2)/F(2m - 3/2)$, $m = 1,2\ldots$.

Thus $a_2(-3/2) = -a_0/2^2 \cdot 1 \cdot (1 - 3/2)$,

$a_4(-3/2) = a_0/2^4 \cdot 2! (1 - 3/2)(2 - 3/2),\ldots$, and

$a_{2m}(-3/2) = (-1)^m a_0/2^{2m} m! (1 - 3/2)(2 - 3/2) \ldots (m - 3/2)$.

Thus a second solution is

$$y_2(x) = x^{-3/2}[1 + \sum_{m=1}^{\infty} \frac{(-1)^m}{m!(1 - 3/2)(2 - 3/2) \ldots (m - 3/2)} (\frac{x}{2})^{2m}].$$

7. Apply the ratio test:

$$\lim_{m \to \infty} \frac{|(-1)^{m+1} x^{2m+2}/2^{2m+2}[(m+1)!]^2|}{|(-1)^m x^{2m}/2^{2m}(m!)^2|} = |x^2| \lim_{m \to \infty} \frac{1}{2^2(m+1)^2} = 0$$

for every x. Thus the series for $J_0(x)$ converges
absolutely for all x.

12. If $\xi = \alpha x^\beta$, then $dy/dx = \frac{1}{2}x^{-1/2}f + x^{1/2}f'\alpha\beta x^{\beta-1}$ where f'

denotes $df/d\xi$. Find d^2y/dx^2 in a similar fashion and use
algebra to show that f satisfies the D.E.
$\xi^2 f'' + \xi f' + [\xi^2 - \upsilon^2]f = 0$, which is the Bessel Equation
of order υ.

13. To compare $y'' - xy = 0$ with the D.E. of Prob.12, we must
multiply by x^2 to get $x^2 y'' - x^3 y = 0$. Thus $2\beta = 3$, $\alpha^2\beta^2$
$= -1$ and $1/4 - \upsilon^2\beta^2 = 0$. Hence $\beta = 3/2$, $\alpha = 2i/3$ and υ^2
$= 1/9$ which yields the desired result.

14. First we verify that $J_0(\lambda_j x)$ satisfies the D.E. We know
that $J_0(t)$ is a solution of the Bessel equation of order
zero:

$$t^2 J_0''(t) + t J_0'(t) + t^2 J_0(t) = 0 \text{ or}$$

$$J_0''(t) + t^{-1} J_0'(t) + J_0(t) = 0.$$

Let $t = \lambda_j x$. Then

$$\frac{d}{dx} J_0(\lambda_j x) = \frac{d}{dt} J_0(t) \frac{dt}{dx} = \lambda_j J_0'(t)$$

$$\frac{d^2}{dx^2} J_0(\lambda_j x) = \lambda_j \frac{d}{dt} [J_0'(t)] \frac{dt}{dx} = \lambda_j^2 J_0''(t).$$

Substituting $y = J_0(\lambda_j x)$ in the given D.E. and making use of these results, we have

$$\lambda_j^2 J_0''(t) + (\lambda_j/t) \lambda_j J_0'(t) + \lambda_j^2 J_0(t) =$$

$$\lambda_j^2 [J_0''(t) + t^{-1} J_0'(t) + J_0(t)] = 0.$$

Thus $y = J_0(\lambda_j x)$ is a solution of the given D.E. For the second part of the problem we follow the hint. First, rewrite the D.E. by multiplying by x to yield
$xy'' + y' + \lambda_j^2 xy = 0$, which can be written as $(xy')' = -\lambda_j^2 xy$.
Now let $y_i(x) = J_0(\lambda_i x)$ and $y_j(x) = J_0(\lambda_j x)$ and we have, respectively:

$$(xy_i')' = -\lambda_i^2 xy_i$$

$$(xy_j')' = -\lambda_j^2 xy_j.$$

Now, multiply the first equation by y_j, the second by y_i, integrate each from 0 to 1, and subtract the second from the first:

$$\int_0^1 [y_j(xy_i')' - y_i(xy_j')'] dx = -(\lambda_i^2 - \lambda_j^2) \int_0^1 xy_i y_j dx.$$

If we integrate each term on the left side once by parts and note that $y_i = y_j = 0$ at $x = 1$, we find that the left side of this equation is identically zero. Hence the right side is identically zero and for $\lambda_i \neq \lambda_j$ this gives the desired result.

CHAPTER 6

Section 6.1, Page 311

1.   The graph of f(t) is shown.
     Since the function is
     continuous on each interval,
     but has a jump discontinuity
     at t = 1, f(t) is piecewise
     continuous.

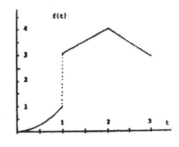

2.   Note that $\lim\limits_{t \to 1^+} (t-1)^{-1} = \infty$.

5b. Since t^2 is continuous for $0 \leq t \leq A$ for any positive A
     and since $t^2 \leq e^{at}$ for any a > 0 and for t sufficiently
     large, it follows from Theorem 6.1.2 that $\pounds\{t^2\}$ exists
     for s > 0. $\pounds\{t^2\} = \int_0^\infty e^{-st} t^2 dt = \lim\limits_{M \to \infty} \int_0^M e^{-st} t^2 dt$

$$= \lim\limits_{M \to \infty} [\frac{-t^2}{s} e^{-st} \big|_0^M + \frac{2}{s} \int_0^M e^{-st} t \, dt]$$

$$= \lim\limits_{M \to \infty} \frac{-M^2}{s} e^{-sM} + \frac{2}{s} \lim\limits_{M \to \infty} [-\frac{1}{s} t e^{-st} \big|_0^M + \frac{1}{s} \int_0^M e^{-st} dt]$$

$$= 0 + 2 \lim\limits_{M \to \infty} \frac{-M}{s^2} e^{-sM} + \frac{2}{s^2} \lim\limits_{M \to \infty} -\frac{1}{s} e^{-st} \big|_0^M = \frac{2}{s^3}.$$

6.   That f(t) = cosat satisfies the hypotheses of Theorem
     6.1.2 can be verified by recalling that $|\cos at| \leq 1$ for
     all t. To determine $\pounds\{\cos at\} = \int_0^\infty e^{-st} \cos at \, dt$ we
     must integrate by parts twice to get

$$\int_0^\infty e^{-st} \cos at \, dt = \lim\limits_{M \to \infty} [(-s^{-1} e^{-st} \cos at + as^{-2} e^{-st} \sin at)\big|_0^M$$

     $- (a^2/s^2) \int_0^M e^{-st} \cos at \, dt]$. Evaluating the first two
     terms and letting $M \to \infty$ yields

$\int_0^\infty e^{-st} \cos at \, dt = \frac{1}{s} - (a^2/s^2) \int_0^\infty e^{-st} \cos at \, dt$ and hence

$[1+a^2/s^2] \int_0^\infty e^{-st} \cos at \, dt = 1/s$, s > 0. Division by $[1+a^2/s^2]$
and simplification yields the desired solution.

9. From the definition for coshbt we have

$\mathcal{L}\{e^{at}coshbt\} = \mathcal{L}\{\frac{1}{2}[e^{(a+b)t} + e^{(a-b)t}]\}$. Using the linearity property of $\mathcal{L}$, Eq.(6), the right side becomes $\frac{1}{2}\mathcal{L}\{e^{(a+b)t}\} + \frac{1}{2}\mathcal{L}\{e^{(a-b)t}\}$ which can be evaluated using the result of Ex. 5 and thus

$$\mathcal{L}\{e^{at}coshbt\} = \frac{1/2}{s-(a+b)} + \frac{1/2}{s-(a-b)}, \text{ for } s-a > |b|$$

$$= \frac{s-a}{(s-a)^2-b^2}.$$

13. We write $sinat = (e^{iat} - e^{-iat})/2i$, then the linearity of the Laplace transform operator allows us to write $\mathcal{L}\{e^{at}sinbt\} = (1/2i)\mathcal{L}\{e^{(a+ib)t}\}-(1/2i)\mathcal{L}\{e^{(a-ib)t}\}$. Each of these two terms can be evaluated by using the result of Ex. 5, where we now have to require s to be greater than the real part of the complex numbers $a \pm ib$ in order for the integrals to converge. Complex algebra then gives the desired result. An alternate method of evaluation would be to use integration on the integral appearing in the definition of $\mathcal{L}\{e^{at}sinbt\}$, but that method requires integration by parts twice.

16. Before starting note that both tsinat and tcosat satisfy Condition 2 of Theorem 6.1.2 and thus the $\lim\limits_{M\to\infty}$ in the follwing are both zero. Using integration by parts twice we have $F(s) = \int_0^{\infty}(tsinat)e^{-st}dt$

$$= \lim_{M\to\infty}(tsinat)|_0^M + \frac{1}{s}\int_0^{\infty}(sinat+acosat)e^{-st}dt$$

$$= \lim_{M\to\infty}(sinat+atcosat)|_0^M + \frac{1}{s^2}\int_0^{\infty}(2atcosat-atsinat)e^{-st}dt$$

$$= \frac{2a}{s^s}\mathcal{L}\{cosat\} - \frac{a}{s^2}F(s) \text{ and thus } (1 + \frac{a}{s^2})F(s) = \frac{2as}{s^2(s^2+a^2)}.$$

Solving for F(s) we find $F(s) = \frac{2as}{(s^2+a^2)^2}$.

21. The integral $\int_0^A (t^2 + 1)^{-1}dt$ can be evaluated in terms of the arctan function and then Eq. (1) can be used. To illustrate Theorem 6.1.1, however, consider that

$$\frac{1}{t^2+1} < \frac{1}{t^2} \text{ for } t \geq 1 \text{ and, from Ex. 3, } \int_1^\infty t^{-2}dt \text{ converges}$$

and hence $\int_1^\infty (t^2 + 1)^{-1}dt$ also converges.

$\int_0^1 (t^2 + 1)^{-1}dt$ is finite and hence does not affect the

convergence of $\int_0^\infty (t^2 + 1)^{-1}dt$ at infinity.

25. If we let $u = f$ and $dv = e^{-st}dt$ then

$$F(s) = \int_0^\infty e^{-st}f(t)dt = \lim_{M \to \infty} -\frac{1}{s}e^{-st}f(t)\Big|_0^M + \frac{1}{s}\int_0^\infty e^{-st}f'(t)dt$$

$$= \frac{1}{s}f(0) + \frac{1}{s}\int_0^\infty e^{-st}f'(t)dt. \text{ This last integral}$$

converges (and is thus finite) using an argument similar to
that given to establish Theoremm 6.1.2. Hence $\lim_{s \to \infty} F(s) = 0$.

27a. let $x = st$ and so that $dx = sdt$. Then use the definition
of $\Gamma(P+1)$ from Prob. 26.

27b. From Part a, $\mathcal{L}\{t^n\} = \frac{1}{s^{n+1}}\int_0^\infty e^{-x}x^n dx = \frac{n}{s^{n+1}}\int_0^\infty e^{-x}x^{n-1}dx$

$$= \frac{n!}{s^{n+1}}\int_0^\infty e^{-x}dx, \text{ using integration by}$$

parts successively. Evaluation of the last integral
yields the desired answer.

27c. From part a, $\mathcal{L}\{t^{-1/2}\} = \frac{1}{\sqrt{s}}\int_0^\infty e^{-x}x^{-1/2}dx$. Let $x = y^2$, then

$2dy = x^{-1/2}dx$ and thus $\mathcal{L}\{t^{-1/2}\} = \frac{2}{\sqrt{s}}\int_0^\infty e^{-y^2}dy$.

27d. Use the definition of $\mathcal{L}\{t^{1/2}\}$ and integrate by parts once
to get $\mathcal{L}\{t^{1/2}\} = (1/2s)\mathcal{L}\{t^{-1/2}\}$. The result follows from
part c.

Section 6.2, Page 320

Problems 1 through 10 are solved by using partial fractions
and algebra to manipulate the given function into a form
matching one of the functions appearing in the middle column
of Table 6.2.1.

2. We have $\dfrac{4}{(s-1)^3} = 2\dfrac{2!}{(s-1)^{2+1}}$ and thus the inverse Laplace transform is $2t^2e^t$, using line 11.

4. We have $\dfrac{3s}{s^2-s-6} = \dfrac{3s}{(s-3)(s+2)} = \dfrac{9/5}{s-3} + \dfrac{6/5}{s+2}$ using partial fractions. Thus $(9/5)e^{3t} + (6/5)e^{-2t}$ is the inverse transform, from line 2.

7. We have $\dfrac{2s+1}{s^2-2s+2} = \dfrac{2s+1}{(s-1)^2+1} = \dfrac{2(s-1)}{(s-1)^2+1} + \dfrac{3}{(s-1)^2+1}$, where we first used the concept of completing the square (in the denominator) and then added and subtracted appropriately to put the numerator in the desired form. Lines 9 and 10 may now be used to find the desired result.

In each of the Problems 11 through 23 it is assumed that the I.V.P. has a solution $y = \phi(t)$ which, with its first two derivatives, satisfies the conditions of the Corollary 6.2.2.

11. Take the Laplace transform of the D.E., using Eq.(1) and Eq.(2), to get
    $s^2Y(s) - sy(0) - y'(0) - [sY(s) - y(0)] - 6Y(s) = 0$.
    Using the I.C. and solving for $Y(s)$ we obtain
    $Y(s) = \dfrac{s-2}{s^2-s-6}$. Following the pattern of Eq.(12) we have
    $\dfrac{s-2}{s^2-s-6} = \dfrac{a}{s+2} + \dfrac{b}{s-3} = \dfrac{a(s-3)+b(s+2)}{(s+2)(s-3)}$. Equating like powers in the numerators we find $a+b = 1$ and $-3a + 2b = -2$. Thus $a = 4/5$ and $b = 1/5$ and
    $Y(s) = \dfrac{4/5}{s+2} + \dfrac{1/5}{s-3}$, which yields the desired solution using Table 6.2.1.

14. Taking the Laplace transform we have
    $s^2Y(s) - sy(0) - y'(0) - 4[sY(s)-y(0)] + 4Y(s) = 0$. Using the I.C. and solving for $Y(s)$ we find $Y(s) = \dfrac{s-3}{s^2-4s+4}$. Since the denominator is a perfect square, the partial fraction form is
    $\dfrac{s-3}{s^2-4s+4} = \dfrac{a}{(s-2)^2} + \dfrac{b}{s-2}$. Solving for a and b, as shown in examples of this section or in Prob. 11, we find $a = -1$ and

$b = 1$. Thus $Y(s) = \dfrac{1}{s-2} - \dfrac{1}{(s-2)^2}$, from which we find

$y(t) = e^{2t} - te^{2t}$ (lines 2 and 11 in Table 6.2.1).

15. Note that $Y(s) = \dfrac{2s-4}{s^2-2s+4} = \dfrac{2s-4}{(s-1)^2+3} = \dfrac{2(s-1)}{(s-1)^2+3} - \dfrac{2}{(s-1)^2+3}$.
Three formulas in Table 6.2.1 are now needed: $F(s-c)$
(with $c = 1$) in line 14 in conjunction with the ones for
cosat and sinat (with $a = \sqrt{3}$), lines 5 and 6.

17. The Laplace transform of the D.E. is
$s^4Y(s) - s^3y(0) - s^2y'(0) - sy''(0) - y'''(0) - 4[s^3Y(s) - s^2y(0)$
$-sy'(0) - y''(0)] + 6[s^2Y(s) - sy(0) - y'(0)] - 4[sY(s) - y(0)]$
$+ Y(s) = 0$. Using the I.C. and solving for $Y(s)$ we find

$Y(s) = \dfrac{s^2 - 4s + 7}{s^4-4s^3+6s^2-4s+1}$. The correct partial fraction

form for this is $\dfrac{a}{(s-1)^4} + \dfrac{b}{(s-1)^3} + \dfrac{c}{(s-1)^2} + \dfrac{d}{s-1}$.
Setting this equal to $Y(s)$ above and equating the
numerators we have $s^2-4s+7 = a + b(s-1) + c(s-1)^2 + d(s-1)^3$.
Solving for $a,b,c,$ and d and use of line 11 in Table 6.2.1
yields the desired solution.

20. The Laplace transform of the D.E. is
$s^2Y(s) - sy(0) - y'(0) + \omega^2Y(s) = s/(s^2+4)$. Applying the
I.C. and solving for $Y(s)$ we get $Y(s) = s/[(s^2+4)(s^2+\omega^2)]$
$+ s/(s^2+\omega^2)$. Decomposing the first term by partial
fractions we have

$Y(s) = \dfrac{s}{(\omega^2-4)(s^2+4)} - \dfrac{s}{(\omega^2-4)(s^2+\omega^2)} + \dfrac{s}{s^2+\omega^2}$

$= (\omega^2-4)^{-1}[\dfrac{(\omega^2-5)s}{s^2+\omega^2} + \dfrac{s}{s^2+4}]$.

Then, using lines 5 and 6 of Table 6.1.2, we have
$y = (\omega^2-4)^{-1}[(\omega^2-5)\cos\omega t + \cos 2t]$.

22. Solving for $Y(s)$ we find
$Y(s) = \dfrac{1}{(s-1)^2+1} + \dfrac{1}{(s+1)[(s-1)^2+1]}$. Using partial

fractions on the second term we obtain
$Y(s) = \dfrac{1}{(s-1)^2+1} + \dfrac{1}{5}[\dfrac{1}{s+1} - \dfrac{s-3}{(s-1)^2+1}]$. Combining the

first and third terms we have

$$Y(s) = \frac{1}{5}[\frac{1}{s+1} - \frac{s-1}{(s-1)^2+1} + \frac{7}{(s-1)^2+1}].$$

Hence, $y = (1/5)(e^{-t} - e^t\cos t + 7e^t\sin t)$.

24. Under the standard assumptions, the Lapace transform of the left side of the D.E. is $s^2Y(s) - sy(0) - y'(0) + 4Y(s)$, or $(s^2 + 4)Y(s) - s$. To transform the right side we must revert to the definition of the Laplace transform to determine $\int_0^\infty e^{-st}f(t)dt$. Since $f(t)$ is piecewise continuous we are able to calculate $\mathcal{L}\{f(t)\}$ by

$$\int_0^\infty e^{-st}f(t)dt = \int_0^\pi e^{-st}\,dt + \lim_{M \to \infty}\int_\pi^M (e^{-st})(0)dt$$

$$= \int_0^\pi e^{-st}dt = (1 - e^{-\pi s})/s.$$

Hence, the Laplace transform $Y(s)$ of the solution is given by $Y(s) = s/(s^2+4) + (1 - e^{-\pi s})/s(s^2+4)$.

27b. The Taylor series for f about $t = 0$ is

$$f(t) = \sum_{n=0}^\infty (-1)^n\frac{t^{2n}}{(2n+1)!},$$ which is obtained from part(a) by

dividing each term of the sine series by t. Also, f is continuous for $t > 0$ since $\lim_{t \to 0+} \frac{\sin t}{t} = 1$. Assuming that we can compute the Laplace transform of f term by term, we obtain $\mathcal{L}\{f(t)\} = \mathcal{L}\{\sum_{n=0}^\infty (-1)^n\frac{t^{2n}}{(2n+1)!}\}$

$$= \sum_{n=0}^\infty \frac{(-1)^n}{(2n+1)!}\mathcal{L}\{t^{2n}\} = \sum_{n=0}^\infty \frac{(-1)^n 2n!}{(2n+1)!s^{2n+1}} = \sum_{n=0}^\infty \frac{(-1)^n}{2n+1}\frac{1}{s^{2n+1}},$$

which converges for $s > 1$. The Taylor series for arctan x is given by $\sum_{n=0}^\infty (-1)^n\frac{x^{2n+1}}{2n+1}$, for $|x| < 1$. Comparing $\mathcal{L}\{f(t)\}$ with the Taylor series for arctanx, we conclude that $\mathcal{L}\{f(t)\} = \arctan(1/s)$, $s > 1$.

30. Setting $n = 2$ in Prob. 28b, we have

$$\mathcal{L}\{t^2\sin bt\} = \frac{d^2}{ds^2}[\frac{b}{s^2+b^2}] = \frac{d}{ds}[\frac{-2bs}{(s^2+b^2)^2}] = \frac{-2b}{(s^2+b^2)^2} + \frac{8bs^2}{(s^2+b^2)^3}$$

$$= \frac{2b(3s^2-b^2)}{(s^2+b^2)^3}.$$

32. Using the result of Prob. 28a, repeatedly, we have

$$\mathcal{L}\{te^{at}\} = -\frac{d}{ds}(s-a)^{-1} = (s-a)^{-2},$$

$$\mathcal{L}\{t^2e^{at}\} = -\frac{d}{ds}(s-a)^{-2} = 2(s-a)^{-3}, \text{ and}$$

$$\mathcal{L}\{t^3e^{at}\} = -\frac{d}{ds}2(s-a)^{-3} = 3!(s-a)^{-4}.$$ Continuing in this

fashion, or using induction, we obtain the desired
result.

36a. Taking the Laplace transform of the D.E. we obtain

$$\mathcal{L}\{y''\} - \mathcal{L}\{ty\} = \mathcal{L}\{y''\} + \mathcal{L}\{-ty\}$$
$$= s^2Y(s) - sy(0) - y'(0) + Y'(s) = 0.$$

Hence, Y satisfies $Y' + s^2Y = s$.

38a. From Eq(i) we have $A_k = \lim_{s\to r_k}(s-r_k)\frac{P(r_k)}{Q(r_k)}$, since Q has

distinct zeros. Thus $A_k = P(r_k)\lim_{s\to r_k}\frac{s-r_k}{Q(r_k)} = \frac{P(r_k)}{Q'(r_k)}$, by

L'Hospital's Rule.

38b. Since $\mathcal{L}^{-1}\left\{\frac{1}{s-r_k}\right\} = e^{r_k t}$, the result follows.

Section 6.3, Page 328

2. From the definition of $u_c(t)$
    we have:
    $$g(t) = (t-3)u_2(t) - (t-2)u_3(t)$$

    $$= \begin{cases} 0 - 0 = 0, & 0 \le t < 2 \\ (t-3) - 0 = t-3, & 2 \le t < 3. \\ (t-3) - (t-2) = -1, & 3 \le t \end{cases}$$

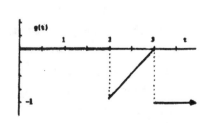

4. As indicated in the discussion
    following Eq.(2), the unit step
    function can be used to
    translate a given function f,
    with domain t≥0, a distance c
    to the right by the
    multiplication $u_c(t)f(t-c)$.
    Hence the required graph of
    $y = u_3(t)f(t-3)$ for $f(t) = \sin t$ is shown.

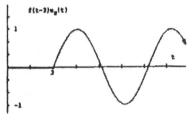

7. There are step changes at
    t = 3,5,7. Thus we use $u_3(t)$,
    $u_5(t)$ and $u_7(t)$ multiplied by
    the appropriate step size to
    obtain
    $f(t) = -2u_3(t)+4u_5(t)-u_7(t)$.

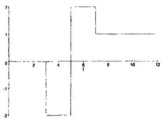

10. There is a discontinuity at
    t = 2 so we use $u_2(t)$
    multiplied by $(1-t^2)$, which
    subtracts the existing t^2
    function. Thus
    $f(t) = t^2 + (1-t^2)u_2(t)$.

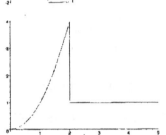

14. In order to use Theorem 6.3.1 we must write f(t) in terms
    of $u_c(t)$. Since $t^2- 2t + 2 = (t-1)^2 + 1$ (by completing
    the square), we can write $f(t) = u_1(t)g(t-1)$, where
    $g(t) = t^2+1$. Now applying Theorem 6.3.1 we have
    $£\{f(t)\} = £\{u_1(t)g(t-1)\} = e^{-s} £\{g(t)\} = e^{-s}(2/s^3 + 1/s)$.

20. Use partial fractions to write
    $F(s) = e^{-2s}\dfrac{1}{3}[\dfrac{1}{s-1} - \dfrac{1}{s+2}]$. For ease in calculations let
    us define $G(s) = (s-1)^{-1}$ and $H(s) = (s+2)^{-1}$. Then
    $F(s) = [e^{-2s} G(s) - e^{-2s} H(s)]/3$. Using the fact that
    $£\{e^{at}\} = (s-a)^{-1}$ and applying Theorem 6.3.1, we have
    $F(s) = [e^{-2s} £\{e^t\} - e^{-2s} £\{e^{-2t}\}]/3$. Thus
    $F(s) = [£\{u_2(t)e^{(t-2)}\} - £\{u_2(t)e^{-2(t-2)}\}]/3$. Using the
    linearity of the Laplace transform, we have
    $£\{f(t)\} = £\{u_2(t)[e^{t-2} - e^{-2(t-2)}]/3\}$. Hence,
    $f(t) = [u_2(t)(e^{t-2} - e^{-2(t-2)})]/3$. An alternate method is
    to complete the square in the denominator:

$$F(s) = \frac{e^{-2s}}{(s+1/2)^2 - 9/4}.$$ From line 7, Table 6.2.1, this

gives $f(t) = (2/3)u_2(t)e^{-(t-2)/2} \sinh\frac{3}{2}(t-2)$, which can be

shown to be the same as that found above.

21. Completing the square in the denominator we have

$$F(s) = \frac{2e^{-2s}(s-1)}{(s-1)^2 + 1}.$$ Since the inverse of $\frac{s-1}{(s-1)^2 + 1}$ is

$e^t \cos t$ we conclude that $f(t) = 2u_2(t)e^{t-2}\cos(t-2)$ since e^{-2s}

causes a shift of 2 units on the t axis (Theorem 6.3.1).

27. By completing the square in the denominator of F we can

write $F(s) = \frac{2s+1}{(2s+1)^2+4}.$ This has the form $G(2s+1)$ where

$G(u) = \frac{u}{u^2+4}.$ We must find $\mathcal{L}^{-1}\{G(2s+1)\}$. Applying the

results of Prob.25(c), with a = 2 and b = 1, we have

$\mathcal{L}^{-1}\{F(s)\} = \frac{1}{2}e^{-t/2}\cos(\frac{2t}{2})$, since $\mathcal{L}^{-1}\{G(s)\} = \cos 2t$.

28. If the approach of Prob.27 is used we find

$f(t) = (1/3)e^{2t/3}\sinh(t/3)$, which is equivalent to the

given answer using the definition of sinh t.

33. Assuming that term-by-term integration of the infinite

series is permissible and recalling that $\mathcal{L}\{u_c(t)\} = e^{-cs}/s$

for s > 0, we have $\mathcal{L}\{f(t)\} = (1/s) + \sum_{k=1}^{\infty}(-1)^k \mathcal{L}\{u_k(t)\}$

$= (1/s) + \sum_{k=1}^{\infty}(-1)^k (e^{-ks}/s) = \frac{1}{s}\sum_{k=0}^{\infty}(-e^{-s})^k.$ We recognize

the last series as the geometric series, $\sum_{k=0}^{\infty}ar^k$, with

a = 1 and $r = -e^{-s}$. This series converges to $[1/(1+e^{-s})]$

if $|r| < 1$ (or s > 0). Hence,

$\mathcal{L}\{f(t)\} = (\frac{1}{s})\frac{1}{1+e^{-s}},$ s > 0.

34. Using the definition of the Laplace transform we have

    $F(s) = \mathcal{L}\{f(t)\} = \int_0^\infty e^{-st}f(t)dt$. Since f is periodic with
    period T, we have $f(t+T) = f(t)$. This suggests that we
    rewrite the improper integral as

    $\int_0^\infty e^{-st}f(t)dt = \sum_{n=0}^\infty \int_{nT}^{(n+1)T} e^{-st}f(t)dt$. The periodicity of f

    also suggests that we make the change of variable
    $t = r + nT$. Hence,

    $$F(s) = \sum_{n=0}^\infty \int_0^T e^{-s(r+nT)}f(r+nT)dr = \sum_{n=0}^\infty (e^{-sT})^n \int_0^T e^{-rs} f(r)dr,$$

    where we have used the fact that
    $f(r+nT) = f(r+(n-1)T) = \ldots = f(r+T) = f(r)$, since f is
    periodic. We recognize this last series as the geometric

    series, $\sum_{n=0}^\infty au^n$, with $a = \int_0^T e^{-rs}f(r)dr$ and $u = e^{-sT}$. The

    geometric series converges to $a/(1-u)$ for $|u| < 1$ and
    consequently we obtain

    $F(s) = (1 - e^{-sT})^{-1}\int_0^T e^{-rs} f(r)dr, \quad s > 0.$

36. The function f is periodic with period 2. The result of
    Prob.34 gives us $\mathcal{L}\{f(t)\} = \int_0^2 e^{-st} f(t)dt/(1-e^{-2s})$.
    Calculating the integral we have

    $$\int_0^2 e^{-st}f(t)dt = \int_0^1 e^{-st}dt - \int_1^2 e^{-st}dt$$
    $$= (1-e^{-s})/s + (e^{-2s}-e^{-s})/s$$
    $$= (e^{-2s}-2e^{-s}+1)/s$$
    $$= (1-e^{-s})^2/s. \text{ Since the denominator of}$$

    $\mathcal{L}\{f(t)\}$, $1 - e^{-2s}$, may be written as $(1-e^{-s})(1+e^{-s})$ we
    obtain the desired answer.

Section 6.4, Page 336

1a. f(t) can be written in the form $f(t) = 1 - u_{3\pi}(t)$ and
    thus the Laplace transform of the D.E. is
    $(s^2+1)Y(s) - sy(0) - y'(0) = (1/s) - e^{-3\pi s}/s$. Using the
    I.C. and solving for Y(s), we obtain
    $Y(s) = (s^2+1)^{-1} + [s(s^2+1)]^{-1} - e^{-3\pi s}/s(s^2+1)$. Using

partial fractions on the second and third terms we find
$Y(s) = (s^2+1)^{-1} + (1/s) - s/(s^2+1) - e^{-3\pi s}/s + e^{-3\pi s}s/(s^2+1)$.
The inverse transform of the first three terms can be
obtained directly from Table 6.2.1. Using Theorem 6.3.1
to find the inverse transform of the last two terms, we
have $\mathcal{L}^{-1}\{e^{-3\pi s}/s\} = u_{3\pi}(t)g(t - 3\pi)$ where

$g(t) = \mathcal{L}^{-1}\{1/s\} = 1$ and
$\mathcal{L}^{-1}\{e^{-3\pi s}s/(s^2+1)\} = u_{3\pi}(t)h(t - 3\pi)$ where

$h(t) = \mathcal{L}^{-1}\{s/(s^2+1)\} = \cos t$. Hence,
$y = 1 + \sin t - \cos t + u_{3\pi}(t)[\cos(t - 3\pi) - 1]$
  $= 1 + \sin t - \cos t - u_{3\pi}(t)[1 + \cos t]$.

1b. The graph of the forcing function is a unit pulse for
$0 \le t < 3\pi$ and 0 thereafter. The graph of the solution is
composed of two segments. The first, for $0 \le t < 3\pi$,
is a sinusoid oscillating about 1, which represents
the system response to a unit forcing function and
the given initial conditions. For $t \ge 3\pi$, the forcing
function, f(t), is
zero and the "initial"
conditions are

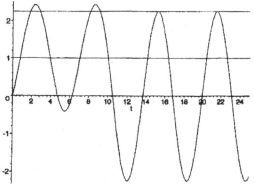

$y(3\pi) = \lim_{t \to 3\pi^-}(1 + \sin t - \cos t) = 2$

and

$y'(3\pi) = \lim_{t \to 3\pi^-}(\cos t + \sin t) = -1$.

For $t \ge 3\pi$ the system
response is
$y(t) = \sin t - 2\cos t$,
which is a sinusoid of
magnitude $\sqrt{5}$ oscillating
about zero.

3a. According to Theorem 6.3.1,
$\mathcal{L}\{u_{2\pi}(t)\sin(t-2\pi)\} = e^{-2\pi s}\mathcal{L}\{\sin t\} = e^{-2\pi s}/(s^2+1)$.
Transforming the D.E., we have
$(s^2+4)Y(s) - sy(0) - y'(0) = 1/(s^2+1) - e^{-2\pi s}/(s^2+1)$.
Using the I.C. and solving for Y(s), we obtain
$Y(s) = (1-e^{-2\pi s})/(s^2+1)(s^2+4)$. We apply partial fractions
to write
$Y(s) = [(s^2+1)^{-1} - (s^2+4)^{-1} - e^{-2\pi s}(s^2+1)^{-1} + e^{-2\pi s}(s^2+4)^{-1}]/3$.
We compute the inverse transform of the first two terms
directly from Table 6.2.1 after noting that
$(s^2+4)^{-1} = (1/2)[2/(s^2+4)]$. We apply Theorem 6.3.1 to the

last two terms to obtain the solution,
y =(1/3){sint-(1/2)sin2t-u_{2π}(t)[sin(t-2π)-(1/2)sin2(t-2π)]}.
This may be simplified, using trigonometric identities,
to y = [(2sint - sin2t)(1-u_{2π}(t))]/6.

3b. Note that the forcing function is sint - sin(t-2π) = 0 for
     t ≥ 2π. The solution is y(t) = 2sint - sin2t for 0 ≤ t < 2π.
     Thus y(2π⁻) = 0 and y'(2π⁻) = 2cos2π - 2cos4π = 0. Hence
     the "initial" value problem for t ≥ 2π is y" + 4y = 0,
     y(2π) = 0, y'(2π) = 0, which has the trivial solution y = 0
     for t ≥ 2π [Note that 1 - u_{2π}(t) = 0 for t ≥ 2π, so this
     agrees with the above solution].

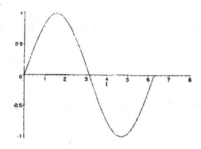   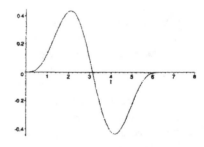

8a. Taking the Laplace transform, applying the I.C. and using
     Theorem 6.3.1 we have $(s^2+s+5/4)Y(s) = (1-e^{-πs/2})/s^2$. Thus

$$Y(s) = \frac{1-e^{-πs/2}}{s^2(s^2+s+5/4)}$$

$$= (1-e^{-πs/2})\left\{\frac{4/5}{s^2} - \frac{16/25}{s} + \frac{(16/25)s-4/25}{(s+1/2)^2+1}\right\}$$

     = (1-e^{-πs/2})H(s), where we have used partial
fractions and completed the square in the denominator of
the last term. Since the numerator of the last term of H
can be written as $\frac{16}{25}[(s+1/2) - 3/4]$, we see that

$£^{-1}{H(s)} = (4/25)(5t - 4 + 4e^{-t/2}cost - 3e^{-t/2}sint)$,
which yields the desired solution.

8b. The graph of the forcing
     function is a ramp (f(t) = t)
     for 0 ≤ t < π/2 and a constant
     (f(t) = π/2) for t ≥ π/2.

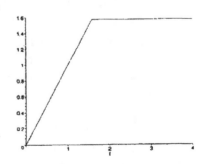

The solution will be a damped sinusoid oscillating about
the "ramp" (20t-16)/25 [which is the first two terms in
the answer of Part a.] for $0 \le t < \pi/2$ and oscillating
about $2\pi/5$ for $t \ge \pi/2$. The first part is shown on the
left over an 'expanded' interval to detect this
behaviour. The graph of the solution is on the right.

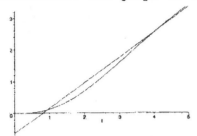 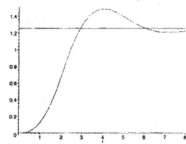

10. Note that $g(t) = \sin t - u_\pi(t)\sin t = \sin t + u_\pi(t)\sin(t-\pi)$.
    Proceeding as in Prob. 8 we find

    $Y(s) = (1+e^{-\pi s}) \dfrac{1}{(s^2+1)(s^2+s+5/4)}$.   The correct partial

    fraction expansion of the quotient is $\dfrac{as+b}{s^2+1} + \dfrac{cs+d}{s^2+s+5/4}$,

    where $a+c = 0$, $a+b+d = 0$, $(5/4)a+b+c = 0$ and $(5/4)b+d = 1$
    by equating coefficients. Solving for a,b,c,d and
    following the steps of Prob. 8 yields the desired
    solution.

16b. Taking the Laplace transform of the D.E. we obtain
     $U(s^2 + s/4 + 1) = k(e^{-3s/2}-e^{-5s/2})/s$, since the I.C. are
     zero. Solving for U and using partial fractions yields

     $U(s) = k(e^{-3s/2}-e^{-5s/2})(\dfrac{1}{s} - \dfrac{s+1/4}{s^2+s/4+1})$.   Thus, if

     $H(s) = (\dfrac{1}{s} - \dfrac{s+1/4}{s^2+s/4+1})$, then, since

     $s^2 + s/4+ 1 = (s+1/8)^2 + 63/64$,

     $h(t) = 1 - e^{-t/8}(\cos\dfrac{3\sqrt{7}}{8}t + \dfrac{\sqrt{7}}{21}\sin\dfrac{3\sqrt{7}}{8}t)$ and

     $u(t) = ku_{3/2}(t)h(t-3/2) - ku_{5/2}(t)h(t-5/2)$.

16c. In all cases the plot will be zero for $0 \le t < 3/2$. For
     $3/2 \le t < 5/2$ the plot will be the system response
     (damped sinusoid) to a step input of magnitude k. For
     $t \ge 5/2$, the plot will be the system response to the
     I.C. $u(5^-/2)$, $u'(5^-/2)$ with no forcing function. The graph

shown is for k = 2.
Varying k will just affect
the amplitude. Note that
the amplitude never
reaches 2, which would be
the steady state response
for the step input $2u_{3/2}(t)$.

Note also that the solution
and its derivative are
continuous at t = 5/2.

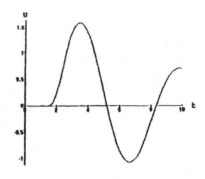

19a. The graph on $0 \le t < 6\pi$ will depend on how large n is.
For instance, if n = 2 then

$$f(t) = \begin{cases} 1 & 0 \le t < \pi, \quad 2\pi \le t \\ -1 & \pi \le t < 2\pi \end{cases}.$$

For n = 4, $f(t) = \begin{cases} 1 & 0 \le t < \pi, \ 2\pi \le t < 3\pi, \ 4\pi \le t \\ -1 & \pi \le t < 2\pi, \ 3\pi \le t < 4\pi \end{cases}.$

19b. Taking the Laplace transform of the D.E. and using the I.C. we

have $Y(s) = \dfrac{1}{s(s^2+1)} [1 + 2\sum\limits_{k=1}^{n} (-1)^k e^{-\pi ks}]$, since

$\mathcal{L}\{u_{\pi k}(t)\} = \dfrac{e^{-\pi ks}}{s}$. By partial fractons $\dfrac{1}{s(s^2+1)} = \dfrac{1}{s} - \dfrac{s}{s^2+1}$,

so $y(t) = 1 - \cos t + 2\sum\limits_{k=1}^{n} (-1)^k u_{\pi k}(t)[1 - \cos(t-\pi k)]$, using

line 13 in Table 6.2.1.

19c. For $0 \le t < \pi$, $y(t)=1-\cos t$,
which peaks at $t=\pi$, just
when the forcing function
changes from +1 to -1.
Thus the forcing function
"reinforces" the natural
motion, creating a
"resonance". This occurs
at each π interval until
t > 15π, at which time the
forcing function no longer
changes and the solution

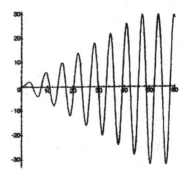

continues oscillating about -1. If n = 16, the
solution would continue to oscillate about +1 for t > 16π.

19d Since $\cos(t-\pi k) = (-1)^k \cos t$, the solution in part b can
be written as

$$y(t) = 1 - \cos t + 2\sum_{k=1}^{n}(-1)^k u_{\pi k}(t) - 2\sum_{k=1}^{n}(-1)^{2k}\cos t$$

$$= 1 - \cos t - 2n\cos t + 2\sum_{k=1}^{n}(-1)^k u_{\pi k}(t)$$ which diverges for

$n\to\infty$, since the nth erm does not approach zero.

20. In this case

$$Y(s) = \frac{1}{s(s^2+.1s+1)}[1 + 2\sum_{k=1}^{n}(-1)^k e^{-\pi ks}]. \quad \text{Using partial}$$

fractions we have

$$H(s) = \frac{1}{s(s^2+.1s+1)} = \frac{1}{s} - \frac{s+.1}{s^2+.1s+1}$$

$$= \frac{1}{s} - \frac{s+.05}{(s+.05)^2+b^2} - \frac{.05}{(s+.05)^2+b^2}, \quad \text{where}$$

$b^2 = [1-(.05)^2] = .9975.$ Now let

$$h(t) = \mathcal{L}^{-1}\{H(s)\} = 1 - e^{-.05t}\cos bt - \frac{.05}{b}e^{-.05t}\sin bt. \quad \text{Hence,}$$

$$y(t) = h(t) + 2\sum_{k=1}^{n}(-1)^k u_{\pi k}(t)\, h(t-\pi k), \text{ and thus, for } t > n\pi \text{ the}$$

solution will be approximated by
$\pm 1 - Ae^{-.05(t-n\pi)}\cos[b(t-n\pi) + \delta]$, and therefore
converges as $t\to\infty$.

20a. y(t) for n = 30   y(t) for n = 31

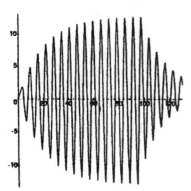

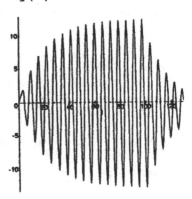

20b. From the graph of part a, $A \cong 12.5$ and the frequency is 2π.

20c. From the graph
     (or analytically)
     $A = 10$ and the
     frequency is 2π.

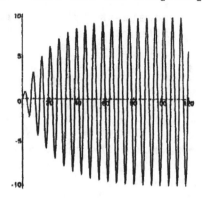

Section 6.5, Page 343

1a. Proceeding as in Ex. 1, we take the Laplace transform
    of the D.E. and apply the I.C.:
    $(s^2 + 2s + 2)Y(s) = s + 2 + e^{-\pi s}$. Thus,

    $Y(s) = (s+2)/[(s+1)^2 + 1] + e^{-\pi s}/[(s+1)^2 + 1]$. We write
    the first term as $(s+1)/[(s+1)^2 + 1] + 1/[(s+1)^2 + 1]$.
    Applying Theorem 6.3.1 and using Table 6.2.1, we obtain
    the solution, $y = e^{-t}\cos t + e^{-t}\sin t + u_\pi(t)e^{-(t-\pi)}\sin(t-\pi)$.

1b.

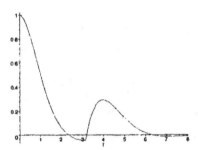

3b.

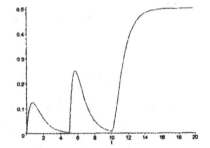

3a. Taking the Laplace transform and using the I.C. we have

    $(s^2+3s+2)Y(s) = \dfrac{1}{2} + e^{-5s} + \dfrac{e^{-10s}}{s}$. Thus

    $Y(s) = \dfrac{1/2}{s^2+3s+2} + \dfrac{e^{-5s}}{s^2+3s+2} + e^{-10s}(\dfrac{1/2}{s} + \dfrac{1/2}{s+2} - \dfrac{1}{s+1})$ and

    hence

    $y(t) = \dfrac{1}{2}h(t) + u_5(t)h(t-5) + u_{10}(t)[\dfrac{1}{2}+\dfrac{1}{2}e^{-2(t-10)}-e^{-(t-10)}]$
    where $h(t) = e^{-t} - e^{-2t}$.

5a. The Laplace transform of the D.E. is

$(s^2+2s+3)Y(s) = \dfrac{1}{s^2+1} + e^{-3\pi s}$, so

$Y(s) = \dfrac{1}{(s^2+1)(s^2+2s+3)} + e^{-3\pi s}[\dfrac{1}{s^2+2s+3}]$. Using partial

fractions or a computer algebra system we
obtain

$y(t) = \dfrac{1}{4}\sin t - \dfrac{1}{4}\cos t + \dfrac{1}{4}e^{-t}\cos\sqrt{2}\,t + \dfrac{1}{\sqrt{2}}u_{3\pi}(t)h(t-3\pi)$,

where $h(t) = e^{-t}\sin\sqrt{2}\,t$.

5b.

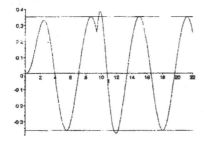

Notice the response to the impulse at $t = 3\pi$ and that the
effect of the impulse is negligible after $t = 15$ or 16.

7a. Taking the Laplace transform of the D.E. yields

$(s^2+1)Y(s) - y'(0) = \displaystyle\int_0^\infty e^{-st}\delta(t-2\pi)\cos t\,dt$. Since

$\delta(t-2\pi) = 0$ for $t \neq 2\pi$ the integral on the right is equal

to $\displaystyle\int_{-\infty}^\infty e^{-st}\delta(t-2\pi)\cos t\,dt$ which equals $e^{-2\pi s}\cos 2\pi$ from

Eq.(16). Substituting for $y'(0)$ and solving for $Y(s)$

gives $Y(s) = \dfrac{1}{s^2+1} + \dfrac{e^{-2\pi s}}{s^2+1}$ and hence

$y(t) = \sin t + u_{2\pi}(t)\sin(t-2\pi) = \begin{cases} \sin t & 0 \le t < 2\pi \\ 2\sin t & 2\pi \le t \end{cases}$.

7b. The effect of the
impulse is barely
seen at $t = 2\pi$.

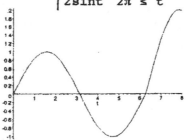

10. Follow the same steps as in the solution for Prob. 7.

13a. From Eq. (22) y(t) will complete one cycle when

$\sqrt{15}$ (t-5)/4 = 2π or T = t − 5 = 8π/$\sqrt{15}$, which is
consistent with the plot in Fig. 6.5.3. Since an impulse
causes a discontinuity in the first derivative, we need
to find the value of y′ at t = 5 and t = 5 + T. From Eq.
(22) we have, for t ≥ 5,

$$y' = e^{-(t-5)/4}[\frac{-1}{2\sqrt{15}}\sin\frac{\sqrt{15}}{4}(t-5) + \frac{1}{2}\cos\frac{\sqrt{15}}{4}(t-5)]. \text{ Thus}$$

$y'(5) = \frac{1}{2}$ and $y'(5+T) = \frac{1}{2}e^{-T/4}$. Since the original

impulse, δ(t−5), caused a discontinuity in y′ of 1/2 at t
= 5, we must choose the impulse at t = 5 + T to be
$-e^{-T/4}$, which is equal and opposite to y′ at 5 + T.

13b. Now consider 2y″ + y′ + 2y = δ(t−5) + kδ(t−5−T) with
y(0) = 0, y′(0) = 0. Using the results of Ex. 1 we have

$$y(t) = \frac{2}{\sqrt{15}}u_5(t)e^{-(t-5)/4}\sin\frac{\sqrt{15}}{4}(t-5)$$

$$+ \frac{2k}{\sqrt{15}}u_{5+T}(t)e^{-(t-5-T)/4}\sin\frac{\sqrt{15}}{4}(t-5-T)$$

$$= \frac{2}{\sqrt{15}}e^{-(t-5)/4}[u_5(t)\sin\frac{\sqrt{15}}{4}(t-5)+ku_{5+T}(t)e^{T/4}\sin\frac{\sqrt{15}}{4}(t-5-T)]$$

$$= \frac{2}{\sqrt{15}}e^{-(t-5)/4}[u_5(t)+ke^{T/4}u_{5+T}(t)]\sin\frac{\sqrt{15}}{4}(t-5), \text{ since}$$

T = 8π/$\sqrt{15}$. If k = $-e^{-T/4}$ then 1 + $ke^{-T/4}$ = 0 for
t > 5 + T and y(t) = 0, which is the desired result.

17b. We have $(s^2+1)Y(s) = \sum_{k=1}^{20} e^{-k\pi s}$ so that $Y(s) = \sum_{k=1}^{20} \frac{e^{-ks}}{s^2+1}$ and

hence $y(t) = \sum_{k=1}^{20} u_{k\pi}(t)\sin(t-k\pi)$

= $u_\pi(t)\sin(t-\pi) + u_{2\pi}(t)\sin(t-2\pi) + ... + u_{20\pi}\sin(t-20\pi)$.
For 0 ≤ t < π, y(t) = 0.
For π ≤ t < 2π, y(t) = sin(t−π) = −sint.   For 2π ≤ t <
3π, y(t) = sin(t−π)+sin(t−2π) = −sint+sint = 0. Due to
the periodicity of sint, the solution will exhibit
this behavior in alternate intervals for 0 ≤ t < 20π.

17c. After $t = 20\pi$ the solution remains at zero.

21b. Taking the transform and using the I.C. we have

$$(s^2+1)Y(s) = \sum_{k=1}^{15} e^{-(2k-1)\pi} \text{ so that } Y(s) = \sum_{k=1}^{15} \frac{e^{-(2k-1)\pi}}{s^2+1}.$$

Thus $y(t) = \sum_{k=1}^{15} u_{(2k-1)\pi}(t)\sin[t-(2k-1)\pi]$.

21c. For $t > 29\pi$, $y(t) = \sin(t-\pi) + \sin(t-3\pi) +...+ \sin(t-29\pi)$
$$= -\sin t - \sin t -....-\sin t$$
$$= -15\sin t.$$

25b. Substituting for $f(t)$ in the integral of Part a, we have

$y = \int_0^t e^{-(t-\tau)}\delta(\tau-\pi)\sin(t-\tau)d\tau$. We know that the integration variable is always less than t (the upper limit) and thus for $t < \pi$ we have $\tau < \pi$ and thus

$\delta(\tau-\pi) = 0$. Hence $y \equiv 0$ for $t < \pi$. For $t > \pi$ utilize Eq.(16).

Section 6.6, Page 350

1c. Using the format of Eqs.(2) and (3) we have

$$f*(g*h) = \int_0^t f(t-\tau)(g*h)(\tau)d\tau$$
$$= \int_0^t f(t-\tau)[\int_0^\tau g(\tau-\eta)h(\eta)d\eta]d\tau$$
$$= \int_0^t [\int_\eta^t f(t-\tau)g(\tau-\eta)d\tau]h(\eta)(d\eta).$$

The last double integral is obtained from the previous line by interchanging the order of the η and τ integrations. Making the change of variable $\omega = \tau - \eta$ on the inside integral yields

$$f*(g*h) = \int_0^t [\int_0^{t-\eta} f(t-\eta-\omega)g(\omega)d\omega]h(\eta)d\eta$$
$$= \int_0^t [(f*g)(t-\eta)]h(\eta)d\eta = (f*g)*h.$$

3. Use the trigonometric identity
    $\sin A\sin B = (1/2)[\cos(A-B)-\cos(A+B)]$ with $A = t-\tau$ and $B = \tau$.

4.   It is possible to determine f(t) explicitly by using
     integration by parts and then find its transform F(s).
     However, it is much more convenient to apply Theorem
     6.6.1. Let us define $g(t) = t^2$ and $h(t) = \cos 2t$. Then,
     $f(t) = \int_0^t g(t-\tau)h(\tau)d\tau$. Using Table 6.2.1, we have

     $G(s) = \mathcal{L}\{g(t)\} = 2/s^3$ and $H(s) = \mathcal{L}\{h(t)\} = s/(s^2+4)$.
     Hence, by Theorem 6.6.1,

     $$\mathcal{L}\{f(t)\} = F(s) = G(s)H(s) = \frac{2}{s^2(s^2+4)}.$$

8.   As was done in Ex. 1 think of F(s) as the product of s^{-4}
     and $(s^2+1)^{-1}$ which, according to Table 6.2.1, are the
     transforms of $t^3/6$ and $\sin t$, respectively. Hence, by
     Theorem 6.6.1, the inverse transform of F(s) is

     $$f(t) = (1/6)\int_0^t (t-\tau)^3 \sin\tau d\tau.$$

14. We take the Laplace transform of the D.E. and apply the
     I.C.: $(s^2 + 2s + 2)Y(s) = \alpha/(s^2 + \alpha^2)$. Solving for Y(s),
     we have $Y(s) = \dfrac{\alpha}{s^2+\alpha^2} \cdot \dfrac{1}{(s+1)^2+1}$, where the second factor
     has been written in a convenient way by completing the
     square. Thus Y(s) is seen to be the product of the
     transforms of $\sin\alpha t$ and $e^{-t}\sin t$ respectively. Hence,
     according to Theorem 6.6.1, $y = \int_0^t e^{-(t-\tau)}\sin(t-\tau)\sin\alpha\tau d\tau.$

16. Proceeding as in Prob.14 we obtain
     $$(s^2+s+5/4)Y(s) - s = \frac{1-e^{-\pi s}}{s} \text{ or}$$

     $$Y(s) = \frac{s}{s^2+s+5/4} + \frac{1-e^{-\pi s}}{s(s^2+s+5/4)}$$

     $$= \frac{(s+1/2) - 1/2}{(s+1/2)^2+1} + \frac{1-e^{-\pi s}}{s} \cdot \frac{1}{(s+1/2)^2+1}$$

     where the first term is obtained by completing the square
     in the denominator and the second term is written as the
     product of two terms whose inverse transforms are known,
     so that Theorem 6.6.1 can be used. Note that
     $\mathcal{L}^{-1}\{(1-e^{-\pi s})/s\} = 1 - u_\pi(t)$. Also note that a different
     form of the same solution would be obtained by writing

the second term as $(1-e^{-\pi s})(\dfrac{a}{s} + \dfrac{bs + c}{(s+1/2)^2+1})$ and solving

for a, b and c. In this case $\mathcal{L}^{-1}\{1-e^{-\pi s}\} = \delta(t) - \delta(t-\pi)$ from Sect.6.5.

18. Taking the Laplace transform, using the I.C. and solving,

we have $Y(s) = \dfrac{s+3}{(s+1)(s+2)} + \dfrac{s}{(s^2+\alpha^2)(s+1)(s+2)}$. As in

Prob. 16, there are several correct ways the second term can be treated in order to use the convolution integral. In order to obtain the desired answer, write the second term as

$\dfrac{s}{s^2+\alpha^2}(\dfrac{a}{s+1} + \dfrac{b}{s+2})$ and solve for a and b.

21. To find $\Phi(s)$ you must recognize the integral that appears in the equation as a convolution integral. Taking the transform of both sides then yields

$\Phi(s) + K(s)\Phi(s) = F(s)$, or $\Phi(s) = \dfrac{F(s)}{1+K(s)}$.

24a Again, we recognize $\displaystyle\int_0^t (t-\xi)\phi(\xi)d\xi$ as the convolution of t

and $\phi(t)$. Thus, taking the transform of both sides, we

get $\Phi(s) - \dfrac{1}{s^2}\Phi(s) = \dfrac{1}{s}$. Solving for Φ we get

$\Phi(s) = \dfrac{s}{s^2-1}$ and hence $\phi(t) = \cosh t$ from line 8 of

Table 6.2.1.

24b. Taking the derivative of the given equation we get

$\phi'(t) - \dfrac{d}{dt}\displaystyle\int_0^t (t-\xi)\phi(\xi)d\xi = 0$, or

$\phi'(t) - \displaystyle\int_0^t \phi(\xi)d\xi - (t-\xi)\phi(\xi)\big|_{\xi=t} = 0$, using Leibnitz's

Rule. Since the last term is zero, we have

$$\phi'(t) - \int_0^t \phi(\xi)d\xi = 0, \quad (i),$$

and then $\phi''(t) - \phi(t) = 0, \qquad (ii)$.

The original equation gives $\phi(0) = 1$ and Eq.(i) gives $\phi'(0) = 0$. Using Eq.(ii) the IVP is $\phi''(t) - \phi(t) = 0$ with $\phi(0) = 1$ and $\phi'(0) = 0$.

24c. The solution to $\phi''(t) - \phi(t) = 0$, $\phi(0) = 1$, $\phi'(0) = 0$ is

also $\phi(t) = \dfrac{e^t + e^{-t}}{2} = \cosh t$.

26a. Taking the transform, we obtain

$$s\Phi(s) - \phi(0) + \frac{1}{s^2}\Phi(s) = \frac{1}{s^2}, \text{ and thus}$$

$$\Phi(s) = \frac{1}{s^3+1} = \frac{1}{(s+1)(s^2-s+1)} = \frac{1/3}{s+1} - \frac{(s-2)/3}{s^2-s+1}.$$

Completing the square in the denominator of the last term and using lines 9 and 10 of Table 6.2.1 we obtain

$$\phi(t) = e^{-t}/3 - (1/3)e^{t/2}\cos(\sqrt{3}\,t/2) + (1/\sqrt{3})e^{t/2}\sin(\sqrt{3}\,t/2).$$

26b. From the given equation $\phi'(0) = 0$. Differentiating we get

$$\phi''(t) + \int_0^t \phi(\xi)d\xi = 1 \text{ and } \phi''(0) = 1. \text{ Another}$$

differentiation gives $\phi'''(t) + \phi(t) = 0$. Thus the IVP is $\phi'''(t) + \phi(t) = 0$, $\phi(0) = 0$, $\phi'(0) = 0$, and $\phi''(0) = 1$.

26c. The characteristic equation is $r^3 + 1 = 0$, so the roots are the cube roots of -1, which are -1 and $(1 \pm \sqrt{3})/2$.

CHAPTER 7

Section 7.1, Page 359

2. As in Ex. 1, let $x_1 = u$ and $x_2 = u'$. Then $x_1' = x_2$ and
    $x_2' = u'' = 3\sin t - .5u' - 2u = -2x_1 - .5x_2 + 3\sin t$.

4. In this case let $x_1 = u$, $x_2 = u'$, $x_3 = u''$, and $x_4 = u'''$.
    The last equation gives $x_4' = x_1$.

6. Let $x_1 = u$ and $x_2 = u'$; then $x_1' = x_2$ is the first of the
    desired pair of equations. The second equation is
    obtained by substituting $u'' = x_2'$, $u' = x_2$, and $u = x_1$ in
    the given D.E. The I.C. become $x_1(0) = u_0$, $x_2(0) = u_0'$.

8a. Follow the steps outlined in Prob.7. Solve the first D.E.
    for x_2 to obtain $x_2 = \dfrac{3}{2}x_1 - \dfrac{1}{2}x_1'$. Substitute this into
    the second D.E. to obtain $x_1'' - x_1' - 2x_1 = 0$.

8b. The solution of the 2nd order ODE of Part a is
    $x_1 = c_1 e^{2t} + c_2 e^{-t}$. Differentiating this and substituting
    into the above equation for x_2 yields $x_2 = \dfrac{1}{2}c_1 e^{2t} + 2c_2 e^{-t}$.
    The I.C. then give
    $c_1 + c_2 = 3$ and $\dfrac{1}{2}c_1 + 2c_2 = \dfrac{1}{2}$, which yield
    $c_1 = \dfrac{11}{3}$, $c_2 = -\dfrac{2}{3}$. Thus $x_1 = \dfrac{11}{3}e^{2t} - \dfrac{2}{3}e^{-t}$ and
    $x_2 = \dfrac{11}{6}e^{2t} - \dfrac{4}{3}e^{-t}$.

8c. Note that for large t, the second term in each solution
    of Part b vanishes and we have $x_1 \cong \dfrac{11}{3}e^{2t}$ and $x_2 \cong \dfrac{11}{6}e^{2t}$,
    so that $x_1 \cong 2x_2$. This says that the graph will be
    asymptotic to the line $x_1 = 2x_2$ for large t.

9a. Solving the first D.E. for x_2 gives $x_2 = \dfrac{4}{3}x_1' - \dfrac{5}{3}x_1$,
    which substituted into the second D.E. yields
    $x_1'' - 2.5x_1' + x_1 = 0$.

9b. From Part a, $x_1 = c_1 e^{t/2} + c_2 e^{2t}$ and $x_2 = -c_1 e^{t/2} + c_2 e^{2t}$.
     Using the I.C. yields $c_1 = -3/2$ and $c_2 = -1/2$. For large
     t, $x_1 \approx (-1/2)e^{2t}$ and $x_2 \approx (-1/2)e^{2t}$ and thus the graph is
     asymptotic to $x_1 = x_2$ in the third quadrant.

9c.   12c.

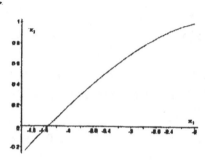   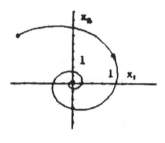

12a. Solving the first D.E. for x_2 gives $x_2 = \dfrac{1}{2}x_1' + \dfrac{1}{4}x_1$ and
     substitution into the second D.E. gives
     $$x_1'' + x_1' + \frac{17}{4}x_1 = 0.$$

12b. Solving the ODE of Part a we find
     $x_1 = e^{-t/2}(c_1\cos 2t + c_2\sin 2t)$ and substitution gives
     $x_2 = e^{-t/2}(c_2\cos 2t - c_1\sin 2t)$. The I.C. yields $c_1 = -2$ and
     $c_2 = 2$.

14. If $a_{12} \neq 0$, then solve the first equation for x_2,
     obtaining $x_2 = [x_1' - a_{11}x_1 - g_1(t)]/a_{12}$. Upon substituting
     this expression into the second equation, we have a
     second order linear O.D.E. for x_1. One I.C. is
     $x_1(0) = x_1^0$. The second I.C. is
     $x_2(0) = [x_1'(0) - a_{11}x_1(0) - g_1(0)]/a_{12} = x_2^0$. Solving for
     $x_1'(0)$ gives $x_1'(0) = a_{12}x_2^0 + a_{11}x_1^0 + g_1(0)$. If $a_{12} = 0$, then
     solve the second equation for x_1 and proceed as above.
     These results hold when $a_{11}, \ldots, a_{22}$ are functions of t
     as long as the derivatives exist and $a_{12}(t)$ and
     $a_{21}(t)$ are not both zero on the interval. The I.C.
     will involve $a_{11}(0)$ and $a_{12}(0)$.

20. Number the nodes 1,2, and 3 clockwise beginning with the top right node in Fig. 7.1.4. Also let I_1, I_2, I_3, and I_4 denote the currents through the resistor $R = 1$, the inductor $L = 1$, the capacitor $C = \dfrac{1}{2}$, and the resistor $R = 2$, respectively. Let V_1, V_2, V_3, and V_4 be the corresponding voltage drops. Kirchhoff's first law applied to nodes 1 and 2, respectively, gives (i) $I_1 - I_2 = 0$ and (ii) $I_2 - I_3 - I_4 = 0$. Kirchhoff's second law applied to each loop gives (iii) $V_1 + V_2 + V_3 = 0$ and (iv) $V_3 - V_4 = 0$. The current-voltage relation through each circuit element yields four more equations: (v) $V_1 = I_1$, (vi) $I_2' = V_2$, (vii) $(1/2)V_3' = I_3$ and (viii) $V_4 = 2I_4$. We thus have a system of eight equations in eight unknowns, and we wish to eliminate all of the variables except I_2 and V_3 from this system of equations. For example, we can use Eqs.(i) and (iv) to eliminate I_1 and V_4 in Eqs.(v) and (viii). Then use the new Eqs.(v) and (viii) to eliminate V_1 and I_4 in Eqs.(ii) and (iii). Finally, use the new Eqs. (ii) and (iii) in Eqs.(vi) and (vii) to obtain $I_2' = - I_2 - V_3$, $V_3' = 2I_2 - V_3$. These equations are identical (when subscripts on the remaining variables are dropped) to the equations given in the text.

22a. Note that the amount of water in each tank remains constant. Thus $Q_1(t)/30$ and $Q_2(t)/20$ represent oz./gal of salt in each tank. We assume the mixture in each tank is well stirred. Then, for the first tank we have $\dfrac{dQ_1}{dt} = 1.5 - 3\dfrac{Q_1(t)}{30} + 1.5\dfrac{Q_2(t)}{20}$, where the first term on the right represents the amount of salt per minute entering the mixture from an external source, the second term represents the loss of salt per minute going to Tank 2 and the third term represents the gain of salt per minute entering from Tank 2. Similarly, we have $\dfrac{dQ_2}{dt} = 3 + 3\dfrac{Q_1(t)}{30} - 4\dfrac{Q_2(t)}{20}$ for Tank 2. We are given that $Q_1(0) = 25$ oz. and $Q_2(0) = 15$ oz.

22b. Solve the second equation for $Q_1(t)$ to obtain $Q_1(t) = 10Q_2' + 2Q_2 - 30$. Substitution into the first

equation then yields $10Q_2'' + 3Q_2' + \dfrac{1}{8}Q_2 = \dfrac{9}{2}$. Equilibrium
is the steady state solution, which is $Q_2^E = 8(9/2) = 36$.
Substituting this value into the equation for Q_1 yields
$Q_1^E = 72 - 30 = 42$. It can be shown that $Q_1(t)$ satisfies
the same second order DE as $Q_2(t)$ (except with the
constant 21/4 on the right side) and thus the
exponentials in the solutions for each are the same.
Hence each tank approaches the equilibrium solution at
then same rate.

22c. Substitute $Q_1 = x_1 + 42$ and $Q_2 = x_2 + 36$ into the
equations found in Part a.

Section 7.2, Page 371

1a. $2\mathbf{A} = \begin{pmatrix} 2 & -4 & 0 \\ 6 & 4 & -2 \\ -4 & 2 & 6 \end{pmatrix}$ so that

$2\mathbf{A} + \mathbf{B} = \begin{pmatrix} 2+4 & -4-2 & 0+3 \\ 6-1 & 4+5 & -2+0 \\ -4+6 & 2+1 & 6+2 \end{pmatrix} = \begin{pmatrix} 6 & -6 & 3 \\ 5 & 9 & -2 \\ 2 & 3 & 8 \end{pmatrix}$

1c. Using Eq.(9) and following Ex. 1 we have

$\mathbf{AB} = \begin{pmatrix} 4+2+0 & -2-10+0 & 3+0+0 \\ 12-2-6 & -6+10-1 & 9+0-2 \\ -8-1+18 & 4+5+3 & -6+0+6 \end{pmatrix}$,

which yields the correct answer.

6a. $\mathbf{AB} = \begin{pmatrix} 6 & -5 & -7 \\ 1 & 9 & 1 \\ -1 & -2 & 8 \end{pmatrix}$ and $\mathbf{BC} = \begin{pmatrix} 5 & 3 & 3 \\ -1 & 7 & 3 \\ 2 & 3 & -2 \end{pmatrix}$ so that

$(\mathbf{AB})\mathbf{C} = \mathbf{A}(\mathbf{BC}) = \begin{pmatrix} 7 & -11 & -3 \\ 11 & 20 & 17 \\ -4 & 3 & -12 \end{pmatrix}$.

In Problems 10 through 19 the method of row reduction, as
illustrated in Ex. 2, can be used to find the inverse matrix
or else to show that none exists. We start with the original

matrix augmented by the identity matrix, describe a suitable
sequence of elementary row operations, and show the result of
applying these operations.

10. Start with the given matrix augmented by the identity

matrix. $\begin{pmatrix} 1 & 4 & . & 1 & 0 \\ & & . & & \\ -2 & 3 & . & 0 & 1 \end{pmatrix}$

Add 2 times the first row to the second row.

$\begin{pmatrix} 1 & 4 & . & 1 & 0 \\ & & . & & \\ 0 & 11 & . & 2 & 1 \end{pmatrix}$

Multiply the second row by (1/11).

$\begin{pmatrix} 1 & 4 & . & 1 & 0 \\ & & . & & \\ 0 & 1 & . & 2/11 & 1/11 \end{pmatrix}$

Add (-4) times the second row to the first row.

$\begin{pmatrix} 1 & 0 & . & 3/11 & -4/11 \\ & & . & & \\ 0 & 1 & . & 2/11 & 1/11 \end{pmatrix}$

Since we have performed the same operation on the given
matrix and the identity matrix, the 2 x 2 matrix
appearing on the right side of this augmented matrix is
the desired inverse matrix. The answer can be checked by
multiplying it by the given matrix; the result should be
the identity matrix.

12. The augmented matrix in this case is:

$\begin{pmatrix} 1 & 2 & 3 & . & 1 & 0 & 0 \\ & & & . & & & \\ 2 & 4 & 5 & . & 0 & 1 & 0 \\ & & & . & & & \\ 3 & 5 & 6 & . & 0 & 0 & 1 \end{pmatrix}$

Add (-2) times the first row to the second row and (-3) times the first row to the third row.

$$\begin{pmatrix} 1 & 2 & 3 & . & 1 & 0 & 0 \\ & & & . & & & \\ 0 & 0 & -1 & .-2 & 1 & 0 \\ & & & . & & & \\ 0 & -1 & -3 & .-3 & 0 & 1 \end{pmatrix}$$

Multiply the second and third rows by (-1) and interchange them.

$$\begin{pmatrix} 1 & 2 & 3 & . & 1 & 0 & 0 \\ & & & . & & & \\ 0 & 1 & 3 & . & 3 & 0 & -1 \\ & & & . & & & \\ 0 & 0 & 1 & . & 2 & -1 & 0 \end{pmatrix}$$

Add (-3) times the third row to the first and second

rows.
$$\begin{pmatrix} 1 & 2 & 0 & . & -5 & 3 & 0 \\ & & & . & & & \\ 0 & 1 & 0 & .-3 & 3 & -1 \\ & & & . & & & \\ 0 & 0 & 1 & . & 2 & -1 & 0 \end{pmatrix}$$

Add (-2) times the second row to the first row.

$$\begin{pmatrix} 1 & 0 & 0 & . & 1 & -3 & 2 \\ & & & . & & & \\ 0 & 1 & 0 & .-3 & 3 & -1 \\ & & & . & & & \\ 0 & 0 & 1 & . & 2 & -1 & 0 \end{pmatrix}$$

The desired answer appears on the right side of this augmented matrix.

14. Again, start with the given matrix augmented by the

identity matrix:
$$\begin{pmatrix} 1 & 2 & 1 & . & 1 & 0 & 0 \\ & & & . & & & \\ -2 & 1 & 8 & . & 0 & 1 & 0 \\ & & & . & & & \\ 1 & -2 & -7 & . & 0 & 0 & 1 \end{pmatrix}$$

Add (2) times the first row to the second row and add(-1) times the first row to the third row.

$$\begin{pmatrix} 1 & 2 & 1 & . & 1 & 0 & 0 \\ & & & . & & & \\ 0 & 5 & 10 & . & 2 & 1 & 0 \\ & & & . & & & \\ 0 & -4 & -8 & . & -1 & 0 & 1 \end{pmatrix}$$

Add (4/5) times the second row to the third row.

$$\begin{pmatrix} 1 & 2 & 1 & . & 1 & 0 & 0 \\ & & & . & & & \\ 0 & 5 & 10 & . & 2 & 1 & 0 \\ & & & . & & & \\ 0 & 0 & 0 & . & 3/5 & 4/5 & 0 \end{pmatrix}$$

Since the third row of the left matrix is all zeros, no further reduction can be performed, and the given matrix is singular.

21c. Differentiate each element of $A(t)$. For instance, $4e^{2t}$ is the derivative of $a_{33}(t)$ and this will then be the element in the 3rd row 3rd column of $A'(t)$.

21d. Integrate each element of $A(t)$ from $t = 0$ to $t = 1$. For instance, $\int_0^1 e^{2t}dt = (1/2)e^{2t}\big|_0^1 = (1/2)(e^2-1) = (e+1)(e-1)/2$ is the integral of $a_{13}(t)$. Thus $(e+1)(e-1)/2$ will be the element in the 1st row 3rd column of $\int_0^1 A(t)dt$

22. $x' = \begin{pmatrix} 4 \\ 2 \end{pmatrix}2e^{2t} = \begin{pmatrix} 8 \\ 4 \end{pmatrix}e^{2t}$; and

$\begin{pmatrix} 3 & -2 \\ 2 & -2 \end{pmatrix} x = \begin{pmatrix} 3 & -2 \\ 2 & -2 \end{pmatrix}\begin{pmatrix} 4 \\ 2 \end{pmatrix}e^{2t} = \begin{pmatrix} 12-4 \\ 8-4 \end{pmatrix}e^{2t} = \begin{pmatrix} 8 \\ 4 \end{pmatrix}e^{2t}.$

25. $\Psi' = \begin{pmatrix} -3e^{-3t} & 2e^{2t} \\ 12e^{-3t} & 2e^{2t} \end{pmatrix} = \begin{pmatrix} 1 & 1 \\ 4 & -2 \end{pmatrix}\begin{pmatrix} e^{-3t} & e^{2t} \\ -4e^{-3t} & e^{2t} \end{pmatrix}.$

Section 7.3, Page 383

1. As in Ex. 1, form the augmented matrix and use row reduction:

$$\begin{pmatrix} 1 & 0 & -1 & . & 0 \\ & & & . & \\ 3 & 1 & 1 & . & 1 \\ & & & . & \\ -1 & 1 & 2 & . & 2 \end{pmatrix}$$

Add (-3) times the first row to the second and add the first row to the third.

$$\begin{pmatrix} 1 & 0 & -1 & . & 0 \\ & & & . & \\ 0 & 1 & 4 & . & 1 \\ & & & . & \\ 0 & 1 & 1 & . & 2 \end{pmatrix}$$

Add (-1) times the second row to the third.

$$\begin{pmatrix} 1 & 0 & -1 & . & 0 \\ & & & . & \\ 0 & 1 & 4 & . & 1 \\ & & & . & \\ 0 & 0 & -3 & . & 1 \end{pmatrix}$$

The third row is equivalent to $- 3x_3 = 1$ or $x_3 = - 1/3$. Likewise the second row is equivalent to $x_2 + 4x_3 = 1$, so $x_2 = 7/3$. Finally, from the first row, $x_1 - x_3 = 0$, so $x_1 = - 1/3$. The answer can be checked by substituting into the original equations.

2. Forming the augmented matrix and using row reduction will yield an augmented mstrix whose third row is equivalent to: $0x_1 + 0x_2 + 0x_3 = 1$, or $0 = 1$. Thus there is no solution.

3. Form the augmented matrix and use row reduction.

$$\begin{pmatrix} 1 & 2 & -1 & . & 2 \\ & & & . & \\ 2 & 1 & 1 & . & 1 \\ & & & . & \\ 1 & -1 & 2 & . & -1 \end{pmatrix}$$

Add (-2) times the first row to the second and add (-1) times the first row to the third.

$$\begin{pmatrix} 1 & 2 & -1 & . & 2 \\ & & & . & \\ 0 & -3 & 3 & . & -3 \\ & & & . & \\ 0 & -3 & 3 & . & -3 \end{pmatrix}$$

Add (-1) times the second row to the third row and then multiply the second row by $(-1/3)$.

$$\begin{pmatrix} 1 & 2 & -1 & . & 2 \\ & & & . & \\ 0 & 1 & -1 & . & 1 \\ & & & . & \\ 0 & 0 & 0 & . & 0 \end{pmatrix}$$

Since the last row has only zero entries, it may be dropped. The second row corresponds to the equation $x_2 - x_3 = 1$. We can assign an arbitrary value to either x_2 or x_3 and use this equation to solve for the other. For example, let $x_3 = c$, where c is arbitrary. Then $x_2 = 1 + c$. The first row corresponds to the equation $x_1 + 2x_2 - x_3 = 2$, so $x_1 = 2 - 2x_2 + x_3 = 2 - 2(1+c) + c = -c$. The solution can be written in vector form as $\mathbf{x} = \begin{pmatrix} -c \\ 1+c \\ c \end{pmatrix} = \begin{pmatrix} 0 \\ 1 \\ 0 \end{pmatrix} + c\begin{pmatrix} -1 \\ 1 \\ 1 \end{pmatrix}$,

where the first vector on the right is a solution of the given nonhomogeneous equation and the second vector is a solution of the related homogeneous equation.

7. To determine whether the given set of vectors is linearly independent we must solve the system $c_1\mathbf{x}^{(1)} + c_2\mathbf{x}^{(2)} + c_3\mathbf{x}^{(3)} = \mathbf{0}$ for c_1, c_2, and c_3. Writing this in scalar form, we have $c_1 \qquad + c_3 = 0$

$$c_1 + c_2 \qquad = 0, \text{ so the}$$
$$c_2 + c_3 = 0$$

augmented matrix is $\begin{pmatrix} 1 & 0 & 1 & . & 0 \\ & & & . & \\ 1 & 1 & 0 & . & 0 \\ & & & . & \\ 0 & 1 & 1 & . & 0 \end{pmatrix}$

Row reduction yields
$$\begin{pmatrix} 1 & 0 & 1 & . & 0 \\ & & & . & \\ 0 & 1 & -1 & . & 0 \\ & & & . & \\ 0 & 0 & 2 & . & 0 \end{pmatrix}.$$

From the third row we have $c_3 = 0$. Then from the second row, $c_2 - c_3 = 0$, so $c_2 = 0$. Finally from the first row $c_1 + c_3 = 0$, so $c_1 = 0$. Since $c_1 = c_2 = c_3 = 0$, we conclude that the given vectors are linearly independent.

9. As in Prob.7 we wish to solve the system
    $c_1\mathbf{x}^{(1)} + c_2\mathbf{x}^{(2)} + c_3\mathbf{x}^{(3)} + c_4\mathbf{x}^{(4)} = \mathbf{0}$ for c_1, c_2, c_3, and
    c_4. Form the augmented matrix and use row reduction:

$$\begin{pmatrix} 1 & -1 & -2 & -3 & . & 0 \\ & & & & . & \\ 2 & 0 & -1 & 0 & . & 0 \\ & & & & . & \\ 2 & 3 & 1 & -1 & . & 0 \\ & & & & . & \\ 3 & 1 & 0 & 3 & . & 0 \end{pmatrix}$$

Add (-2) times the first row to the second, add (-2) times the first row to the third, and add (-3) times the first row to the fourth.

$$\begin{pmatrix} 1 & -1 & -2 & -3 & . & 0 \\ & & & & . & \\ 0 & 2 & 3 & 6 & . & 0 \\ & & & & . & \\ 0 & 5 & 5 & 5 & . & 0 \\ & & & & . & \\ 0 & 4 & 6 & 12 & . & 0 \end{pmatrix}$$

Multiply the second row by $(1/2)$ and then add (-5) times the second row to the third and add (-4) times the second row to the fourth.

$$\begin{pmatrix} 1 & -1 & -2 & -3 & . & 0 \\ & & & & . & \\ 0 & 1 & 3/2 & 3 & . & 0 \\ & & & & . & \\ 0 & 0 & -5/2 & -10 & . & 0 \\ & & & & . & \\ 0 & 0 & 0 & 0 & . & 0 \end{pmatrix}$$

The third row is equivalent to the equation $c_3 + 4c_4 = 0$.
One way to satisfy this equation is by choosing $c_4 = -1$;
then $c_3 = 4$. From the second row we then have
$c_2 = - (3/2)c_3 - 3c_4 = - 6 + 3 = -3$. Then, from the first
row, $c_1 = c_2 + 2c_3 + 3c_4 = -3 + 8 - 3 = 2$. Hence the
given vectors are linearly dependent, and satisfy
$2\mathbf{x}^{(1)} - 3\mathbf{x}^{(2)} + 4\mathbf{x}^{(3)} - \mathbf{x}^{(4)} = \mathbf{0}$.

14. Consider $c_1\mathbf{x}^{(1)}(t) + c_2\mathbf{x}^{(2)}(t) = \mathbf{0}$, which has the

augmented matrix $\begin{pmatrix} 2\sin t & \sin t & . & 0 \\ & & . & \\ \sin t & 2\sin t & . & 0 \end{pmatrix}$. Multiplying the

first row by $-1/2$ and adding to the second row yields
$\begin{pmatrix} 2\sin t & \sin t & . & 0 \\ 0 & (3/2)\sin t & . & 0 \end{pmatrix}$. Since $\sin t$ is not identically zero,

we conclude, from the last row, that $c_2 = 0$. Using this
in the first row gives $c_1 = 0$ and thus $\mathbf{x}^{(1)}(t)$ and $\mathbf{x}^{(2)}(t)$
are linearly independent for $-\infty < t < \infty$.

15. Let $t = t_0$ be a fixed value of t in the interval
$0 \le t \le 1$. To determine whether $\mathbf{x}^{(1)}(t_0)$ and $\mathbf{x}^{(2)}(t_0)$ are
linearly dependent we must solve $c_1\mathbf{x}^{(1)}(t_0)+c_2\mathbf{x}^{(2)}(t_0)=\mathbf{0}$.
We have the augmented matrix

$$\begin{pmatrix} e^{t_0} & 1 & . & 0 \\ & & & \\ t_0 e^{t_0} & t_0 & . & 0 \end{pmatrix}.$$

Multiply the first row by $(-t_0)$ and add to the second row to obtain $\begin{pmatrix} e^{t_0} & 1 & . & 0 \\ 0 & 0 & . & 0 \end{pmatrix}$.

Thus, for example, we can choose $c_1 = 1$ and $c_2 = -e^{t_0}$, and hence the given vectors are linearly dependent at t_0. Since t_0 is arbitrary the vectors are linearly dependent at each point in the interval. However, there is no linear relation between $\mathbf{x}^{(1)}$ and $\mathbf{x}^{(2)}$ that is valid throughout the interval $0 \le t \le 1$. For example, if $t_1 \ne t_0$, and if c_1 and c_2 are chosen as above, then

$c_1\mathbf{x}^{(1)}(t_1) + c_2\mathbf{x}^{(2)}(t_1)$

$= \begin{pmatrix} e^{t_1} \\ t_1 e^{t_1} \end{pmatrix} + -e^{t_0}\begin{pmatrix} 1 \\ t_1 \end{pmatrix} = \begin{pmatrix} e^{t_1} - e^{t_0} \\ t_1 e^{t_1} - t_1 e^{t_0} \end{pmatrix} \ne \begin{pmatrix} 0 \\ 0 \end{pmatrix}$.

Hence the given vectors must be linearly independent on $0 \le t \le 1$. In fact, the same argument applies to any interval.

16. To find the eigenvalues and eigenvectors of the given matrix we must solve $\begin{pmatrix} 5-\lambda & -1 \\ 3 & 1-\lambda \end{pmatrix}\begin{pmatrix} x_1 \\ x_2 \end{pmatrix} = \begin{pmatrix} 0 \\ 0 \end{pmatrix}$. The determinant of coefficients is $(5-\lambda)(1-\lambda) - (-1)(3) = 0$, or $\lambda^2 - 6\lambda + 8 = 0$. Hence $\lambda_1 = 2$ and $\lambda_2 = 4$ are the eigenvalues. The eigenvector corresponding to λ_1 must satisfy $\begin{pmatrix} 3 & -1 \\ 3 & -1 \end{pmatrix}\begin{pmatrix} x_1 \\ x_2 \end{pmatrix} = \begin{pmatrix} 0 \\ 0 \end{pmatrix}$, or $3x_1 - x_2 = 0$. If we let $x_1 = 1$, then $x_2 = 3$ and the eigenvector is $\mathbf{x}^{(1)} = \begin{pmatrix} 1 \\ 3 \end{pmatrix}$, or any constant multiple of this vector. Similarly, the eigenvector corresponding to λ_2 must satisfy $\begin{pmatrix} 1 & -1 \\ 3 & -3 \end{pmatrix}\begin{pmatrix} x_1 \\ x_2 \end{pmatrix} = \begin{pmatrix} 0 \\ 0 \end{pmatrix}$, or $x_1 - x_2 = 0$. Hence $\mathbf{x}^{(2)} = \begin{pmatrix} 1 \\ 1 \end{pmatrix}$, or a multiple thereof.

19. Since $\bar{a}_{12} = a_{21}$, the given matrix is Hermitian and we know in advance that its eigenvalues are real. To find the eigenvalues and eigenvectors we must solve

$$\begin{pmatrix} 1-\lambda & i \\ -i & 1-\lambda \end{pmatrix} \begin{pmatrix} x_1 \\ x_2 \end{pmatrix} = \begin{pmatrix} 0 \\ 0 \end{pmatrix}.$$ The determinant of coefficients

is $(1-\lambda)^2 - i(-i) = \lambda^2 - 2\lambda$, so the eigenvalues are
$\lambda_1 = 0$ and $\lambda_2 = 2$; observe that they are indeed real even
though the given matrix has imaginary entries.   The
eigenvector corresponding to λ_1 must satisfy

$$\begin{pmatrix} 1 & i \\ -i & 1 \end{pmatrix} \begin{pmatrix} x_1 \\ x_2 \end{pmatrix} = \begin{pmatrix} 0 \\ 0 \end{pmatrix},$$ or $x_1 + ix_2 = 0$.   Note that the second

equation $-ix_1 + x_2 = 0$ is a multiple of the first.
If $x_1 = 1$, then $x_2 = i$, and the eigenvector is

$$\mathbf{x}^{(1)} = \begin{pmatrix} 1 \\ i \end{pmatrix}.$$ In a similar way we find that the

eigenvector associated with λ_2 is $\mathbf{x}^{(2)} = \begin{pmatrix} 1 \\ -i \end{pmatrix}.$

22. The eigenvalues and eigenvectors satisfy
$$\begin{pmatrix} 1-\lambda & 0 & 0 \\ 2 & 1-\lambda & -2 \\ 3 & 2 & 1-\lambda \end{pmatrix} \begin{pmatrix} x_1 \\ x_2 \\ x_3 \end{pmatrix} = \begin{pmatrix} 0 \\ 0 \\ 0 \end{pmatrix}.$$ The determinant of coefficients is

$(1-\lambda)[(1-\lambda)^2 + 4] = 0$, which has roots $\lambda = 1, 1 \pm 2i$. For
$\lambda = 1$, we then have $2x_1 - 2x_3 = 0$ and $3x_1 + 2x_2 = 0$.   Choosing

$x_1 = 2$ then yields $\begin{pmatrix} 2 \\ -3 \\ 2 \end{pmatrix}$ as the eigenvector corresponding to

$\lambda = 1$. For $\lambda = 1 + 2i$ we have
$-2ix_1 = 0$, $2x_1 - 2ix_2 - 2x_3 = 0$ and $3x_1 + 2x_2 - 2ix_3 = 0$,

yielding $x_1 = 0$ and $x_3 = -ix_2$.   Thus $\begin{pmatrix} 0 \\ 1 \\ -i \end{pmatrix}$ is the eigenvector

corresponding to $\lambda = 1 + 2i$.   A similar calculation shows that
$\begin{pmatrix} 0 \\ 1 \\ i \end{pmatrix}$ is the eigenvector corresponding to $\lambda = 1 - 2i$.

25. Since the given matrix is real and symmetric, we know
    that the eigenvalues are real. Further, even if there
    are repeated eigenvalues, there will be a full set of
    three linearly independent eigenvectors. To find the
    eigenvalues and eigenvectors we must solve

$$\begin{pmatrix} 3-\lambda & 2 & 4 \\ 2 & -\lambda & 2 \\ 4 & 2 & 3-\lambda \end{pmatrix} \begin{pmatrix} x_1 \\ x_2 \\ x_3 \end{pmatrix} = \begin{pmatrix} 0 \\ 0 \\ 0 \end{pmatrix}.$$ The determinant of

coefficients is $(3-\lambda)[-\lambda(3-\lambda)-4] - 2[2(3-\lambda)-8] + 4[4+4\lambda]$
$= -\lambda^3 + 6\lambda^2 + 15\lambda + 8$. Setting this equal to zero and
solving we find $\lambda_1 = \lambda_2 = -1$, $\lambda_3 = 8$. The eigenvectors
corresponding to λ_1 and λ_2 must satisfy

$$\begin{pmatrix} 4 & 2 & 4 \\ 2 & 1 & 2 \\ 4 & 2 & 4 \end{pmatrix} \begin{pmatrix} x_1 \\ x_2 \\ x_3 \end{pmatrix} = \begin{pmatrix} 0 \\ 0 \\ 0 \end{pmatrix};$$ hence there is only the single

relation $2x_1 + x_2 + 2x_3 = 0$ to be satisfied.
Consequently, two of the variables can be selected
arbitrarily and the third is then determined by this
equation. For example, if $x_1 = 1$ and $x_3 = 1$, then $x_2 = -4$,

and we obtain the eigenvector $\mathbf{x}^{(1)} = \begin{pmatrix} 1 \\ -4 \\ 1 \end{pmatrix}$. Similarly, if

$x_1 = 1$ and $x_2 = 0$, then $x_3 = -1$, and we have the

eigenvector $\mathbf{x}^{(2)} = \begin{pmatrix} 1 \\ 0 \\ -1 \end{pmatrix}$, which is linearly independent of

$\mathbf{x}^{(1)}$. There are many other choices that could have been
made; however, by Eq.(38) there can be no
more than two linearly independent eigenvectors
corresponding to the eigenvalue -1. To find the
eigenvector corresponding to λ_3 we must solve

$$\begin{pmatrix} -5 & 2 & 4 \\ 2 & -8 & 2 \\ 4 & 2 & -5 \end{pmatrix} \begin{pmatrix} x_1 \\ x_2 \\ x_3 \end{pmatrix} = \begin{pmatrix} 0 \\ 0 \\ 0 \end{pmatrix}.$$ Interchange the first and

second rows and use row reduction to obtain the
equivalent system $x_1 - 4x_2 + x_3 = 0$, $2x_2 - x_3 = 0$. Since

there are two equations to satisfy only one variable can be assigned an arbitrary value. If we let $x_2 = 1$, then

$x_3 = 2$ and $x_1 = 2$, so we find that $\mathbf{x}^{(3)} = \begin{pmatrix} 2 \\ 1 \\ 2 \end{pmatrix}$.

28. We are given that $\mathbf{Ax} = \mathbf{b}$ has solutions and thus we have $(\mathbf{Ax}, \mathbf{y}) = (\mathbf{b}, \mathbf{y})$. Using $\mathbf{A}^*\mathbf{y} = \mathbf{0}$ and the result of Prob.26b, we have $(\mathbf{Ax}, \mathbf{y}) = (\mathbf{x}, \mathbf{A}^*\mathbf{y}) = 0$. Thus $(\mathbf{b}, \mathbf{y}) = 0$. For Ex. 2, since $\mathbf{A}$ is real,

$\mathbf{A}^* = \mathbf{A}^T = \begin{pmatrix} 1 & -1 & 2 \\ -2 & 1 & -1 \\ 3 & -2 & 3 \end{pmatrix}$ and, using row reduction, the

augmeted matrix for $\mathbf{A}^*\mathbf{y} = \mathbf{0}$ becomes $\begin{pmatrix} 1 & -1 & 2 & 0 \\ 0 & 1 & -3 & 0 \\ 0 & 0 & 0 & 0 \end{pmatrix}$.

Thus $\mathbf{y} = c\begin{pmatrix} 1 \\ 3 \\ 1 \end{pmatrix}$ and hence $(\mathbf{b}, \mathbf{y}) = b_1 + 3b_2 + b_3 = 0$.

Section 7.4, Page 389

1. Use Mathematical Induction. It has already been proven that if $x^{(1)}$ and $x^{(2)}$ are solutions, then so is $c_1 x^{(1)} + c_2 x^{(2)}$. Assume that if $x^{(1)}$, $x^{(2)}$, ..., $x^{(k)}$ are solutions, then $x = c_1 x^{(1)} + \cdots + c_k x^{(k)}$ is a solution. Then use Theorem 7.4.1 to conclude that $x + c_{k+1} x^{(k+1)}$ is also a solution and thus $c_1 x^{(1)} + \cdots + c_{k+1} x^{(k+1)}$ is a solution if $x^{(1)}, ..., x^{(k+1)}$ are solutions.

2a. From Eq.(10) we have
$W = \begin{vmatrix} x_1^{(1)} & x_1^{(2)} \\ x_2^{(1)} & x_2^{(2)} \end{vmatrix} = x_1^{(1)} x_2^{(2)} - x_2^{(1)} x_1^{(2)}$. Taking the

derivative of these two products yields four terms which may be written as
$$\frac{dW}{dt} = [\frac{dx_1^{(1)}}{dt} x_2^{(2)} - x_2^{(1)} \frac{dx_1^{(2)}}{dt}] + [x_1^{(1)} \frac{dx_2^{(2)}}{dt} - \frac{dx_2^{(1)}}{dt} x_1^{(2)}].$$

The terms in the square brackets can now be recognized as

the respective determinants appearing in the desired
result. A similar result was obtained in Prob. 20 of
Sect. 4.1.

2b. If $x^{(1)}$ is substituted into Eq.(3) we have

$$\frac{dx_1^{(1)}}{dt} = p_{11}\, x_1^{(1)} + p_{12}\, x_2^{(1)}$$

$$\frac{dx_2^{(1)}}{dt} = p_{21}\, x_1^{(1)} + p_{22}\, x_2^{(1)}.$$

Substituting the first equation above and its counterpart
for $x^{(2)}$ into the first determinant appearing in dW/dt
and evaluating the result yields $p_{11}\begin{vmatrix} x_1^{(1)} & x_1^{(2)} \\ x_2^{(1)} & x_2^{(2)} \end{vmatrix} = p_{11}W.$

Similarly, the second determinant in dW/dt is evaluated
as $p_{22}W$, yielding the desired result.

2c. From Part(b) we have $\dfrac{dW}{W} = [p_{11}(t) + p_{22}(t)]dt$ which gives

$$W(t) = c\ \exp\!\int [p_{11}(t) + p_{22}(t)]dt.$$

6a. $W = \begin{vmatrix} t & t^2 \\ 1 & 2t \end{vmatrix} = 2t^2 - t^2 = t^2.$

6b. Pick $t = t_0$, then $c_1 x^{(1)}(t_0) + c_2 x^{(2)}(t_0) = 0$ implies

$c_1 \begin{pmatrix} t_0 \\ 1 \end{pmatrix} + c_2 \begin{pmatrix} t_0^2 \\ 2t_0 \end{pmatrix} = \begin{pmatrix} 0 \\ 0 \end{pmatrix}$, which has a non-zero solution

for c_1 and c_2 if and only if $\begin{vmatrix} t_0 & t_0^2 \\ 1 & 2t_0 \end{vmatrix} = 2t_0^2 - t_0^2 = t_0^2 = 0.$

Thus $x^{(1)}(t)$ and $x^{(2)}(t)$ are linearly independent at each
point except $t = 0$. Thus they are linearly independent
on every interval.

6c. From Part(a) we see that the Wronskian vanishes at $t = 0$,
but not at any other point. By Theorem 7.4.3, if $P(t)$,
from Eq.(3), is continuous, then the Wronskian is either
identically zero or else never vanishes. Hence, we
conclude that the D.E. satisfied by $x^{(1)}(t)$ and $x^{(2)}(t)$ must
have at least one discontinuous coefficient at $t = 0$.

6d. To obtain the system satisfied by $x^{(1)}$ and $x^{(2)}$ we
consider

$$\mathbf{x} = c_1\mathbf{x}^{(1)} + c_2\mathbf{x}^{(2)}, \text{ or } \begin{pmatrix} x_1 \\ x_2 \end{pmatrix} = c_1 \begin{pmatrix} t \\ 1 \end{pmatrix} + c_2 \begin{pmatrix} t^2 \\ 2t \end{pmatrix}.$$

Taking the derivative we obtain $\begin{pmatrix} x_1' \\ x_2' \end{pmatrix} = c_1 \begin{pmatrix} 1 \\ 0 \end{pmatrix} + c_2 \begin{pmatrix} 2t \\ 2 \end{pmatrix}.$

Solving this last system for c_1 and c_2 we find
$c_1 = x_1' - tx_2'$ and $c_2 = x_2'/2$. Thus

$$\begin{pmatrix} x_1 \\ x_2 \end{pmatrix} = (x_1' - tx_2') \begin{pmatrix} t \\ 1 \end{pmatrix} + \frac{x_2'}{2} \begin{pmatrix} t^2 \\ 2t \end{pmatrix}, \text{ which yields}$$

$x_1 = tx_1' - \dfrac{t^2}{2} x_2'$ and $x_2 = x_1'$. Writing this system in

matrix form we have $\mathbf{x} = \begin{pmatrix} t & -t^2/2 \\ 1 & 0 \end{pmatrix}\mathbf{x}'$. Finding the

inverse of the matrix multiplying $\mathbf{x}'$ yields the desired
solution. Note that at $t = 0$ two of the elements in $P(t)$
are discontinuous.

Section 7.5, Page 398

1. Assuming that there are solutions of the form $\mathbf{x} = \xi e^{rt}$,
    we substitute into the D.E. to find

    $r\xi e^{rt} = \begin{pmatrix} 3 & -2 \\ 2 & -2 \end{pmatrix} \xi e^{rt}.$ Since $\xi = I\xi = \begin{pmatrix} 1 & 0 \\ 0 & 1 \end{pmatrix} \xi$, we can

    write this equation as $\begin{pmatrix} 3 & -2 \\ 2 & -2 \end{pmatrix} \xi - r \begin{pmatrix} 1 & 0 \\ 0 & 1 \end{pmatrix} \xi = 0$ and

    thus we must solve $\begin{pmatrix} 3-r & -2 \\ 2 & -2-r \end{pmatrix} \begin{pmatrix} \xi_1 \\ \xi_2 \end{pmatrix} = \begin{pmatrix} 0 \\ 0 \end{pmatrix}$ for r, ξ_1, ξ_2.

    The determinant of the coefficients is

    $(3-r)(-2-r) + 4 = r^2 - r - 2$, so the eigenvalues are

    $r = -1, 2$. The eigenvector corresponding to $r = -1$

    satisfies $\begin{pmatrix} 4 & -2 \\ 2 & -1 \end{pmatrix} \begin{pmatrix} \xi_1 \\ \xi_2 \end{pmatrix} = \begin{pmatrix} 0 \\ 0 \end{pmatrix}$, which yields $2\xi_1 - \xi_2 = 0$.

    Thus $\mathbf{x}^{(1)}(t) = \xi^{(1)}e^{-t} = \begin{pmatrix} 1 \\ 2 \end{pmatrix} e^{-t}$, where we have set $\xi_1 = 1$.

    (Any other non zero choice would also work). In a

similar fashion, for r = 2, we have $\begin{pmatrix} 1 & -2 \\ 2 & -4 \end{pmatrix} \begin{pmatrix} \xi_1 \\ \xi_2 \end{pmatrix} = \begin{pmatrix} 0 \\ 0 \end{pmatrix}$,

or $\xi_1 - 2\xi_2 = 0$. Hence $x^{(2)}(t) = \xi^{(2)} e^{2t} = \begin{pmatrix} 2 \\ 1 \end{pmatrix} e^{2t}$ by

setting $\xi_2 = 1$. The general solution is then

$x = c_1 x^{(1)}(t) + c_2 x^{(2)}(t)$. To sketch the trajectories we
follow the steps illustrated in Exs. 1 and 2. Setting

$c_2 = 0$ we have $x = \begin{pmatrix} x_1 \\ x_2 \end{pmatrix} = c_1 \begin{pmatrix} 1 \\ 2 \end{pmatrix} e^{-t}$ or $x_1 = c_1 e^{-t}$ and

$x_2 = 2c_1 e^{-t}$ and thus one asymptote is given by
$x_2 = 2x_1$. In a similar
fashion $c_1 = 0$ gives
$x_2 = (1/2)x_1$ as a second
asymptote. Since the
roots differ in sign,
the trajectories for
this problem are similar
in nature to those in
Ex. 1. For $c_2 \neq 0$, all
solutions will be

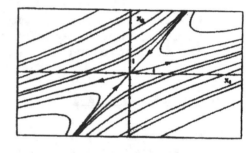

asymptotic to $x_2 = (1/2)x_1$ as $t \to \infty$. For $c_2 = 0$, the
solution approaches the origin along the line $x_2 = 2x_1$.

5.   Proceeding as in Prob. 1 we assume a solution of the form
     $x = \xi e^{rt}$, where r, ξ_1, ξ_2 must now satisfy
     $\begin{pmatrix} -2-r & 1 \\ 1 & -2-r \end{pmatrix} \begin{pmatrix} \xi_1 \\ \xi_2 \end{pmatrix} = \begin{pmatrix} 0 \\ 0 \end{pmatrix}$. Evaluating the determinant of the
     coefficients set equal to zero yields $r = -1, -3$ as the
     eigenvalues. For $r = -1$ we find $\xi_1 = \xi_2$ and thus

     $\xi^{(1)} = \begin{pmatrix} 1 \\ 1 \end{pmatrix}$ and for $r = -3$ we find $\xi_2 = -\xi_1$ and hence

     $\xi^{(2)} = \begin{pmatrix} 1 \\ -1 \end{pmatrix}$. The general solution is then

     $x = c_1 \begin{pmatrix} 1 \\ 1 \end{pmatrix} e^{-t} + c_2 \begin{pmatrix} 1 \\ -1 \end{pmatrix} e^{-3t}$. Since there are two negative

     eigenvalues, we would expect the trajectories to be
     similar to those of Ex. 2. Setting $c_2 = 0$ and eliminating
     t (as in Prob. 1) we find

that $\begin{pmatrix} 1 \\ 1 \end{pmatrix} e^{-t}$ approaches the

origin along the line

$x_2 = x_1$. Similarly $\begin{pmatrix} 1 \\ -1 \end{pmatrix} e^{-3t}$

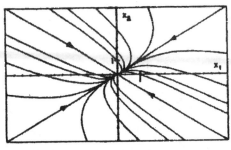

approaches the origin along
the line $x_2 = -x_1$. As long
as $c_1 \neq 0$ (since e^{-t} is the dominnt term as $t \to 0$). all
trajectories approach the origin asymptotic to $x_2 = x_1$.
For $c_1 = 0$, the trajectory approaches the origin along
$x_2 = -x_1$, as shown in the graph.

6. The characteristic equation is $(5/4 - r)^2 - 9/16 = 0$, so
    $r = 2, 1/2$. Since the roots are of the same sign, the
    behavior of the solutions is similar to Prob. 5, except
    the trajectories are reversed, since the roots are
    positive.

7. Again assuming $x = \xi e^{rt}$ we find that r, ξ_1, ξ_2 must
    satisfy $\begin{pmatrix} 4-r & -3 \\ 8 & -6-r \end{pmatrix} \begin{pmatrix} \xi_1 \\ \xi_2 \end{pmatrix} = \begin{pmatrix} 0 \\ 0 \end{pmatrix}$. The determinant of the

    coefficients set equal to zero yields $r = 0, -2$. For
    $r = 0$ we find $4\xi_1 = 3\xi_2$. Choosing $\xi_2 = 4$ we find $\xi_1 = 3$
    and thus $\xi^{(1)} = \begin{pmatrix} 3 \\ 4 \end{pmatrix}$. Similarly for $r = -2$ we have

    $\xi^{(2)} = \begin{pmatrix} 1 \\ 2 \end{pmatrix}$ and thus $x = c_1 \begin{pmatrix} 3 \\ 4 \end{pmatrix} + c_2 \begin{pmatrix} 1 \\ 2 \end{pmatrix} e^{-2t}$. To sketch the

    trajectories, note that the general solution is
    equivalent to the simultaneous equations $x_1 = 3c_1 + c_2 e^{-2t}$
    and $x_2 = 4c_1 + 2c_2 e^{-2t}$. Solving the first equation for
    $c_2 e^{-2t}$ (assuming $c_2 \neq 0$) and substituting into the second
    yields $x_2 = 2x_1 - 2c_1$ and thus the trajectories are
    parallel straight lines. If $c_2 = 0$, the solution is fixed
    at a point.

9. The eigvalues are given by
    $\begin{vmatrix} 1-r & i \\ -i & 1-r \end{vmatrix} = (1-r)^2 + i^2 = r(r-2) = 0$. For $r = 0$ we have

$$\begin{pmatrix} 1 & i \\ -i & 1 \end{pmatrix}\begin{pmatrix} \xi_1 \\ \xi_2 \end{pmatrix} = 0 \text{ or } -i\xi_1 + \xi_2 = 0 \text{ and thus } \begin{pmatrix} 1 \\ i \end{pmatrix} \text{ is one}$$

eigenvector. Similarly $\begin{pmatrix} 1 \\ -i \end{pmatrix}$ is the eigenvector for $r = 2$.

14. The eigenvalues and eigenvectors of the coefficient

matrix satisfy $\begin{pmatrix} 1-r & -1 & 4 \\ 3 & 2-r & -1 \\ 2 & 1 & -1-r \end{pmatrix}\begin{pmatrix} \xi_1 \\ \xi_2 \\ \xi_3 \end{pmatrix} = \begin{pmatrix} 0 \\ 0 \\ 0 \end{pmatrix}$. The determinant

of coefficients set equal to zero reduces to

$r^3 - 2r^2 - 5r + 6 = 0$, so the eigenvalues are

$r_1 = 1$, $r_2 = -2$, and $r_3 = 3$. The eigenvector

corresponding to r_1 must satisfy $\begin{pmatrix} 0 & -1 & 4 \\ 3 & 1 & -1 \\ 2 & 1 & -2 \end{pmatrix}\begin{pmatrix} \xi_1 \\ \xi_2 \\ \xi_3 \end{pmatrix} = \begin{pmatrix} 0 \\ 0 \\ 0 \end{pmatrix}$.

Using row reduction we obtain the equivalent system

$\xi_1 + \xi_3 = 0$, $\xi_2 - 4\xi_3 = 0$. Letting $\xi_1 = 1$, it follows that

$\xi_3 = -1$ and $\xi_2 = -4$, so $\xi^{(1)} = \begin{pmatrix} 1 \\ -4 \\ -1 \end{pmatrix}$. In a similar way the

eigenvectors corresponding to r_2 and r_3 are found to be

$\xi^{(2)} = \begin{pmatrix} 1 \\ -1 \\ -1 \end{pmatrix}$ and $\xi^{(3)} = \begin{pmatrix} 1 \\ 2 \\ 1 \end{pmatrix}$, respectively. Thus the

general solution of the given D.E. is

$$x = c_1\begin{pmatrix} 1 \\ -4 \\ -1 \end{pmatrix} e^t + c_2\begin{pmatrix} 1 \\ -1 \\ -1 \end{pmatrix} e^{-2t} + c_3\begin{pmatrix} 1 \\ 2 \\ 1 \end{pmatrix} e^{3t}.$$ Notice that the

"trajectories" of this solution would lie in the x_1 x_2 x_3 three dimensional space.

16. The eigenvalues and eigenvectors of the coefficient

matrix are found to be $r_1 = -1$, $\xi^{(1)} = \begin{pmatrix} 1 \\ 1 \end{pmatrix}$ and $r_2 = 3$,

$\xi^{(2)} = \begin{pmatrix} 1 \\ 5 \end{pmatrix}$. Thus the general solution of the given D.E.

is $x = c_1 \begin{pmatrix} 1 \\ 1 \end{pmatrix} e^{-t} + c_2 \begin{pmatrix} 1 \\ 5 \end{pmatrix} e^{3t}$. The I.C. yields the

system of equations $c_1 \begin{pmatrix} 1 \\ 1 \end{pmatrix} + c_2 \begin{pmatrix} 1 \\ 5 \end{pmatrix} = \begin{pmatrix} 1 \\ 3 \end{pmatrix}$. The augmented

matrix of this system is $\begin{pmatrix} 1 & 1 & . & 1 \\ & & . & \\ 1 & 5 & . & 3 \end{pmatrix}$ and by row reduction

we obtain $\begin{pmatrix} 1 & 1 & . & 1 \\ & & . & \\ 0 & 1 & . & 1/2 \end{pmatrix}$. Thus $c_2 = 1/2$ and $c_1 = 1/2$.

Substituting these values in the general solution gives
the solution of the I.V.P. As $t \to \infty$, the solution

becomes asymptotic to $x = \dfrac{1}{2} \begin{pmatrix} 1 \\ 5 \end{pmatrix} e^{3t}$, or $x_2 = 5x_1$.

20. Substituting $x = \xi t^r$ into the D.E. we obtain

$r\xi t^r = \begin{pmatrix} 2 & -1 \\ 3 & -2 \end{pmatrix} \xi t^r$. For $t \neq 0$ this equation can be

written as $\begin{pmatrix} 2-r & -1 \\ 3 & -2-r \end{pmatrix} \begin{pmatrix} \xi_1 \\ \xi_2 \end{pmatrix} = \begin{pmatrix} 0 \\ 0 \end{pmatrix}$. The eigenvalues and

eigenvectors are $r_1 = 1$, $\xi^{(1)} = \begin{pmatrix} 1 \\ 1 \end{pmatrix}$ and $r_2 = -1$,

$\xi^{(2)} = \begin{pmatrix} 1 \\ 3 \end{pmatrix}$. Substituting these in the assumed form we

obtain the general solution $x = c_1 \begin{pmatrix} 1 \\ 1 \end{pmatrix} t + c_2 \begin{pmatrix} 1 \\ 3 \end{pmatrix} t^{-1}$.

25.

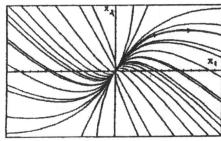
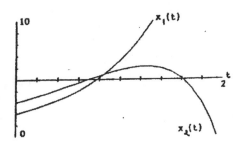

27.
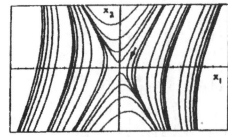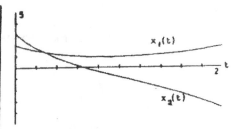

31c. The eigevalues are given by

$$\begin{vmatrix} -1-r & -1 \\ -\alpha & -1-r \end{vmatrix} = r^2 + 2r + 1 - \alpha = 0. \quad \text{Thus } r_{1,2} = -1\pm\sqrt{\alpha}.$$

Note that in Part (a) the eigenvalues are both negative while in Part (b) they differ in sign. Thus, in this part, if we choose $\alpha = 1$, then one eigenvalue is zero, which is the transition of the one root from negative to positive. This is the desired bifurcation point.

Section 7.6, Page 409

1a. We assume a solution of the form $x = \xi e^{rt}$ thus r and ξ are solutions of $\begin{pmatrix} 3-r & -2 \\ 4 & -1-r \end{pmatrix} \begin{pmatrix} \xi_1 \\ \xi_2 \end{pmatrix} = \begin{pmatrix} 0 \\ 0 \end{pmatrix}$. The determinant of coefficients is $(r^2-2r-3) + 8 = r^2 - 2r + 5$, so the eigenvalues are $r = 1 \pm 2i$. The eigenvector corresponding to $1 + 2i$ satisfies $\begin{pmatrix} 2-2i & -2 \\ 4 & -2-2i \end{pmatrix} \begin{pmatrix} \xi_1 \\ \xi_2 \end{pmatrix} = \begin{pmatrix} 0 \\ 0 \end{pmatrix}$, or $(2-2i)\xi_1 - 2\xi_2 = 0$. If $\xi_1 = 1$, then $\xi_2 = 1-i$ and $\xi^{(1)} = \begin{pmatrix} 1 \\ 1-i \end{pmatrix}$ and thus one complex-valued solution of the D.E. is $x^{(1)}(t) = \begin{pmatrix} 1 \\ 1-i \end{pmatrix} e^{(1+2i)t}$. To find real-valued solutions (see Eqs.10 and 11) we take the real and imaginary parts, respectively of $x^{(1)}(t)$.

Thus $x^{(1)}(t) = \begin{pmatrix} 1 \\ 1-i \end{pmatrix} e^t(\cos 2t + i\sin 2t)$

$$= e^t \begin{pmatrix} \cos 2t + i\sin 2t \\ \cos 2t + \sin 2t + i(\sin 2t - \cos 2t) \end{pmatrix}$$

$$= e^t \begin{pmatrix} \cos 2t \\ \cos 2t + \sin 2t \end{pmatrix} + ie^t \begin{pmatrix} \sin 2t \\ \sin 2t - \cos 2t \end{pmatrix}.$$

Hence the general solution of the D.E. is

$$x = c_1 e^t \begin{pmatrix} \cos 2t \\ \cos 2t + \sin 2t \end{pmatrix} + c_2 e^t \begin{pmatrix} \sin 2t \\ \sin 2t - \cos 2t \end{pmatrix}.$$ The

solutions spiral to ∞ as $t \to \infty$ due to the e^t terms.

1b.

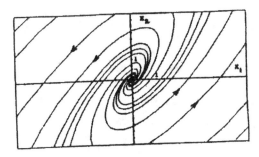

7. The eigenvalues and eigenvectors of the coefficient

matrix satisfy $\begin{pmatrix} 1-r & 0 & 0 \\ 2 & 1-r & -2 \\ 3 & 2 & 1-r \end{pmatrix} \begin{pmatrix} \xi_1 \\ \xi_2 \\ \xi_3 \end{pmatrix} = \begin{pmatrix} 0 \\ 0 \\ 0 \end{pmatrix}.$ The

determinant of coefficients reduces to $(1-r)(r^2 - 2r + 5)$
so the eigenvalues are $r_1 = 1$, $r_2 = 1 + 2i$, and
$r_3 = 1 - 2i$. The eigenvector corresponding to r_1
satisfies

$$\begin{pmatrix} 0 & 0 & 0 \\ 2 & 0 & -2 \\ 3 & 2 & 0 \end{pmatrix} \begin{pmatrix} \xi_1 \\ \xi_2 \\ \xi_3 \end{pmatrix} = \begin{pmatrix} 0 \\ 0 \\ 0 \end{pmatrix};$$ hence $\xi_1 - \xi_3 = 0$ and

$3\xi_1 + 2\xi_2 = 0$. If we let $\xi_2 = -3$ then $\xi_1 = 2$ and $\xi_3 = 2$,

so one solution of the D.E. is $\begin{pmatrix} 2 \\ -3 \\ 2 \end{pmatrix} e^t$. The eigenvector

corresponding to r_2 satisfies $\begin{pmatrix} -2i & 0 & 0 \\ 2 & -2i & -2 \\ 3 & 2 & -2i \end{pmatrix} \begin{pmatrix} \xi_1 \\ \xi_2 \\ \xi_3 \end{pmatrix} = \begin{pmatrix} 0 \\ 0 \\ 0 \end{pmatrix}.$

Hence $\xi_1 = 0$ and $i\xi_2 + \xi_3 = 0$. If we let $\xi_2 = 1$, then
$\xi_3 = -i$. Thus a complex-valued solution is

$$\begin{pmatrix} 0 \\ 1 \\ -i \end{pmatrix} e^t(\cos 2t + i \sin 2t).$$ Taking the real and imaginary

parts, see Prob. 1, we obtain $\begin{pmatrix} 0 \\ \cos 2t \\ \sin 2t \end{pmatrix} e^t$ and $\begin{pmatrix} 0 \\ \sin 2t \\ -\cos 2t \end{pmatrix} e^t$,

respectively. Thus the general solution is

$$\mathbf{x} = c_1 \begin{pmatrix} 2 \\ -3 \\ 2 \end{pmatrix} e^t + c_2 e^t \begin{pmatrix} 0 \\ \cos 2t \\ \sin 2t \end{pmatrix} + c_3 e^t \begin{pmatrix} 0 \\ \sin 2t \\ -\cos 2t \end{pmatrix}, \text{ which spirals}$$

to ∞ about the x_1 axis in the $x_1 x_2 x_3$ space as $t \to \infty$.

9. The eigenvalues and eigenvectors of the coefficient

matrix satisfy $\begin{pmatrix} 1-r & -5 \\ 1 & -3-r \end{pmatrix}\begin{pmatrix} \xi_1 \\ \xi_2 \end{pmatrix} = \begin{pmatrix} 0 \\ 0 \end{pmatrix}$. The determinant of

coefficients is $r^2 + 2r + 2$ so that the eigenvalues are $r = -1 \pm i$. The eigenvector corresponding to $r = -1 + i$

is given by $\begin{pmatrix} 2-i & -5 \\ 1 & -2-i \end{pmatrix}\begin{pmatrix} \xi_1 \\ \xi_2 \end{pmatrix} = 0$ so that $\xi_1 = (2+i)\xi_2$ and

thus one complex-valued solution is

$$\mathbf{x}^{(1)}(t) = \begin{pmatrix} 2+i \\ 1 \end{pmatrix} e^{(-1+i)t}.$$ Finding the real and complex

parts of $\mathbf{x}^{(1)}$, as in Prob.1, leads to the general

solution $\mathbf{x} = c_1 e^{-t} \begin{pmatrix} 2\cos t - \sin t \\ \cos t \end{pmatrix} + c_2 e^{-t} \begin{pmatrix} 2\sin t + \cos t \\ \sin t \end{pmatrix}$.

Setting $t = 0$ we find $\mathbf{x}(0) = \begin{pmatrix} 1 \\ 1 \end{pmatrix} = c_1 \begin{pmatrix} 2 \\ 1 \end{pmatrix} + c_2 \begin{pmatrix} 1 \\ 0 \end{pmatrix}$, which

is equivalent to the system $\begin{aligned} 2c_1 + c_2 &= 1 \\ c_1 + 0 &= 1 \end{aligned}$. Thus $c_1 = 1$,

$c_2 = -1$ and $\mathbf{x}(t) = e^{-t}\begin{pmatrix} 2\cos t - \sin t \\ \cos t \end{pmatrix} - e^{-t}\begin{pmatrix} 2\sin t + \cos t \\ \sin t \end{pmatrix}$

$$= e^{-t}\begin{pmatrix} \cos t - 3\sin t \\ \cos t - \sin t \end{pmatrix}, \text{ which spirals to}$$

zero as $t \to \infty$, due to the e^{-t} term.

11a. The eigenvalues are given by

$$\begin{vmatrix} 3/4-r & -2 \\ 1 & -5/4-r \end{vmatrix} = r^2 + r/2 + 17/16 = 0, \text{ so } r = -1/4 \pm i.$$

11d. Choose $x(0) = \begin{pmatrix} 5 \\ 5 \end{pmatrix}$, then

the trajectory starts at
(5,5) in the $x_1 x_2$ plane
and spirals around the
t-axis and converges
to the t axis as $t \to \infty$.

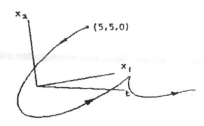

15a. The eigenvalues satisfy $\begin{vmatrix} 2-r & -5 \\ \alpha & -2-r \end{vmatrix} = r^2 - 4 + 5\alpha = 0$, so

$r_1, r_2 = \pm\sqrt{4-5\alpha}$.

15b. The qualitative nature of the phase portrait changes when
r goes from real to complex. Thus $\alpha = 4/5$ is the
critical value and $r_1 = r_2 = 0$.

15c.

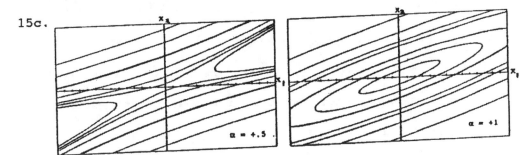

16a. $\begin{vmatrix} 5/4-r & 3/4 \\ \alpha & 5/4-r \end{vmatrix} = r^2 - 5r/2 + (25/16 - 3\alpha/4) = 0$, so

$r_{1,2} = 5/4 \pm \sqrt{3\alpha}/2$.

16b. There are two critical values of α. For $\alpha < 0$ the
eigenvalues are complex, while for $\alpha > 0$ they are real.
There will be a second critical value of α when $r_2 = 0$,
or $\alpha = 25/12$. In this case the second real eigenvalue
goes from positive to negative.

16c.

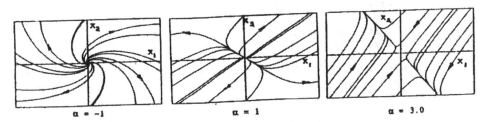

18a. We have $\begin{vmatrix} 3-r & \alpha \\ -6 & -4-r \end{vmatrix} = r^2 + r - 12 + 6\alpha = 0$, so

$r_1, r_2 = -1/2 \pm \sqrt{49-24\alpha}\,/2$.

18b. The critical values occur when $49 - 24\alpha = 1$ (in which case

$r_2 = 0$) and when $49 - 24\alpha = 0$, in which case $r_1 = r_2 = -1/2$.

Thus $\alpha = 2$ and $\alpha = 49/24 \approx 2.04$.

18c.

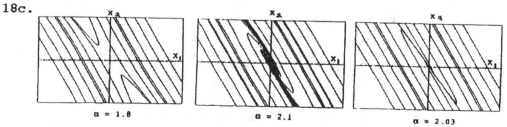

21. If we seek solutions of the form $\mathbf{x} = \xi t^r$, then

$\xi r t^r = \begin{pmatrix} -1 & -1 \\ 2 & -1 \end{pmatrix} \xi t^r$. Thus r must be an eigenvalue and ξ a

corresponding eigenvector of the coefficient matrix.

Thus r and ξ satisfy $\begin{pmatrix} -1-r & -1 \\ 2 & -1-r \end{pmatrix}\begin{pmatrix} \xi_1 \\ \xi_2 \end{pmatrix} = \begin{pmatrix} 0 \\ 0 \end{pmatrix}$. The

determinant of coefficients is $(-1-r)^2 + 2 = r^2 + 2r + 3$,

so the eigenvalues are $r = -1 \pm \sqrt{2}\,i$. The eigenvector

corresponding to $-1 + \sqrt{2}\,i$ satisfies

$\begin{pmatrix} -\sqrt{2}\,i & -1 \\ 2 & -\sqrt{2}\,i \end{pmatrix}\begin{pmatrix} \xi_1 \\ \xi_2 \end{pmatrix} = \begin{pmatrix} 0 \\ 0 \end{pmatrix}$ or $\sqrt{2}\,i\xi_1 + \xi_2 = 0$. If we let

$\xi_1 = 1$, then $\xi_2 = -\sqrt{2}\,i$, and $\xi^{(1)} = \begin{pmatrix} 1 \\ -\sqrt{2}\,i \end{pmatrix}$. Thus a

complex-valued solution of the given D.E. is

$\begin{pmatrix} 1 \\ -\sqrt{2}\,i \end{pmatrix} t^{-1+\sqrt{2}\,i}$. From Eq.(16) of Sect. 5.4

we have (since $t^{\sqrt{2}\,i} = e^{\ln t^{\sqrt{2}\,i}} = e^{\sqrt{2}\,i\ln t}$)

$t^{-1+\sqrt{2}\,i} = t^{-1}[\cos(\sqrt{2}\,\ln t) + i\sin(\sqrt{2}\,\ln t)]$ for $t > 0$.

Separating the complex valued solution into real and

imaginary parts, we obtain the two real-valued solutions

$\mathbf{u} = t^{-1}\begin{pmatrix} \cos(\sqrt{2}\,\ln t) \\ \sqrt{2}\sin(\sqrt{2}\,\ln t) \end{pmatrix}$ and $\mathbf{v} = t^{-1}\begin{pmatrix} \sin(\sqrt{2}\,\ln t) \\ -\sqrt{2}\cos(\sqrt{2}\,\ln t) \end{pmatrix}$.

23a. The eigenvalues are given by $(r+1/4)[(r+1/4)^2 + 1] = 0$.

23b.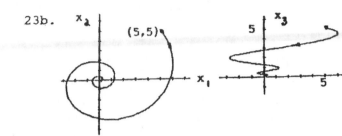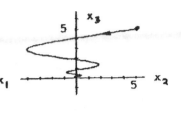

23c. Graph starts in the first octant and spirals around the x_3 axis, converging to zero.

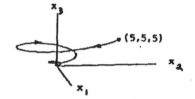

30a. Following the steps leading to Eq.(24), and using the given values for the m's and k's, we obtain $y_1' = y_3$,

$y_2' = y_4$, $y_3' = -4y_1 + 3y_2$, and $y_4' = (9/4)y_1 - (13/4)y_2$.

Thus $\mathbf{Y}' = \mathbf{AY}$, where $\mathbf{A} = \begin{pmatrix} 0 & 0 & 1 & 0 \\ 0 & 0 & 0 & 1 \\ -4 & 3 & 0 & 0 \\ 9/4 & -13/4 & 0 & 0 \end{pmatrix}$.

30b. The eigenvalues of $\mathbf{A}$ are given by $\det(\mathbf{A} - r\mathbf{I}) = r^4 + (29/4)r^2 + 25/4 = 0$, and thus $r_{1,2} = \pm i$ and $r_{3,4} = \pm(5/2)i$. The eigenvector corresponding to $r_1 = i$ satisfies

$\begin{pmatrix} -i & 0 & 1 & 0 \\ 0 & -i & 0 & 1 \\ -4 & 3 & -i & 0 \\ 9/4 & -13/4 & 0 & -i \end{pmatrix} \begin{pmatrix} \xi_1 \\ \xi_2 \\ \xi_3 \\ \xi_4 \end{pmatrix} = \begin{pmatrix} 0 \\ 0 \\ 0 \\ 0 \end{pmatrix}$, or $\xi_3 = i\xi_1$, $\xi_4 = i\xi_2$,

$-4\xi_1 + 3\xi_2 = i\xi_3$, and $(9/4)\xi_1 - (13/4)\xi_2 = i\xi_4$. Setting

$\xi_1 = 1$, the first three equations yield $\xi^{(1)} = \begin{pmatrix} 1 \\ 1 \\ i \\ i \end{pmatrix}$ and thus $\xi^{(2)} = \begin{pmatrix} 1 \\ 1 \\ -i \\ -i \end{pmatrix}$ by Eq.(13). It should be noted that the

fourth equation is also satisfied by this choice.

Similarly, $\xi^{(3)}$ satisfies $\begin{pmatrix} -5i/2 & 0 & 1 & 0 \\ 0 & -5i/2 & 0 & 1 \\ -4 & 3 & -5i/2 & 0 \\ 9/4 & -13/4 & 0 & -5i/2 \end{pmatrix} \begin{pmatrix} \xi_1 \\ \xi_2 \\ \xi_3 \\ \xi_4 \end{pmatrix} = \begin{pmatrix} 0 \\ 0 \\ 0 \\ 0 \end{pmatrix}$,

which yields $\xi^{(3)} = \begin{pmatrix} 4 \\ -3 \\ 10i \\ -15i/2 \end{pmatrix}$ and $\xi^{(4)} = \begin{pmatrix} 4 \\ -3 \\ -10i \\ 15i/2 \end{pmatrix}$.

30c. Taking the real and imaginary parts of $e^{it} \begin{pmatrix} 1 \\ 1 \\ i \\ i \end{pmatrix}$ yields

$w_1(t) = (\cos t, \cos t, -\sin t, -\sin t)^T$ and
$w_2(t) = (\sin t, \sin t, \cos t, \cos t)^T$ as the corresponding two

real valued solutions. Similarly, $e^{(5/2)it} \begin{pmatrix} 4 \\ -3 \\ 10i \\ -15i/2 \end{pmatrix}$ yields

the other two real valued solutions, denoted as $w_3(t)$ and
$w_4(t)$. The general solution of the system is then

$y(t) = c_1 w_1(t) + c_2 w_2(t) + c_3 w_3(t) + c_4 w_4(t).$

30d. There are two fundamental modes, one represented by
$\cos(t - \delta_1)$, of frequency 1, and the other represented by
$\cos(5t/2 - \delta_2)$, of frequency 5/2 (see Sect. 3.7).

30e. From Part(c),

$$\mathbf{y}(0) = c_1\begin{pmatrix} 1 \\ 1 \\ 0 \\ 0 \end{pmatrix} + c_2\begin{pmatrix} 0 \\ 0 \\ 1 \\ 1 \end{pmatrix} + c_3\begin{pmatrix} 4 \\ -3 \\ 0 \\ 0 \end{pmatrix} + c_4\begin{pmatrix} 0 \\ 0 \\ 10 \\ -15/2 \end{pmatrix} = \begin{pmatrix} 2 \\ 1 \\ 0 \\ 0 \end{pmatrix}, \text{ which}$$

yields $c_2 = c_4 = 0$, $c_1 = 10/7$, and $c_3 = 1/7$.

Section 7.7, Page 420

Each of the Problems 1 through 10, except 2 and 8, has been solved in one of the previous sections. Thus a fundamental matrix for the given systems can be readily written down. The fundamental matrix $\Phi(t)$ satisfying $\Phi(0) = I$ can then be found, as shown in the following problems.

2a. The characteristic equation is given by

$$\begin{vmatrix} -3/4-r & 1/2 \\ 1/8 & -3/4-r \end{vmatrix} = r^2 + 3r/2 + 1/2 = 0, \text{ so } r = -1, -1/2. \text{ For}$$

$r = -1$ we have $\begin{pmatrix} 1/4 & 1/2 \\ 1/8 & 1/4 \end{pmatrix} \begin{pmatrix} \xi_1 \\ \xi_2 \end{pmatrix} = \begin{pmatrix} 0 \\ 0 \end{pmatrix}$, and $\xi^{(1)} = \begin{pmatrix} -2 \\ 1 \end{pmatrix}$.

Likewise $\xi^{(2)} = \begin{pmatrix} 2 \\ 1 \end{pmatrix}$. Thus $\mathbf{x}^{(1)}(t) = \begin{pmatrix} -2 \\ 1 \end{pmatrix}e^{-t}$ and

$\mathbf{x}^{(2)}(t) = \begin{pmatrix} 2 \\ 1 \end{pmatrix}e^{-t/2}$ so a fundamental matrix Ψ is $\begin{pmatrix} -2e^{-t} & 2e^{-t/2} \\ e^{-t} & e^{-t/2} \end{pmatrix}$.

2b. To find the first column of Φ we choose c_1 and c_2 so that

$c_1\mathbf{x}^{(1)}(0) + c_2\mathbf{x}^{(2)}(0) = \begin{pmatrix} 1 \\ 0 \end{pmatrix}$, which yields $-2c_1 + 2c_2 = 1$

and $c_1 + c_2 = 0$. Thus $c_1 = -1/4$ and $c_2 = 1/4$ and the first

column of Φ is $\begin{pmatrix} e^{-t}/2 + e^{-t/2}/2 \\ -e^{-t}/4 + e^{-t/2}/4 \end{pmatrix}$. The second column of Φ

is determined by $d_1\mathbf{x}^{(1)}(0) + d_2\mathbf{x}^{(2)}(0) = \begin{pmatrix} 0 \\ 1 \end{pmatrix}$ which yields

$d_1 = d_2 = 1/2$ and thus the second column of Φ is

$\begin{pmatrix} -e^{-t} + e^{-t/2} \\ e^{-t}/2 + e^{-t/2}/2 \end{pmatrix}$.

4a. From Prob. 4 of Sect. 7.5 we have the two linearly

independent solutions $\mathbf{x}^{(1)}(t) = \begin{pmatrix} 1 \\ -4 \end{pmatrix} e^{-3t}$ and

$\mathbf{x}^{(2)}(t) = \begin{pmatrix} 1 \\ 1 \end{pmatrix} e^{2t}$. Hence a fundamental matrix Ψ is given

by $\Psi(t) = \begin{pmatrix} e^{-3t} & e^{2t} \\ -4e^{-3t} & e^{2t} \end{pmatrix}$.

4b. To find the fundamental matrix $\Phi(t)$ satisfying the I.C.
$\Phi(0) = I$ we can proceed in either of two ways. One way
is to find $\Psi(0)$, invert it to obtain $\Psi^{-1}(0)$, and then to
form the product $\Psi(t)\Psi^{-1}(0)$, which is $\Phi(t)$.
Alternatively, we can find the first column of Φ by
determining the linear combination

$c_1\mathbf{x}^{(1)}(t) + c_2\mathbf{x}^{(2)}(t)$ that satisfies the I.C. $\begin{pmatrix} 1 \\ 0 \end{pmatrix}$. This

requires that $c_1 + c_2 = 1$, $-4c_1 + c_2 = 0$, so we obtain
$c_1 = 1/5$ and $c_2 = 4/5$. Thus the first column of $\Phi(t)$ is
$\begin{pmatrix} (1/5)e^{-3t} + (4/5)e^{2t} \\ -(4/5)e^{-3t} + (4/5)e^{2t} \end{pmatrix}$. Similarly, the second column of
Φ is that linear combination of $\mathbf{x}^{(1)}(t)$ and $\mathbf{x}^{(2)}(t)$ that

satisfies the I.C. $\begin{pmatrix} 0 \\ 1 \end{pmatrix}$. Thus we must have

$c_1 + c_2 = 0$, $-4c_1 + c_2 = 1$; therefore $c_1 = -1/5$ and
$c_2 = 1/5$. Hence the second column of $\Phi(t)$ is
$\begin{pmatrix} -(1/5)e^{-3t} + (1/5)e^{2t} \\ (4/5)e^{-3t} + (1/5)e^{2t} \end{pmatrix}$.

6a. Two linearly independent real-valued solutions of the
given D.E. were found in Prob. 2 of Sect. 7.6. Using the
result of that problem, we have
$$\Psi(t) = \begin{pmatrix} -2e^{-t}\sin 2t & 2e^{-t}\cos 2t \\ e^{-t}\cos 2t & e^{-t}\sin 2t \end{pmatrix}.$$

6b. To find $\Phi(t)$ we determine the linear combinations of the

columns of $\Psi(t)$ that satisfy the I.C. $\begin{pmatrix} 1 \\ 0 \end{pmatrix}$ and $\begin{pmatrix} 0 \\ 1 \end{pmatrix}$,

respectively. In the first case c_1 and c_2 satisfy
$0c_1 + 2c_2 = 1$ and $c_1 + 0c_2 = 0$. Thus $c_1 = 0$ and $c_2 = 1/2$.
In the second case we have $0c_1 + 2c_2 = 0$ and $c_1 + 0c_2 = 1$,

so $c_1 = 1$ and $c_2 = 0$. Using these values of c_1 and c_2 to form the first and second columns of $\Phi(t)$ respectively, we obtain $\Phi(t) = \begin{pmatrix} e^{-t}\cos 2t & -2e^{-t}\sin 2t \\ e^{-t}\sin 2t/2 & e^{-t}\cos 2t \end{pmatrix}$.

10b. From Prob. 14 Sect. 7.5 we have $x^{(1)} = \begin{pmatrix} 1 \\ -4 \\ -1 \end{pmatrix} e^t$,

$x^{(2)} = \begin{pmatrix} 1 \\ -1 \\ -1 \end{pmatrix} e^{-2t}$ and $x^{(3)} = \begin{pmatrix} 1 \\ 2 \\ 1 \end{pmatrix} e^{3t}$. For the first column of Φ we want to choose c_1, c_2, c_3 such that

$c_1 x^{(1)}(0) + c_2 x^{(2)}(0) + c_3 x^{(3)}(0) = \begin{pmatrix} 1 \\ 0 \\ 0 \end{pmatrix}$. Thus

$c_1 + c_2 + c_3 = 1$, $-4c_1 - c_2 + 2c_3 = 0$ and $-c_1 - c_2 + c_3 = 0$, which yield $c_1 = 1/6$, $c_2 = 1/3$ and $c_3 = 1/2$. The first column of Φ is then
$(e^t/6 + e^{-2t}/3 + e^{3t}/2,\ -2e^t/3 - e^{-2t}/3 + e^{3t},$
$$-e^t/6 - e^{-2t}/3 + e^{3t}/2)^T.$$
Likewise, for the second column we have

$d_1 x^{(1)}(0) + d_2 x^{(2)}(0) + d_3 x^{(3)}(0) = \begin{pmatrix} 0 \\ 1 \\ 0 \end{pmatrix}$, which yields

$d_1 = -1/3$, $d_2 = 1/3$ and $d_3 = 0$ and thus
$(-e^t/3 + e^{-2t}/3,\ 4e^t/3 - e^{-2t}/3,\ e^t/3 - e^{-2t}/3)^T$ is the second column of $\Phi(t)$. Finally, for the third column we have $e_1 x^{(1)}(0) + e_2 x^{(2)}(0) + e_3 x^{(3)}(0) = \begin{pmatrix} 0 \\ 0 \\ 1 \end{pmatrix}$, which gives

$e_1 = 1/2$, $e_2 = -1$ and $e_3 = 1/2$ and hence
$(e^t/2 - e^{-2t} + e^{3t}/2,\ -2e^t + e^{-2t} + e^{3t},\ -e^t/2 + e^{-2t} + e^{3t}/2)^T$ is the third column of $\Phi(t)$.

11. From Eq. (14) the solution is given by $\Phi(t)x^0$. Thus
$$x = \begin{pmatrix} 3e^t/2 - e^{-t}/2 & -e^t/2 + e^{-t}/2 \\ 3e^t/2 - 3e^{-t}/2 & -e^t/2 + 3e^{-t}/2 \end{pmatrix} \begin{pmatrix} 2 \\ -1 \end{pmatrix}$$
$$= \begin{pmatrix} 7e^t/2 - 3e^{-t}/2 \\ 7e^t/2 - 9e^{-t}/2 \end{pmatrix} = \frac{7}{2}\begin{pmatrix} 1 \\ 1 \end{pmatrix} e^t - \frac{3}{2}\begin{pmatrix} 1 \\ 3 \end{pmatrix} e^{-t}.$$

Section 7.8, Page 428

1a.

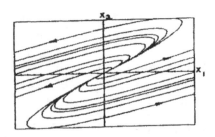

1b. From the general solution we have $\dfrac{x_2}{x_1} = \dfrac{c_1 + c_2 t}{2c_1 + 2c_2 t + c_2}$,

so that $\lim\limits_{t \to \infty} \dfrac{x_2}{x_1} = \dfrac{1}{2}$. Thus all solutions diverge to

infinity along lines of slope $\dfrac{1}{2}$ which can be seen in the

trajectories shown in Part(a).

1c. The eigenvalues and eigenvectors of the given coefficient

matrix satisfy $\begin{pmatrix} 3-r & -4 \\ 1 & -1-r \end{pmatrix} \begin{pmatrix} \xi_1 \\ \xi_2 \end{pmatrix} = \begin{pmatrix} 0 \\ 0 \end{pmatrix}$. The determinant of

coefficients is $(3-r)(-1-r) + 4 = r^2 - 2r + 1 = (r-1)^2$ so

$r_1 = 1$ and $r_2 = 1$. The eigenvectors corresponding to

this double eigenvalue satisfy $\begin{pmatrix} 2 & -4 \\ 1 & -2 \end{pmatrix} \begin{pmatrix} \xi_1 \\ \xi_2 \end{pmatrix} = \begin{pmatrix} 0 \\ 0 \end{pmatrix}$, or

$\xi_1 - 2\xi_2 = 0$. Thus the only eigenvectors are multiples

of $\xi^{(1)} = \begin{pmatrix} 2 \\ 1 \end{pmatrix}$. One solution of the given D.E. is

$x^{(1)}(t) = \begin{pmatrix} 2 \\ 1 \end{pmatrix} e^t$, but there is no second solution of this

form. To find a second solution we assume, as in

Eq. (13), that $x = \xi t e^t + \eta e^t$ and substitute this

expression into the D.E. As in Ex. 2 we find that ξ is an

eigenvector, so we choose $\xi = \begin{pmatrix} 2 \\ 1 \end{pmatrix}$. Then η must satisfy

Eq.(24): $(A - rI)\eta = \xi$, or $\begin{pmatrix} 2 & -4 \\ 1 & -2 \end{pmatrix} \begin{pmatrix} \eta_1 \\ \eta_2 \end{pmatrix} = \begin{pmatrix} 2 \\ 1 \end{pmatrix}$ for this

problem. Solving these equations yields $\eta_1 - 2\eta_2 = 1$.

If $\eta_2 = k$, where k is an arbitrary constant, then

$\eta_1 = 1 + 2k$. Hence the second solution that we obtain is

$x^{(2)}(t) = \begin{pmatrix} 2 \\ 1 \end{pmatrix} t e^t + \begin{pmatrix} 1 + 2k \\ k \end{pmatrix} e^t = \begin{pmatrix} 2 \\ 1 \end{pmatrix} t e^t + \begin{pmatrix} 1 \\ 0 \end{pmatrix} e^t + k \begin{pmatrix} 2 \\ 1 \end{pmatrix} e^t$.

The last term is a multiple of the first solution $\mathbf{x}^{(1)}(t)$ and may be neglected, that is, we may set $k = 0$. Thus

$$\mathbf{x}^{(2)}(t) = \begin{pmatrix} 2 \\ 1 \end{pmatrix} te^t + \begin{pmatrix} 1 \\ 0 \end{pmatrix} e^t \text{ and the general solution is}$$

$$\mathbf{x} = c_1 \mathbf{x}^{(1)}(t) + c_2 \mathbf{x}^{(2)}(t).$$

3b. The origin is attracting.
     That is, as $t \to \infty$
     the solution approaches
     the origin tangent to
     the line $x_2 = x_1/2$, which
     is obtained by taking the

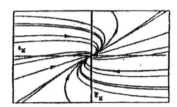

$$\lim_{t \to 0} \frac{x_2}{x_1} \text{ similar to Prob.1.}$$

5. Substituting $\mathbf{x} = \xi e^{rt}$ into the given system, we find that the eigenvalues and eigenvectors satisfy

$$\begin{pmatrix} 1-r & 1 & 1 \\ 2 & 1-r & -1 \\ 0 & -1 & 1-r \end{pmatrix} \begin{pmatrix} \xi_1 \\ \xi_2 \\ \xi_3 \end{pmatrix} = \begin{pmatrix} 0 \\ 0 \\ 0 \end{pmatrix}. \text{ The determinant of coefficients}$$

is $-r^3 + 3r^2 - 4$ and thus $r_1 = -1$, $r_2 = 2$ and $r_3 = 2$. The eigenvector corresponding to r_1 satisfies

$$\begin{pmatrix} 2 & 1 & 1 \\ 2 & 2 & -1 \\ 0 & -1 & 2 \end{pmatrix} \begin{pmatrix} \xi_1 \\ \xi_2 \\ \xi_3 \end{pmatrix} = \begin{pmatrix} 0 \\ 0 \\ 0 \end{pmatrix} \text{ which yields } \xi^{(1)} = \begin{pmatrix} -3 \\ 4 \\ 2 \end{pmatrix} \text{ and}$$

$$\mathbf{x}^{(1)} = \begin{pmatrix} -3 \\ 4 \\ 2 \end{pmatrix} e^{-t}. \text{ The eigenvectors corresponding to the}$$

double eigenvalue must satsify $\begin{pmatrix} -1 & 1 & 1 \\ 2 & -1 & -1 \\ 0 & -1 & -1 \end{pmatrix} \begin{pmatrix} \xi_1 \\ \xi_2 \\ \xi_3 \end{pmatrix} = \begin{pmatrix} 0 \\ 0 \\ 0 \end{pmatrix},$

which yields the single eigenvector $\xi^{(2)} = \begin{pmatrix} 0 \\ 1 \\ -1 \end{pmatrix}$ and hence

$$\mathbf{x}^{(2)}(t) = \begin{pmatrix} 0 \\ 1 \\ -1 \end{pmatrix} e^{2t}. \text{ The second solution corresponding to}$$

the double eigenvalue will have the form specified by

Eq.(13), which yields $\mathbf{x}^{(3)} = \begin{pmatrix} 0 \\ 1 \\ -1 \end{pmatrix} te^{2t} + \eta e^{2t}$.

Substituting this into the given system, or using

Eq.(24), we find that η satisfies $\begin{pmatrix} -1 & 1 & 1 \\ 2 & -1 & -1 \\ 0 & -1 & -1 \end{pmatrix} \begin{pmatrix} \eta_1 \\ \eta_2 \\ \eta_3 \end{pmatrix} = \begin{pmatrix} 0 \\ 1 \\ -1 \end{pmatrix}$.

Using row reduction we find that $\eta_1 = 1$ and $\eta_2 + \eta_3 = 1$, where either η_2 or η_3 is arbitrary. If we choose $\eta_2 = 0$,

then $\eta = \begin{pmatrix} 1 \\ 0 \\ 1 \end{pmatrix}$ and thus $\mathbf{x}^{(3)} = \begin{pmatrix} 0 \\ 1 \\ -1 \end{pmatrix} te^{2t} + \begin{pmatrix} 1 \\ 0 \\ 1 \end{pmatrix} e^{2t}$. The

general solution is then $\mathbf{x} = c_1\mathbf{x}^{(1)} + c_2\mathbf{x}^{(2)} + c_3\mathbf{x}^{(3)}$.

9a. We have $\begin{vmatrix} 2-r & 3/2 \\ -3/2 & -1-r \end{vmatrix} = (r-1/2)^2 = 0$. For $r = 1/2$, the

eigenvector is given by $\begin{pmatrix} 3/2 & 3/2 \\ -3/2 & -3/2 \end{pmatrix} \begin{pmatrix} \xi_1 \\ \xi_2 \end{pmatrix} = 0$, so $\xi = \begin{pmatrix} 1 \\ -1 \end{pmatrix}$

and $\begin{pmatrix} 1 \\ -1 \end{pmatrix} e^{t/2}$ is one solution. For the second solution we

have $\mathbf{x} = \xi te^{t/2} + \eta e^{t/2}$, where $(A - \frac{1}{2}I)\eta = \xi$, A being

the coefficient matrix for this problem. This last
equation reduces to $3\eta_1/2 + 3\eta_2/2 = 1$ and
$-3\eta_1/2 - 3\eta_2/2 = -1$. Choosing $\eta_2 = 0$ yields $\eta_1 = 2/3$
and hence the general solution is

$$\mathbf{x} = c_1\begin{pmatrix} 1 \\ -1 \end{pmatrix} e^{t/2} + c_2\begin{pmatrix} 2/3 \\ 0 \end{pmatrix} e^{t/2} + c_2\begin{pmatrix} 1 \\ -1 \end{pmatrix} te^{t/2}. \quad \mathbf{x}(0) = \begin{pmatrix} 3 \\ -2 \end{pmatrix}$$

gives $c_1 + 2c_2/3 = 3$ and $-c_1 = -2$, and hence $c_1 = 2$,
$c_2 = 3/2$. The graphs are shown for $-10 \le t \le 1$.

9b.

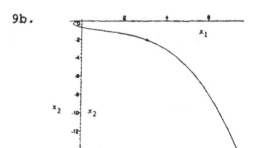
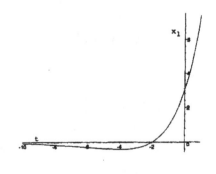

11a. Since the coefficient matrix is lower triangular, the eigenvalues are easily found to be $r = 1, 1, 2$. For $r = 2$,

we have $\begin{pmatrix} -1 & 0 & 0 \\ -4 & -1 & 0 \\ 3 & 6 & 0 \end{pmatrix} \begin{pmatrix} \xi_1 \\ \xi_2 \\ \xi_3 \end{pmatrix} = \begin{pmatrix} 0 \\ 0 \\ 0 \end{pmatrix}$, which yields $\xi = \begin{pmatrix} 0 \\ 0 \\ 1 \end{pmatrix}$, so one

solution is $\mathbf{x}^{(1)} = \begin{pmatrix} 0 \\ 0 \\ 1 \end{pmatrix} e^{2t}$. For $r = 1$, we have $\begin{pmatrix} 0 & 0 & 0 \\ -4 & 0 & 0 \\ 3 & 6 & 1 \end{pmatrix}$

$\begin{pmatrix} \xi_1 \\ \xi_2 \\ \xi_3 \end{pmatrix} = \begin{pmatrix} 0 \\ 0 \\ 0 \end{pmatrix}$, which yields the second solution

$\mathbf{x}^{(2)} = \begin{pmatrix} 0 \\ 1 \\ -6 \end{pmatrix} e^t$. The third solution is of the form

$\mathbf{x}^{(3)} = \begin{pmatrix} 0 \\ 1 \\ -6 \end{pmatrix} te^t + \eta e^t$, where $\begin{pmatrix} 0 & 0 & 0 \\ -4 & 0 & 0 \\ 3 & 6 & 1 \end{pmatrix} \eta = \begin{pmatrix} 0 \\ 1 \\ -6 \end{pmatrix}$ and thus

$\eta_1 = -1/4$ and $6\eta_2 + \eta_3 = -21/4$. Choosing $\eta_2 = 0$ gives $\eta_3 = -21/4$ and hence

$\mathbf{x}(t) = c_1 \begin{pmatrix} 0 \\ 0 \\ 1 \end{pmatrix} e^{2t} + c_2 \begin{pmatrix} 0 \\ 1 \\ -6 \end{pmatrix} e^t + c_3 \left[\begin{pmatrix} -1/4 \\ 0 \\ -21/4 \end{pmatrix} e^t + \begin{pmatrix} 0 \\ 1 \\ -6 \end{pmatrix} te^t \right]$. The

I.C. then yield $c_1 = 2$, $c_2 = 4$ and $c_3 = 3$ and hence

$\mathbf{x} = \begin{pmatrix} -1 \\ 2 \\ -33 \end{pmatrix} e^t + 4 \begin{pmatrix} 0 \\ 1 \\ -6 \end{pmatrix} te^t + 3 \begin{pmatrix} 0 \\ 0 \\ 1 \end{pmatrix} e^{2t}$, which becomes unbounded

as $t \to \infty$.

12b.

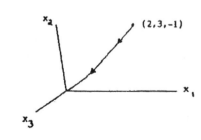

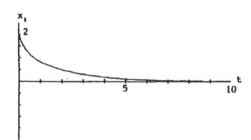

14. Assuming $x = \xi t^r$ and substituting into the given system, we find r and ξ must satisfy $\begin{pmatrix} 1-r & -4 \\ 4 & -7-r \end{pmatrix} \begin{pmatrix} \xi_1 \\ \xi_2 \end{pmatrix} = \begin{pmatrix} 0 \\ 0 \end{pmatrix}$, which has the double eigenvalue r = -3 and single eigenvector $\begin{pmatrix} 1 \\ 1 \end{pmatrix}$. Hence one solution of the given D.E. is

$x^{(1)}(t) = \begin{pmatrix} 1 \\ 1 \end{pmatrix} t^{-3}$. By analogy with the scalar case considered in Sect. 5.4 and Ex. 2 of this section, we seek a second solution of the form $x = \eta t^{-3} \ln t + \zeta t^{-3}$. Substituting this expression into the D.E. we find that η and ζ satisfy the equations $(A + 3I)\eta = 0$ and

$(A + 3I)\zeta = \eta$, where $A = \begin{pmatrix} 1 & -4 \\ 4 & -7 \end{pmatrix}$ and I is the identity

matrix. Thus $\eta = \begin{pmatrix} 1 \\ 1 \end{pmatrix}$, from above, and ζ then satisfies

$\begin{pmatrix} 4 & -4 \\ 4 & -4 \end{pmatrix}\begin{pmatrix} \zeta_1 \\ \zeta_2 \end{pmatrix} = \begin{pmatrix} 1 \\ 1 \end{pmatrix}$. Choosing $\zeta_1 = 0$ we obtain $\zeta_2 = -1/4$ and

hence a second solution is $x^{(2)}(t) = \begin{pmatrix} 1 \\ 1 \end{pmatrix} t^{-3} \ln t + \begin{pmatrix} 0 \\ -1/4 \end{pmatrix} t^{-3}$.

15. The eigenvalues are given by $r^2 - (a+d)r + ad-bc = 0$. Use the quadratic formula to find the roots. Then show that the roots are either real and negative or else are complex with negative real part when a+d < 0 and ad-bc>0. In both these cases the solution approaches zero as t → ∞.

17a. The eigenvalues and eigenvectors of the coefficient matrix satisfy $\begin{pmatrix} 1-r & 1 & 1 \\ 2 & 1-r & -1 \\ -3 & 2 & 4-r \end{pmatrix} \begin{pmatrix} \xi_1 \\ \xi_2 \\ \xi_3 \end{pmatrix} = \begin{pmatrix} 0 \\ 0 \\ 0 \end{pmatrix}$. The determinant

of coefficients is $8 - 12r + 6r^2 - r^3 = (2-r)^3$, so the eigenvalues are $r_1 = r_2 = r_3 = 2$. The eigenvectors corresponding to this triple eigenvalue satisfy

$\begin{pmatrix} -1 & 1 & 1 \\ 2 & -1 & -1 \\ -3 & 2 & 2 \end{pmatrix} \begin{pmatrix} \xi_1 \\ \xi_2 \\ \xi_3 \end{pmatrix} = \begin{pmatrix} 0 \\ 0 \\ 0 \end{pmatrix}$. Using row reduction we can reduce this to the equivalent system $\xi_1 - \xi_2 - \xi_3 = 0$, and $\xi_2 + \xi_3 = 0$. If we let $\xi_2 = 1$, then $\xi_1 = 0$ and $\xi_3 = -1$,

so the only eigenvectors are multiples of $\xi = \begin{pmatrix} 0 \\ 1 \\ -1 \end{pmatrix}$.

17b. From Part(a), one solution of the given D.E. is

$x^{(1)}(t) = \begin{pmatrix} 0 \\ 1 \\ -1 \end{pmatrix} e^{2t}$, but there are no other linearly

independent solutions of this form.

17c. We now seek a second solution of the form
$x = \xi t e^{2t} + \eta e^{2t}$. Thus $Ax = A\xi t e^{2t} + A\eta e^{2t}$ and
$x' = 2\xi t e^{2t} + \xi e^{2t} + 2\eta e^{2t}$. Equating like terms, we then
have $(A-2I)\xi = 0$ and $(A-2I)\eta = \xi$. Thus ξ is the same as
in Part(a) and the second equation yields

$\begin{pmatrix} -1 & 1 & 1 \\ 2 & -1 & -1 \\ -3 & 2 & 2 \end{pmatrix} \begin{pmatrix} \eta_1 \\ \eta_2 \\ \eta_3 \end{pmatrix} = \begin{pmatrix} 0 \\ 1 \\ -1 \end{pmatrix}$. By row reduction this is

equivalent to the system $\begin{pmatrix} 1 & -1 & -1 \\ 0 & 1 & 1 \\ 0 & 0 & 0 \end{pmatrix} \begin{pmatrix} \eta_1 \\ \eta_2 \\ \eta_3 \end{pmatrix} = \begin{pmatrix} 0 \\ 1 \\ 0 \end{pmatrix}$. If we

choose $\eta_3 = 0$, then $\eta_2 = 1$ and $\eta_1 = 1$, so $\eta = \begin{pmatrix} 1 \\ 1 \\ 0 \end{pmatrix}$. Hence

a second solution of the D.E. is

$x^{(2)}(t) = \begin{pmatrix} 0 \\ 1 \\ -1 \end{pmatrix} te^{2t} + \begin{pmatrix} 1 \\ 1 \\ 0 \end{pmatrix} e^{2t}$.

17d. Assuming $x = \xi(t^2/2)e^{2t} + \eta t e^{2t} + \zeta e^{2t}$, we have
$Ax = A\xi(t^2/2)e^{2t} + A\eta t e^{2t} + A\xi e^{2t}$ and
$x' = \xi t e^{2t} + 2\xi(t^2/2)e^{2t} + \eta e^{2t} + 2\eta t e^{2t} + 2\zeta e^{2t}$ and thus
$(A-2I)\xi = 0$, $(A-2I)\eta = \xi$ and $(A-2I)\zeta = \eta$. Again, ξ and η
are as found previously and the last equation is
equivalent to

$$\begin{pmatrix} -1 & 1 & 1 \\ 2 & -1 & -1 \\ -3 & 2 & 2 \end{pmatrix} \begin{pmatrix} \zeta_1 \\ \zeta_2 \\ \zeta_3 \end{pmatrix} = \begin{pmatrix} 1 \\ 1 \\ 0 \end{pmatrix}.$$ By row reduction we find the

equivalent system $\begin{pmatrix} 1 & -1 & -1 \\ 0 & 1 & 1 \\ 0 & 0 & 0 \end{pmatrix} \begin{pmatrix} \zeta_1 \\ \zeta_2 \\ \zeta_3 \end{pmatrix} = \begin{pmatrix} -1 \\ 3 \\ 0 \end{pmatrix}.$ If we let

$\zeta_2 = 0$, then $\zeta_3 = 3$ and $\zeta_1 = 2$, so $\zeta = \begin{pmatrix} 2 \\ 0 \\ 3 \end{pmatrix}$ and

$$\mathbf{x}^{(3)}(t) = \begin{pmatrix} 0 \\ 1 \\ -1 \end{pmatrix} (t^2/2)e^{2t} + \begin{pmatrix} 1 \\ 1 \\ 0 \end{pmatrix} te^{2t} + \begin{pmatrix} 2 \\ 0 \\ 3 \end{pmatrix} e^{2t}.$$

17e. Ψ is the matrix with $\mathbf{x}^{(1)}$ as the first column, $\mathbf{x}^{(2)}$ as the second column and $\mathbf{x}^{(3)}$ as the third column.

17f. $\mathbf{T} = \begin{pmatrix} 0 & 1 & 2 \\ 1 & 1 & 0 \\ -1 & 0 & 3 \end{pmatrix}$ and using row operations on $\mathbf{T}$ and $\mathbf{I}$, or a

computer algebra system, $\mathbf{T}^{-1} = \begin{pmatrix} -3 & 3 & 2 \\ 3 & -2 & -2 \\ -1 & 1 & 1 \end{pmatrix}$ and thus

$\mathbf{T}^{-1}\mathbf{A}\mathbf{T} = \begin{pmatrix} 2 & 1 & 0 \\ 0 & 2 & 1 \\ 0 & 0 & 2 \end{pmatrix} = \mathbf{J}$, which is equivalent to Eq.(29) for

this problem.

19a. $\mathbf{J}^2 = \mathbf{J}\mathbf{J} = \begin{pmatrix} \lambda & 1 \\ 0 & \lambda \end{pmatrix} \begin{pmatrix} \lambda & 1 \\ 0 & \lambda \end{pmatrix} = \begin{pmatrix} \lambda^2 & 2\lambda \\ 0 & \lambda^2 \end{pmatrix}$

$\mathbf{J}^3 = \mathbf{J}\mathbf{J}^2 = \begin{pmatrix} \lambda & 1 \\ 0 & \lambda \end{pmatrix} \begin{pmatrix} \lambda^2 & 2\lambda \\ 0 & \lambda^2 \end{pmatrix} = \begin{pmatrix} \lambda^3 & 3\lambda^2 \\ 0 & \lambda^3 \end{pmatrix}$

19b. Based upon the results of Part(a), assume
$\mathbf{J}^n = \begin{pmatrix} \lambda^n & n\lambda^{n-1} \\ 0 & \lambda^n \end{pmatrix}$, then

$$J^{n+1} = JJ^n = \begin{pmatrix} \lambda & 1 \\ 0 & \lambda \end{pmatrix} \begin{pmatrix} \lambda^n & n\lambda^{n-1} \\ 0 & \lambda^n \end{pmatrix}$$

$$= \begin{pmatrix} \lambda^{n+1} & (n+1)\lambda^n \\ 0 & \lambda^{n+1} \end{pmatrix}, \text{ which is the same as } J^n \text{ with } n$$

replaced by n+1. Thus, by mathematical induction, J^n has the desired form.

19c. From Eq.(23), Sect. 7.7, we have

$$\exp(Jt) = I + \sum_{n=1}^{\infty} \frac{J^n t^n}{n!}$$

$$= I + \sum_{n=1}^{\infty} \begin{pmatrix} \dfrac{\lambda^n t^n}{n!} & \dfrac{n\lambda^{n-1}t^n}{n!} \\ 0 & \dfrac{\lambda^n t^n}{n!} \end{pmatrix}$$

$$= \begin{pmatrix} 1 + \sum_{n=1}^{\infty} \dfrac{\lambda^n t^n}{n!} & \sum_{n=1}^{\infty} \dfrac{\lambda^{n-1}t^n}{(n-1)!} \\ 0 & 1 + \sum_{n=1}^{\infty} \dfrac{\lambda^n t^n}{n!} \end{pmatrix}$$

$$= \begin{pmatrix} e^{\lambda t} & te^{\lambda t} \\ 0 & e^{\lambda t} \end{pmatrix}, \text{ since}$$

$$\sum_{n=1}^{\infty} \frac{\lambda^{n-1}t^n}{(n-1)!} = t(1 + \sum_{n=1}^{\infty} \frac{\lambda^n t^n}{n!}) = te^{\lambda t}.$$

19d. From Eq.(28), Sect. 7.7, we have

$$x = \exp(Jt)x^0 = \begin{pmatrix} e^{\lambda t} & te^{\lambda t} \\ 0 & e^{\lambda t} \end{pmatrix} \begin{pmatrix} x_1^0 \\ x_2^0 \end{pmatrix} = \begin{pmatrix} x_1^0 e^{\lambda t} + x_2^0 te^{\lambda t} \\ x_2^0 e^{\lambda t} \end{pmatrix}$$

$$= \begin{pmatrix} x_1^0 \\ x_2^0 \end{pmatrix} e^{\lambda t} + \begin{pmatrix} x_2^0 \\ 0 \end{pmatrix} te^{\lambda t}.$$

Section 7.9, Page 439

1. From Sect. 7.5 Prob. 3 we have

$$x^{(c)} = c_1 \begin{pmatrix} 1 \\ 1 \end{pmatrix} e^t + c_2 \begin{pmatrix} 1 \\ 3 \end{pmatrix} e^{-t}. \text{ Note that}$$

$\mathbf{g}(t) = \begin{pmatrix} 1 \\ 0 \end{pmatrix} e^t + \begin{pmatrix} 0 \\ 1 \end{pmatrix} t$ and that $r = 1$ is an eigenvalue of the coefficient matrix. Thus if the method of undetermined coefficients is used, the assumed form is given by Eq.(18). Following the steps of Ex.2 then yields the desired solution.

2. Using methods of previous sections, we find that the eigenvalues are $r_1 = 2$ and $r_2 = -2$, with corresponding eigenvectors $\begin{pmatrix} \sqrt{3} \\ 1 \end{pmatrix}$ and $\begin{pmatrix} 1 \\ -\sqrt{3} \end{pmatrix}$. Thus

$\mathbf{x}^{(c)} = c_1 \begin{pmatrix} \sqrt{3} \\ 1 \end{pmatrix} e^{2t} + c_2 \begin{pmatrix} 1 \\ -\sqrt{3} \end{pmatrix} e^{-2t}$. Writing the

nonhomogeneous term as $\begin{pmatrix} 1 \\ 0 \end{pmatrix} e^t + \begin{pmatrix} 0 \\ \sqrt{3} \end{pmatrix} e^{-t}$ we see that we

can assume $\mathbf{v}(t) = \mathbf{a}e^t + \mathbf{b}e^{-t}$ as the particlar solution. Substituting this in the D.E., we obtain

$\mathbf{a}e^t - \mathbf{b}e^{-t} = \mathbf{A}\mathbf{a}e^t + \mathbf{A}\mathbf{b}e^{-t} + \begin{pmatrix} 1 \\ 0 \end{pmatrix} e^t + \begin{pmatrix} 0 \\ \sqrt{3} \end{pmatrix} e^{-t}$, where A

is the given coefficient matrix. All the terms involving e^t must add to zero and thus we have $\mathbf{A}\mathbf{a} - \mathbf{a} + \begin{pmatrix} 1 \\ 0 \end{pmatrix} = \begin{pmatrix} 0 \\ 0 \end{pmatrix}$.

This is equivalent to the system

$\sqrt{3}\, a_2 = -1$ and $\sqrt{3}\, a_1 - 2a_2 = 0$, or $a_1 = -2/3$ and $a_2 = -1/\sqrt{3}$. Likewise the terms involving e^{-t} must add

to zero, which yields $\mathbf{A}\mathbf{b} + \mathbf{b} + \begin{pmatrix} 0 \\ \sqrt{3} \end{pmatrix} = \begin{pmatrix} 0 \\ 0 \end{pmatrix}$. This is

equilvalent to $2b_1 + \sqrt{3}\, b_2 = 0$ and $\sqrt{3}\, b_1 = -\sqrt{3}$ and thus $b_1 = -1$ and $b_2 = 2/\sqrt{3}$. Substituting these values for $\mathbf{a}$ and $\mathbf{b}$ into $\mathbf{v}(t)$ and adding $\mathbf{v}(t)$ to $\mathbf{x}^{(c)}$ yields the desired solution.

3. From Prob.3 of Sect.7.6 we find that a fundamental matrix is

$\Psi(t) = \begin{pmatrix} 5\cos t & 5\sin t \\ 2\cos t + \sin t & -\cos t + 2\sin t \end{pmatrix}$. The inverse

matrix is

$$\Psi^{-1}(t) = \begin{pmatrix} \dfrac{\cos t - 2\sin t}{5} & \sin t \\ \dfrac{2\cos t + \sin t}{5} & -\cos t \end{pmatrix}, \text{ which may be found as}$$

in Sect. 7.2 or, more efficiently, by using a computer algebra system. Thus we may use the method of variation of parameters where $\mathbf{x} = \Psi(t)\mathbf{u}(t)$ and $\mathbf{u}(t)$ is given by $\mathbf{u}'(t) = \Psi^{-1}(t)\mathbf{g}(t)$ from Eq.(27). For this problem

$$\mathbf{g}(t) = \begin{pmatrix} -\cos t \\ \sin t \end{pmatrix} \text{ and thus}$$

$$\mathbf{u}'(t) = \begin{pmatrix} \dfrac{\cos t - 2\sin t}{5} & \sin t \\ \dfrac{2\cos t + \sin t}{5} & -\cos t \end{pmatrix} \begin{pmatrix} -\cos t \\ \sin t \end{pmatrix}$$

$$= \frac{1}{5}\begin{pmatrix} 2 - 3\cos 2t + \sin 2t \\ -1 - \cos 2t - 3\sin 2t \end{pmatrix},$$

after multiplying and using appropriate trigonometric identities. Integration and multiplication by Ψ yields the desired solution, using trigonometric identities.

This problem may also be solved using undetermined coefficients. However it is not straight forward, since the assumed form of $\mathbf{v}(t) = \mathbf{a}\cos t + \mathbf{b}\sin t$ leads to singular equations for $\mathbf{a}$ and $\mathbf{b}$. The assumed form requires that $t\cos t$ and $t\sin t$ be used, as illustrated in Eq.(18) and the subsequent discussion.

4.   In this problem we use the method illustrated in Ex.1. From Prob 4 of Sect.7.5 we have the transformation matrix

$$\mathbf{T} = \begin{pmatrix} 1 & 1 \\ -4 & 1 \end{pmatrix}. \text{ Inverting } \mathbf{T} \text{ we find that } \mathbf{T}^{-1} = \frac{1}{5}\begin{pmatrix} 1 & -1 \\ 4 & 1 \end{pmatrix}.$$

If we let $\mathbf{x} = \mathbf{Ty}$ and substitute into the D.E., we obtain

$$\mathbf{y}' = \frac{1}{5}\begin{pmatrix} 1 & -1 \\ 4 & 1 \end{pmatrix}\begin{pmatrix} 1 & 1 \\ 4 & -2 \end{pmatrix}\begin{pmatrix} 1 & 1 \\ -4 & 1 \end{pmatrix}\mathbf{y} + \frac{1}{5}\begin{pmatrix} 1 & -1 \\ 4 & 1 \end{pmatrix}\begin{pmatrix} e^{-2t} \\ -2e^{t} \end{pmatrix}$$

$$= \begin{pmatrix} -3 & 0 \\ 0 & 2 \end{pmatrix}\mathbf{y} + \frac{1}{5}\begin{pmatrix} e^{-2t} + 2e^{t} \\ 4e^{-2t} - 2e^{t} \end{pmatrix}. \text{ This corresponds to}$$

the two scalar equations

$$y_1' + 3y_1 = (1/5)e^{-2t} + (2/5)e^{t},$$
$$y_2' - 2y_2 = (4/5)e^{-2t} - (2/5)e^{t},$$

which may be solved by the methods of Sect. 2.1. For the

first equation the integrating factor is e^{3t} and we
obtain $(e^{3t}y_1)' = (1/5)e^t + (2/5)e^{4t}$, so
$e^{3t}y_1 = (1/5)e^t + (1/10)e^{4t} + c_1$. For the second equation
the integrating factor is e^{-2t}, so
$(e^{-2t}y_2)' = (4/5)e^{-4t} - (2/5)e^{-t}$. Hence
$e^{-2t}y_2 = -(1/5)e^{-4t} + (2/5)e^{-t} + c_2$. Thus

$$\mathbf{y} = \begin{pmatrix} 1/5 \\ -1/5 \end{pmatrix} e^{-2t} + \begin{pmatrix} 1/10 \\ 2/5 \end{pmatrix} e^t + \begin{pmatrix} c_1 e^{-3t} \\ c_2 e^{2t} \end{pmatrix}. \quad \text{Finally,}$$

multiplying by $\mathbf{T}$, we obtain

$$\mathbf{x} = \mathbf{Ty} = \begin{pmatrix} 0 \\ -1 \end{pmatrix} e^{-2t} + \begin{pmatrix} 1/2 \\ 0 \end{pmatrix} e^t + c_1 \begin{pmatrix} 1 \\ -4 \end{pmatrix} e^{-3t} + c_2 \begin{pmatrix} 1 \\ 1 \end{pmatrix} e^{2t}.$$

The last two terms are the general solution of the
corresponding homogeneous system, while the first two
terms constitute a particular solution of the
nonhomogeneous system.

8. For this problem we illustrate the use of Laplace

Transforms. As in Eq.(43), $(s\mathbf{I} - \mathbf{A})\mathbf{X} = \begin{pmatrix} \dfrac{1}{s-1} \\ \dfrac{-1}{s-1} \end{pmatrix}$, where

$\mathbf{A} = \begin{pmatrix} 2 & -1 \\ 3 & -2 \end{pmatrix}$ (and we have assumed zero I.C. in order to

find a particular solution), thus $\mathbf{X} = (s\mathbf{I} - \mathbf{A})^{-1} \begin{pmatrix} \dfrac{1}{s-1} \\ \dfrac{-1}{s-1} \end{pmatrix}$.

$(s\mathbf{I} - \mathbf{A})^{-1}$ is found to be $\dfrac{1}{s^2-1} \begin{pmatrix} s+2 & -1 \\ 3 & s-2 \end{pmatrix}$ and hence

$$\mathbf{X}(s) = \frac{1}{(s^2-1)(s-1)} \begin{pmatrix} s+3 \\ 5-s \end{pmatrix} = \begin{pmatrix} \dfrac{2}{(s-1)^2} - \dfrac{1/2}{s-1} + \dfrac{1/2}{s+1} \\ \dfrac{2}{(s-1)^2} - \dfrac{3/2}{s-1} + \dfrac{3/2}{s+1} \end{pmatrix}, \text{ using}$$

partial fractions. The inverse transform gives

$$\mathbf{x}(t) = 2\begin{pmatrix} 1 \\ 1 \end{pmatrix} te^t - \frac{1}{2}\begin{pmatrix} 1 \\ 3 \end{pmatrix} e^t + \frac{1}{2}\begin{pmatrix} 1 \\ 3 \end{pmatrix} e^{-t}. \text{ Note that this}$$

particular solution differs from the one shown in the text by a multiple of the homogeneous solution.

12. Since the coefficient matrix is the same as that of Prob.3, use the same procedure as done in that problem, including the Ψ^{-1} found there. In the interval $\pi/2 < t < \pi$ sint > 0 and cost < 0; hence $|\text{sint}| = \text{sint}$, but $|\text{cost}| = -\text{cost}$.

14. To verify that $\mathbf{x}^{(c)}$ is the general solution of the corresponding homogeneous system it is sufficient to substitute $\mathbf{x}_1(t) = \begin{pmatrix} 1 \\ 1 \end{pmatrix} t$ and $\mathbf{x}_2(t) = \begin{pmatrix} 1 \\ 3 \end{pmatrix} t^{-1}$ individually into $t\mathbf{x}' = \begin{pmatrix} 2 & -1 \\ 3 & -2 \end{pmatrix} \mathbf{x}$, since $\mathbf{x}_1$ and $\mathbf{x}_2$ are linearly independent. For the nonhomogeous solution, substitute $\mathbf{x} = \Psi(t)\mathbf{u}(t)$, where $\Psi(t) = \begin{pmatrix} t & 1/t \\ t & 3/t \end{pmatrix}$, into the given D.E. to obtain $t\Psi'\mathbf{u} + t\Psi\mathbf{u}' = A\Psi\mathbf{u} + \mathbf{g}(t)$. Here A is the coefficeint matrix and $\mathbf{g}(t) = \begin{pmatrix} 1-t^2 \\ 2t \end{pmatrix}$. Since $t\Psi' = A\Psi$, we then have $\mathbf{u}' = (1/t)\Psi^{-1}(t)\mathbf{g}(t)$. Using a computer algebra system or row operations on Ψ and I, we find that $\Psi^{-1} = \begin{pmatrix} 3/2t & -1/2t \\ -t/2 & t/2 \end{pmatrix}$ and hence
$u_1' = \dfrac{3}{2t^2} - \dfrac{3}{2} - \dfrac{1}{t}$ and $u_2' = \dfrac{-1}{2} + \dfrac{t^2}{2} + t$, which yields
$u_1 = \dfrac{-3}{2t} - \dfrac{3t}{2} - \ln t + c_1$ and $u_2 = -\dfrac{1}{2}t + \dfrac{t^3}{6} + \dfrac{t^2}{2} + c_2$.
Multiplication of $\mathbf{u}$ by $\Psi(t)$ yields the desired solution.

CHAPTER 8

Section 8.1, Page 451

In the following problems that ask for a large number of
numerical calculations the first few steps are shown. It is
then necessary to use these samples as a model to format a
computer program or calculator to find the remaining values.

1a. The Euler formulas is $y_{n+1} = y_n + h(3 + t_n - y_n)$ for
    $n = 0,1,2,3...$ and with $t_0 = 0$ and $y_0 = 1$. Thus
    $y_1 = 1 + .05(3 + 0 - 1) = 1.1$
    $y_2 = 1.1 + .05(3 + .05 - 1.1) = 1.1975 \cong y(.1)$
    $y_3 = 1.1975 + .05(3 + .1 - 1.1975) = 1.29263$
    $y_4 = 1.29263 + .05(3 + .15 - 1.29263) = 1.38549 \cong y(.2)$.

1c. The backward Euler formula is $y_{n+1} = y_n + h(3 + t_{n+1} - y_{n+1})$.
    Solving this for y_{n+1} we find $y_{n+1} = [y_n + h(3 + t_{n+1})]/(1+h)$.
    Thus $y_1 = \dfrac{1 + .05(3.05)}{1.05} = 1.097619$,

    $y_2 = \dfrac{1.097619 + .05(3.1)}{1.05} = 1.192971 \cong y(.1)$,

    $y_3 = \dfrac{1.192971 + .05(3.15)}{1.05} = 1.286162$, and

    $y_4 = \dfrac{1.286162 + .05(3.2)}{1.05} = 1.377298 \cong y(.2)$.

5a. $y_1 = y_0 + h\dfrac{y_0^2 + 2t_0 y_0}{3 + t_0^2} = .5 + .05\dfrac{(.5)^2 + 0}{3 + 0} = .504167$

    $y_2 = .504167 + .05\dfrac{(.504167)^2 + 2(.05)(.504167)}{3 + (.05)^2} = .509239$

5c. $y_1 = .5 + .05\dfrac{y_1^2 + 2(.05)y_1}{3 + (.05)^2}$, which is a quadratic
    equation in y_1. Using the quadratic formula, or an
    equation solver, we obtain $y_1 = .5050895$. Thus

    $y_2 = .5050895 + .05\dfrac{y_2^2 + 2(.1)y_2}{3 + (.1)^2}$ which is again quadratic

    in y_2, yielding $y_2 = .5111273$.

7a. For Part(a) eighty steps must be taken, that is,
    $n = 0,1,...79$ and for Part(b) 160 steps must taken with
    $n = 0,1,...159$. Thus use of a programmable calculator or
    a computer is required.

7c. We have $y_{n+1} = y_n + h(.5 - t_{n+1} + 2y_{n+1})$, which is linear in y_{n+1} and thus we have $y_{n+1} = \dfrac{y_n + .5h - ht_{n+1}}{1 - 2h}$. Again, 80 steps are needed here and 160 steps in Part(d). In This case a spreadsheet is very useful. The first few lines, the middle three lines and the last two lines are shown for h = .025:

n	y_n	t_n	y_{n+1}
0	1	0	1.06513
1	1.06513	.025	1.13303
2	1.13303	.050	1.20381
⋮			
38	7.49768	.950	7.87980
39	7.87980	.975	8.28137
40	8.28137	1.000	8.70341
⋮			
78	55.62105	1.950	58.50966
79	58.50966	1.975	61.54964
80	61.54964	2.000	

At least eight decimal places were used in all calculations.

9c. The backward Euler formula gives $y_{n+1} = y_n + h\sqrt{t_{n+1} + y_{n+1}}$. Subtracting y_n from both sides, squaring both sides, and solving for y_{n+1} yields $y_{n+1} = y_n + \dfrac{h^2}{2} + h\sqrt{y_n + t_{n+1} + h^2/4}$, where the positive root is chosen since y' is always positive and thus y is increasing. Alternately, an equation solver can be used to solve $y_{n+1} = y_n + h\sqrt{t_{n+1} + y_{n+1}}$ for y_{n+1}. The first few values, for h = 0.25, are $y_1 = 3.043795$, $y_2 = 2.088082$, $y_3 = 3.132858$ and $y_4 = 3.178122 \cong y(.1)$.

15. If $y' = 1 - t + 4y$ then differentiation of both sides yields $y'' = -1 + 4y' = -1 + 4(1-t+4y) = 3 - 4t + 16y$. In Eq.(12) we let y_n, y_n' and y_n'' denote the approximate values of $\phi(t_n)$, $\phi'(t_n)$, and $\phi''(t_n)$, respectively. Keeping the first three terms in the Taylor series we have
$$y_{n+1} = y_n + y_n'h + y_n'' h^2/2$$
$$= y_n + (1 - t_n + 4y_n)h + (3 - 4t_n + 16y_n)h^2/2.$$ For n = 0, $t_0 = 0$ and $y_0 = 1$ we have

$$y_1 = 1 + (1 - 0 + 4)(.1) + (3 - 0 + 16)\frac{(.1)^2}{2} = 1.595 \cong y(.1) \text{ and}$$

$$y_2 = 1.595 + [1 - .1 + 4(1.595)](.1) + [3 - .4 + 16(1.595)]\frac{(.1)^2}{2} = 2.4636 \cong y(.2).$$

These compare to 1.609 and 2.5053, respectively, in Table 8.1.2

16. If $y = \phi(t)$ is the exact solution of the I.V.P., then
$\phi'(t) = 2\phi(t) - 1$ and $\phi''(t) = 2\phi'(t) = 4\phi(t) - 2$. From
Eq.(21), $e_{n+1} = [2\phi(\bar{t}_n) - 1]h^2$, $t_n < \bar{t}_n < t_n + h$. Thus a
bound for e_{n+1} is $|e_{n+1}| \le [1 + 2\max_{0 \le t \le 1}|\phi(t)|]h^2$. Since the
exact solution is $y = \phi(t) = [1 + \exp(2t)]/2$,
$e_{n+1} = h^2\exp(2\bar{t}_n)$. Therefore
$|e_1| \le (0.1)^2\exp(0.2) = 0.012$ and
$|e_4| \le (0.1)^2\exp(0.8) = 0.022$, since the maximum value of
$\exp(2\bar{t}_n)$ occurs at $t = .1$ and $t = .4$ respectively. From
Prob. 2 of Sect. 2.7, the actual error in the first step
is .0107.

19. The local truncation error is $e_{n+1} = \phi''(\bar{t}_n)h^2/2$. For this
problem $\phi'(t) = 5t - 3\phi^{1/2}(t)$ and thus
$\phi''(t) = 5 - (3/2)\phi^{-1/2}\phi' = 19/2 - (15/2)t\phi^{-1/2}$.
Substituting this last expression into e_{n+1} yields the
desired answer.

22d. Since $y'' = -5\pi\sin5\pi t$, Eq.(21) gives $e_{n+1} = -(5\pi/2)\sin(5\pi\bar{t}_n)h^2$.
Thus $|e_{n+1}| < \frac{5\pi}{2}h^2$, since $|\sin5\pi\bar{t}_n| < 1$. For $|e_{n+1}| < .05$ we
must then have $h < \dfrac{1}{\sqrt{50\pi}} \cong .08$.

23a. From Eq.(14) we have $E_n = \phi(t_n) - y_n$. Using this in
Eq.(20) we obtain

$E_{n+1} = E_n + h\{f[t_n,\phi(t_n)] - f(t_n,y_n)\} + \phi''(\bar{t}_n)h^2/2$. Using
the given inequality involving L we have
$|f[t_n,\phi(t_n)] - f(t_n,y_n)| \le L|\phi(t_n) - y_n| = L|E_n|$ and thus
$|E_{n+1}| \le |E_n| + hL|E_n| + \max_{t_0 \le t \le t_n}|\phi''(t)|h^2/2 = \alpha|E_n| + \beta h^2$.

23b. Since $\alpha = 1 + hL$, $\alpha - 1 = hL$. Hence $\beta h^2(\alpha^n - 1)/(\alpha - 1) = \beta h^2[(1+hL)^n - 1]/hL = \beta h[(1+hL)^n - 1]/L$.

23c. $(1+hL)^n \le \exp(nhL)$ follows from the observation that
$\exp(nhL) = [\exp(nL)]^n = (1 + hL + h^2L^2/2! + \dots)^n$ and
that all the neglected terms are positive. Noting that
$nh = t_n - t_0$, the rest follows from Eq.(ii).

24. The Taylor series for $\phi(t)$ about $t = t_{n+1}$ is

$$\phi(t) = \phi(t_{n+1}) + \phi'(t_{n+1})(t-t_{n+1}) + \phi''(t_{n+1})\frac{(t-t_{n+1})^2}{2} + \ldots.$$

Letting $\phi'(t) = f(t,\phi(t))$, $t = t_n$ and $h = t_{n+1} - t_n$ we

have $\phi(t_n) = \phi(t_{n+1}) - f(t_{n+1},\phi(t_{n+1}))h + \phi''(\bar{t}_n)h^2/2$, where

$t_n < \bar{t}_n < t_{n+1}$. Thus

$\phi(t_{n+1}) = \phi(t_n) + f(t_{n+1},\phi(t_{n+1}))h - \phi''(\bar{t}_n)h^2/2$. Comparing

this to Eq. 13 we then have $e_{n+1} = -\phi''(\bar{t}_n)h^2/2$.

25b. From Prob. 1 we have $y_{n+1} = y_n + h(3 + t_n - y_n)$, so
$y_1 = 1 + .05(3 + 0 - 1) = 1.1$
$y_2 = 1.1 + .05(3 + .05 - 1.1) = 1.20 \approx y(.1)$
$y_3 = 1.20 + .05(3 + .1 - 1.20) = 1.30$
$y_4 = 1.30 + .05(3 + .15 - 1.30) = 1.39 \approx y(.2)$.

Section 8.2, Page 458

1a. The improved Euler formula is given by Eq.(5):
$y_{n+1} = y_n + [y_n' + f(t_n + h, y_n + hy_n')]h/2$ where
$y' = f(t,y) = 3 + t - y$. Hence $y_n' = 3 + t_n - y_n$ and
$f(t_n + h, y_n + hy_n') = 3 + t_{n+1} - (y_n + hy_n')$. Thus we obtain

$$y_{n+1} = y_n + (3 + t_n - y_n)h + \frac{h^2}{2}(1 - y_n')$$

$$= y_n + (3 + t_n - y_n)h + \frac{h^2}{2}(-2 - t_n + y_n). \text{ Thus}$$

$$y_1 = 1 + (3-1)(.05) + \frac{(.05)^2}{2}(-2+1) = 1.098750 \text{ and}$$

$$y_2 = y_1 + (3 +.05 - y_1)(.05) + \frac{(.05)^2}{2}(-2 -.05 + y_1) = 1.19512$$

are the first two steps. In this case, the equation
specifying y_{n+1} is somewhat more complicated when
$y_n' = 3 + t_n - y_n$ was substituted. When designing the steps to
calculate y_{n+1} on a computer, y_n' can be calculated first and
thus the simpler formula for y_{n+1} can be used. The exact
solution is $y(t) = 2 + t - e^{-t}$, so $y(.1) = 1.19516$,
$y(.2) = 1.38127$, $y(.3) = 1.55918$ and $y(.04) = 1.72968$, so the
approximations using $h = .0125$ are quite accurate to five
decimal places.

4.   In this case $y_n' = 2t_n + e^{-t_n y_n}$ and thus the improved Euler
formula is

$$y_{n+1} = y_n + \frac{[(2t_n + e^{-t_n y_n}) + 2t_{n+1} + e^{-t_{n+1}(y_n + hy_n')}]h}{2}.$$ For

n = 0, 1, 2 we get $y_1 = 1.05122$, $y_2 = 1.10483$ and $y_3 = 1.16072$
for h = .05.

10. See Problem 4.

11. The improved Euler formula is
$$y_{n+1} = y_n + \frac{f(t_n, y_n) + f(t_{n+1}, y_n + hf(t_n, y_n))}{2} h.$$ As suggested in
the text, it's best to perform the following steps when
implementing this formula: let $k_1 = (4 - t_n y_n)/(1 + y_n^2)$,
$k_2 = y_n + hk_1$ and $k_3 = (4 - t_{n+1}k_2)/(1 + k_2^2)$. Then
$y_{n+1} = y_n + (k_1 + k_3)h/2$.

14a. Since $\phi(t_n + h) = \phi(t_{n+1})$ we have, using the first part of
Eq.(5) and the given equation,
$$e_{n+1} = \phi(t_{n+1}) - y_{n+1}$$
$$= [\phi(t_n) - y_n] + [\phi'(t_n) - \frac{y_n' + f(t_n + h, \ y_n + hy_n')}{2}]h$$
$$+ \phi''(t_n)h^2/2! + \phi'''(\bar{t}_n)h^3/3!.$$
Since $y_n = \phi(t_n)$ and $y_n' = \phi'(t_n) = f(t_n, y_n)$ this reduces
to $e_{n+1} = \phi''(t_n)h^2/2! - \{f[t_n + h, \ y_n + hf(t_n, y_n)]$
$$- f(t_n, y_n)\}h/2! + \phi'''(\bar{t}_n)h^3/3!,$$
which can be written in the form of Eq.(i).

14b. First observe that $y' = f(t, y)$ and $y'' = f_t(t, y) +$
$f_y(t, y)y'$. Hence $\phi''(t_n) = f_t(t_n, y_n) + f_y(t_n, y_n)f(t_n, y_n)$.
Using the given Taylor series, with $a = t_n$, $h = h$, $b = y_n$
and $k = hf(t_n, y_n)$ we have

$$f[t_n + h, y_n + hf(t_n, y_n)] = f(t_n, y_n) + f_t(t_n, y_n)h + f_y(t_n, y_n)hf(t_n, y_n)$$
$$+ [f_{tt}(\xi, \eta)h^2 + 2f_{ty}(\xi, \eta)h^2 f(t_n, y_n) + f_{yy}(\xi, \eta)h^2 f^2(t_n, y_n)]/2!$$

where $t_n < \xi < t_n + h$ and $|\eta - y_n| < h|f(t_n, y_n)|$.
Substituting this in Eq.(i) and using the earlier
expression for $\phi''(t_n)$ we find that the first term on the
right side of Eq.(i) reduces to
$-[f_{tt}(\xi, \eta) + 2f_{ty}(\xi, \eta)f(t_n, y_n) + f_{yy}(\xi, \eta)f^2(t_n, y_n)]h^3/4$,
which is proportional to h^3 plus, possibly, higher order
terms. The reason that there may be higher order terms
is because ξ and η will, in general, depend upon h.

14c. If $f(t,y)$ is linear in t and y, then $f_{tt} = f_{ty} = f_{yy} = 0$
and the terms appearing in the last formula of Part(b)
are all zero.

15. Since $\phi(t) = [4t - 3 + 19\exp(4t)]/16$ we have
$\phi'''(t) = 76\exp(4t)$ and thus from Prob. 14c, since f is
linear in t and y, we find $e_{n+1} = 38[\exp(4t_n)]h^3/3$.
Thus, since e^{4t} is increasing on [0,2],
$|e_{n+1}| \le (38h^3/3)\exp(8) = 37,758.8h^3$ on $0 \le t \le 2$.
For $n = 1$, we have
$|e_1| = |\phi(t_1) - y_1| \le (38/3)\exp(0.2)(.05)^3 = .001934$,
which is approximately 1/15 of the error indicated in
Eq.(27) of Sect.8.1.

19. The Euler method gives
$y_1 = y_0 + h(5t_0 - 3\sqrt{y_0}) = 2 + .1(-3\sqrt{2}) = 1.57574$ and the
improved Euler method gives

$$y_1 = y_0 + \frac{f(t_0,y_0) + f(t_1,y_1)}{2} h$$
$$= 2 + [-3\sqrt{2} + (.5 - 3\sqrt{1.57574})].05 = 1.62458.$$

Thus, the estimated error in using the Euler method is
$1.62458 - 1.57574 = .04884$. Since we want our error tolerance
to be no greater than .0025 we need to adjust the step size
(see the text discussion on the variation of the step size)
downward by a factor of $\sqrt{.0025/.04884} \cong .226$. Thus a step
size of $h = (.1)(.23) = .023$ would be needed for the required
local truncation error bound of .0025.

24. The modified Euler formula is
$y_{n+1} = y_n + hf[t_n + h/2, y_n + (h/2)f(t_n,y_n)]$ where
$f(t,y) = 5t - 3\sqrt{y}$. Thus
$y_1 = 2 + .05[5(t_0 + .025) - 3\mathrm{sqrt}(2 + .025(5t_0 - 3\sqrt{2}))]$
$= 1.79982$ for $t_0 = 0$. Likewise
$y_2 = y_1 + .05[5(.075) - 3\mathrm{sqrt}\{y_1 + .025(5(.05) - 3\sqrt{2})\}]$
$= 1.62268$, which is between the values for $h = .05$ and
for $h = .025$ using the improved Euler method in Prob.2.

Section 8.3, Page 463

4. The Runge-Kutta formula is
$y_{n+1} = y_n + h(k_{n1} + 2k_{n2} + 2k_{n3} + k_{n4})/6$ where k_{n1}, k_{n2}
etc. are given by Eqs.(3). Thus for

$f(t,y) = 2t + e^{-ty}$, $(t_0,y_0) = (0,1)$ and $h = .1$ we have
$k_{01} = 0 + e^0 = 1$
$k_{02} = 2(0 + .05) + e^{-(0+.05)(1+.05k_{01})} = 1.048854$
$k_{03} = 2(.05) + e^{-(.05)(1+.05k_{02})} = 1.048738$
$k_{04} = 2(.1) + e^{-(.1)(1+.1k_{03})} = 1.095398$ and hence
$y(.1) \cong y_1 = 1 + .1(k_{01} + 2k_{02} + 2k_{03} + k_{04})/6 = 1.104843$.

For $y(.2)$ we have
$k_{11} = .2 + e^{-.1(1.104843)} = 1.095400$
$k_{12} = .3 + e^{-.15(1.104843 + .05k_{11})} = 1.140346$
$k_{13} = .3 + e^{-.15(1.104843 + .05k_{12})} = 1.140062$
$k_{14} = .4 + e^{-.2(1.104843 + .1k_{13})} = 1.183668$ and hence
$y(.2) \cong y_2 = 1.104843 + .1(k_{11}+2k_{12}+2k_{13}+k_{14})/6 = 1.218841$.

Note that for $h = .05$ we get the same result (after two steps) for $y(.1)$ above, indicating the high accuracy of the Runge-Kutta formula. For comparison to the Euler and improved Euler results see Prob.4 in Sects, 8.1 and 8.2.

11a. We have $f(t_n,y_n) = (4 - t_ny_n)/(1 + y_n^2)$. Thus for $t_0 = 0$, $y_0 = -2$ and $h = .1$ we have
$k_{01} = f(0,-2) = .8$
$k_{02} = f(.05,-2+.05(.8)) = f(.05,-1.96) = .846414$,
$k_{03} = f(.05,-2+.05k_{02}) = f(.05,-1.957679) = .847983$,
$k_{04} = f(.1,-2+.1k_{03}) = f(.1,-1.915202) = .897927$, and
$y_1 = -2 + .1(k_{01} + 2k_{02} + 2k_{03} + k_{04})/6 = -1.915221$. For comparison, see Prob. 11 in Sects. 8.1 and 8.2.

14a.

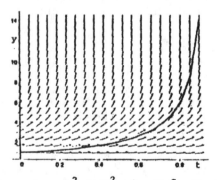

14b. We have $f(t_n,y_n) = t_n^2 + y_n^2$, $t_0 = 0$, $y_0 = 1$ and $h = .1$ so
$k_{01} = 0^2 + 1^2 = 1$
$k_{02} = (.05)^2 + [1 + .05(1)]^2 = 1.105$
$k_{03} = (.05)^2 + [1 + .05(1.105)]^2 = 1.11605$
$k_{04} = (.1)^2 + [1 + .1(1.11605)]^2 = 1.245666$ and thus
$y_1 = 1 + .1(k_{01} + 2k_{02} + 2k_{03} + k_{04})/6 = 1.111463$. Using these steps in a computer program, we obtain the following values

for y:

t	h =.1	h = .05	h = .025	h = 0.0125
.8	5.842	5.8481	5.8483	5.8486
.9	14.0218	14.2712	14.3021	14.3046
.95		46.578	49.757	50.3935

14c. No accurate solution can be obtained for $y(1)$, as the values at $t = .975$ for $h = .025$ and $h = .0125$ are 1218 and 23,279 respectively. These are caused by the slope field becoming vertical as $t \to 1$.

Section 8.4, Page 469

4a. The predictor formula, Eq.(6), is
$y_{n+1} = y_n + h(55f_n - 59f_{n-1} + 37f_{n-2} - 9f_{n-3})/24$
and the corrector formula, Eq.(10), is
$y_{n+1} = y_n + h(9f_{n+1} + 19f_n - 5f_{n-1} + f_{n-2})/24$, where
$f_n = 2t_n + \exp(-t_n y_n)$. Using the Runge-Kutta method, from
Sect. 8.3, Prob. 4a, we have for $t_0 = 0$ and $y_0 = 1$,
$y_1 = 1.1048431$, $y_2 = 1.2188411$ and $y_3 = 1.3414680$. Thus the
predictor formula gives $y_4 = 1.4725974$, so $f_4 = 1.3548603$ and
the corrector formula then gives $y_4 = 1.4726173$, which is the
desired value. These results, and the next step, are
summarized in the following table:

n	y_n	f_n	y_{n+1} Predicted	f_{n+1}	y_{n+1} Corrected
0	1	1			
1	1.1048431	1.0954004			
2	1.2188411	1.1836692			
3	1.3414680	1.2686862	1.4725974	1.3548603	1.4726173
4	1.4726173	1.3548559	1.6126246	1.4465016	1.6126215
5	1.6126215				

Note that the value for f_4 on the line for $n = 4$ uses the
corrected value for y_4, and differs slightly from f_4 on
the line for $n = 3$, which uses the predicted value for y_4.

4b. The fourth order Adams-Moulton method is given by
Eq.(10): $y_{n+1} = y_n + (h/24)(9f_{n+1} + 19f_n - 5f_{n-1} + f_{n-2})$.
Substituting $h = .1$ we obtain
$y_{n+1} = y_n + (.1)(19f_n - 5f_{n-1} + f_{n-2})/24 + .0375f_{n+1}$.
For $n = 2$ we then have

$y_3 = y_2 + (.1)(19f_2 - 5f_1 + f_0)/24 + .0375f_3$
   $= 1.293894103 + .0375(.6 + e^{-.3y_3})$, using values for
y_2, f_0, f_1, f_2 from Part(a). An equation solver then yields
$y_3 = 1.341469821$. Likewise

$y_4 = y_3 + (.1)(19f_3 - 5f_2 + f_1)/24 + .0375f_4$

$\qquad = 1.421811841 + .0375(.8 + e^{-.4y_4})$, where f_3 is calculated
using the y_3 found above. This last equation yields
$y_4 = 1.472618922$. Finally

$y_5 = y_4 + (.1)(19f_4 - 5f_3 + f_2)/24 + .0375f_5$

$\qquad = 1.558379316 + .0375(1.0 + e^{-.5y_5})$, which gives
$y_5 = 1.612623138$.

4c. We use Eq.(16):
$\quad y_{n+1} = (1/25)(48y_n - 36y_{n-1} + 16y_{n-2} - 3y_{n-3} + 12hf_{n+1})$.
Thus $y_4 = .04(48y_3 - 36y_2 + 16y_1 - 3y_0) + .048f_4$

$\qquad\qquad = 1.40758686 + .048(.8 + e^{-.4y_4})$, using values
for y_0, y_1. y_2, y_3 from Part(a). An equation solver then
yields $y_4 = 1.472619913$. Likewise
$y_5 = .04(48y_4 - 36y_3 + 16y_2 - 3y_1) + .048f_5$

$\qquad = 1.54319349 + .048(1 + e^{-.5y_5})$, which gives
$y_5 = 1.612625556$.

All three methods give the same result through five
decimal places for y_5.

7a. Using the predictor and corrector formulas (Eqs.6 and 10)
with $f_n = .5 - t_n + 2y_n$ and using the Runge-Kutta method
to calculate y_1, y_2 and y_3, we obtain the following table
for $h = .05$, $t_0 = 0$, $y_0 = 1$:

n	y_n	f_n	y_{n+1} Predicted	f_{n+1}	y_{n+1} Corrected
0	1	2.5			
1	1.130171	2.710342			
2	1.271403	2.9420805			
3	1.424858	3.199717	1.591820	3.483640	1.591825
4	1.591825	3.483649	1.773716	3.797433	1.773721
5	1.773721	3.797443	1.972114	4.144227	1.972119
6	1.972119	4.144238	2.188747	4.527495	2.188753
7	2.188753	4.527507	2.425535	4.951070	2.425542
8	2.425542	4.951084	2.684597	5.419194	2.684604
9	2.684604	5.419209	2.968276	5.936551	2.968284
10	2.968284				

Thus $y(.5) \cong y_{10} = 2.968284$.

7b. From Eq.(10) we have

$$y_{n+1} = y_n + \frac{h}{24}(9f_{n+1} + 19f_n - 5f_{n-1} + f_{n-2})$$

$$= y_n + \frac{h}{24}[9(.5 - t_{n+1} + 2y_{n+1}) + 19f_n - 5f_{n-1} + f_{n-2}].$$

Solving for y_{n+1} we obtain

$$y_{n+1} = [y_n + \frac{h}{24}(19f_n - 5f_{n-1} + f_{n-2} + 4.5 - 9t_{n+1})]/(1-.75h).$$

For $h = .05$, $t_0 = 0$, $y_0 = 1$ and using y_1 and y_2 as calculated using the Runge-Kutta formula, we obtain the following table:

n	y_n	f_n	y_{n+1}
0	1	2.5	
1	1.130171	2.710342	
2	1.271403	2.942805	1.424859
3	1.424859	3.199718	1.591825
4	1.591825	3.483650	1.773722
5	1.773722	3.797444	1.972120
6	1.972120	4.144241	2.188755
7	2.188755	4.527510	2.425544
8	2.425544	4.951088	2.684607
9	2.684607	5.419214	2.968287
10	2.968287		

Again $y(.5) \cong y_{10} = 2.968287$

7c. From Eq.(16) we have

$$y_{n+1} = (48y_n - 36y_{n-1} + 16y_{n-2} - 3y_{n-3} + 12hf_{n+1})/25$$
$$= [48y_n - 36y_{n-1} + 16y_{n-2} - 3y_{n-3} + 12h(.5-t_{n+1})]/25+(24/25)hy_{n+1}.$$

Solving for y_{n+1} we have

$$y_{n+1} = [48y_n - 36y_{n-1} + 16y_{n-2} - 3y_{n-3} + 12h(.5-t_{n+1})]/(25-24h).$$

Again, using Runge-Kutta to find y_1 and y_2, we then obtain the following table:

n	y_n	y_{n+1}
0	1	
1	1.130170833	
2	1.271402571	
3	1.424858497	1.591825573
4	1.591825573	1.773724801
5	1.773724801	1.972125968
6	1.972125968	2.188764173
7	2.188764173	2.425557376
8	2.425557376	2.684625416
9	2.684625416	2.968311063
10	2.968311063	

The exact solution is $y(t) = e^{2t} + t/2$ so $y(.5) = 2.9682818$ and $y(2) = 55.59815$, so we see that the predictor-corrector method in Part(a) is accurate at $t = .5$ through five decimal places and at $t = 2$ through three decimal places.

16. Let $P_2(t) = At^2 + Bt + C$. As in Eqs. (12) and (13) let $P_2(t_{n-1}) = y_{n-1}$, $P_2(t_n) = y_n$, $P_2(t_{n+1}) = y_{n+1}$ and $P_2'(t_{n+1}) = y_{n+1}' = f(t_{n+1},y_{n+1}) = f_{n+1}$. Recall that $t_{n-1} = t_n - h$ and $t_{n+1} = t_n + h$ and thus we have the four equations:

$$A(t_n-h)^2 + B(t_n-h) + C = y_{n-1} \qquad (i)$$
$$At_n^2 \qquad\quad + Bt_n \qquad + C = y_n \qquad (ii)$$
$$A(t_n+h)^2 + B(t_n+h) + C = y_{n+1} \qquad (iii)$$
$$\qquad\qquad 2A(t_n+h) + B = f_{n+1} \qquad (iv)$$

Subtracting Eq.(i) from Eq.(ii) to get Eq.(v) (not shown) and subtracting Eq.(ii) from Eq.(iii) to get Eq.(vi) (not shown), then subtracting Eq.(v) from Eq.(vi) yields $y_{n+1} - 2y_n + y_{n-1} = 2Ah^2$, which can be solved for A. Now $B = f_{n+1} - 2A(t_n+h)$ [from Eq.(iv)] and $C = y_n - t_n f_{n+1} + At_n^2 + 2At_n h$ [from Eq.(ii)]. Using these values for A, B and C in Eq.(iii) yields $y_{n+1} = (1/3)(4y_n - y_{n-1} + 2hf_{n+1})$, which is Eq.(15).

Section 8.5, Page 479

1b. The D.E. is linear and thus $(e^{-t}y)' = (t-3)e^{-t}$. Integrating by parts and applying the I.C. yields $\phi_2(t) = (2-t) + (.001)e^t$. Thus $\phi_1(1) = 1$, $\phi_2(1) = 1.0027$ and $\phi_2(t) - \phi_1(t) = (.001)e^t$.

2a. If $0 \le t \le 1$ then we know $0 \le t^2 \le 1$ and hence $e^y \le t^2 + e^y \le 1 + e^y$. Since each of these terms represents a slope, we may conclude that the solution of Eq.(i) is bounded above by the solution of Eq.(ii) and is bounded below by the solution of Eq.(iii).

2b. $\phi_1(t)$ and $\phi_2(t)$ can each be found by separation of variables. For $\phi_1(t)$ we have $\dfrac{1}{1+e^y}dy = dt$, or

$\dfrac{e^{-y}}{e^{-y}+1}dy = dt$. Integrating both sides yields

$-\ln(e^{-y}+1) = t + c$. Solving for y we find
$y = \ln[1/(c_1e^{-t}-1)]$. Setting $t = 0$ and $y = 0$, we obtain
$c_1 = 2$ and thus $\phi_1(t) = \ln[e^t/(2-e^t)]$. As $t \to \ln 2$, we
see that $\phi_1(t) \to \infty$. A similar analysis shows that
$\phi_2(t) = \ln[1/(c_2-t)]$, where $c_2 = 1$ when the I.C. are
used. Thus $\phi_2(t) \to \infty$ as $t \to 1$ and thus we conclude
that $\phi(t) \to \infty$ for some t such that $\ln 2 \le t \le 1$.

2c. From Part(b): $\phi_1(.9) = \ln[1/(c_1e^{-.9}-1)] = 3.4298$ yields
$c_1 = 2.5393$ and thus $\phi_1(t) \to \infty$ when $t \cong .9319$.
Similarly for $\phi_2(t)$ we have $c_2 = .9324$ and thus
$\phi_2(t) \to \infty$ when $t \cong .932$.

4a. The D.E. is $y' + 10y = 2.5t^2 + .5t$. So $y_h = ce^{-10t}$ is the
solution of the related homogeneous equation and the
particular solution, using undetermined coefficients, is

$y_p = At^2 + Bt + C$.
Substituting this
into the D.E. yields
$A = 1/4$, $B = C = 0$.
To satisfy the I.C.,
$c = 4$, so
$y(t) = 4e^{-10t} + (t^2/4)$,
which is shown in the
graph.

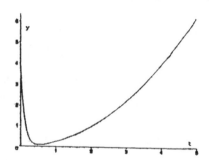

4b. From the discussion following Eq.(15), we see that h must
be less than $\dfrac{2}{|r|}$ for the Euler method to be stable.
Thus, for $r = 10$, $h < .2$. For $h = .2$ we obtain the
following values:

$t =$	4	4.2	4.4	4.6	4.8	5.0
$y =$	8	.4	8.84	1.28	9.76	2.24

and for $h = .18$ we obtain:

$t =$	4.14	4.32	4.50	4.68	4.86	5.04
$y =$	4.26	4.68	5.04	5.48	5.89	6.35.

Clearly the second set of values is stable, although not
very accurate.

4c. For a step size of .25 we find

$$t = 4 \qquad 4.25 \quad 4.75 \quad 5.00$$
$$y = 4.018 \; 4.533 \quad 5.656 \quad 6.205,$$

for a step size of .28 we find

$$t = \; 4.2 \quad\; 4.48 \quad 4.76 \qquad 5.00$$
$$y = 10.14 \quad 10.89 \quad 11.68 \qquad 12.51,$$

and for a step size of .3 we find

$$t = 4.2 \qquad 4.5 \qquad 4.8 \qquad 5.1$$
$$y = 353 \qquad 484 \qquad 664 \qquad 912.$$

Thus instability appears to occur about h = .28 and
certainly by h = .3. Note that the exact solution for
t = 5 is y = 6.2500, so for h = .25 we do obtain a good
approximation.

4d. For h = .5 the error at t = 5 is .013, while for h = .385,
the error at t = 5.005 is .01.

5a. The general solution of the D.E. is $y(t) = t + ce^{\lambda t}$,
where $y(0) = 0 \Rightarrow c = 0$ and thus $y(t) = t$, which is
independent of λ.

5c. Your result in Part(b) will depend upon the particular
computer system and software that you use. If there is
sufficient accuracy, you will obtain the solution y = t
for t on $0 \le t \le 1$ for each value of λ that is given,
since there is no discretization error. If there is not
sufficient accuracy, then round-off error will affect
your calculations. For the larger values of λ, the
numerical solution will quickly diverge from the exact
solution, y = t, to the general solution $y = t + ce^{\lambda t}$,
where the value of c depends upon the round-off error.
If the latter case does not occur, you may simulate it by
computing the numerical solution to the I.V.P.
$y' - \lambda y = 1 - \lambda t$, y(.1) = .10000001. Here we have
assumed that the numerical solution, with a step size of
.01, is exact up to the point t = .09 [i.e. y(.09) = .09]
and that at t = .1 round-off error has occurred as
indicated by the slight error in the I.C. It has also
been found that a larger step size (h = .05 or h = .1)
may also lead to round-off error.

Section 8.6, Page 483

2a. The Euler formula is $x_{n+1} = x_n + hf_n$, where

$$f_n = \begin{pmatrix} 2x_n+t_ny_n \\ x_ny_n \end{pmatrix}, \quad x_0 = 1 \text{ and } y_0 = 1. \text{ Thus}$$

$$f_0 = \begin{pmatrix} 2+0 \\ (1)(1) \end{pmatrix} = \begin{pmatrix} 2 \\ 1 \end{pmatrix}, \quad x_1 = \begin{pmatrix} 1+.1(2) \\ 1+.1(1) \end{pmatrix} = \begin{pmatrix} 1.2 \\ 1.1 \end{pmatrix},$$

$$f_1 = \begin{pmatrix} 2.4+.1(1.1) \\ (1.2)(1.1) \end{pmatrix} = \begin{pmatrix} 2.51 \\ 1.32 \end{pmatrix}$$

$$\text{and } x_2 = \begin{pmatrix} 1.2+.1(2.51) \\ 1.1+.1(1.32) \end{pmatrix} = \begin{pmatrix} 1.451 \\ 1.232 \end{pmatrix} \cong \begin{pmatrix} \phi(.2) \\ \psi(.2) \end{pmatrix}$$

2b. Eqs.(7) give:

$$k_{01} = \begin{pmatrix} f(0,1,1) \\ g(0,1,1) \end{pmatrix} = \begin{pmatrix} 2+0 \\ (1)(1) \end{pmatrix} = \begin{pmatrix} 2 \\ 1 \end{pmatrix}$$

$$k_{02} = \begin{pmatrix} 2.4+.1(1.1) \\ (1.2)(1.1) \end{pmatrix} = \begin{pmatrix} 2.51 \\ 1.32 \end{pmatrix}$$

$$k_{03} = \begin{pmatrix} 2.502+.1(1.132) \\ (1.251)(1.132) \end{pmatrix} = \begin{pmatrix} 2.6152 \\ 1.41613 \end{pmatrix}$$

$$k_{04} = \begin{pmatrix} 3.04608+.2(1.28323) \\ (1.52304)(1.28323) \end{pmatrix} = \begin{pmatrix} 3.30273 \\ 1.95441 \end{pmatrix}$$

Using Eq (6) in scalar form, we then have
$x_1 = 1+(.2/6)[2+2(2.51)+2(2.6152)+3.30273] = 1.51844$
$y_1 = 1+(.2/6)[1+2(1.32)+2(1.41613)+1.95441] = 1.28089$,
which are approximations to $\phi(.2)$ and $\psi(.2)$
respectively.

7.   Write a computer program to do this problem as there are
     twenty steps or more for $h \le .05$.

8.   If we let $y = x'$, then $y' = x''$ and we obtain the system
     $x' = y$ and $y' = t-3x-t^2y$, with $x(0) = 1$ and
     $y(0) = x'(0) = 2$. Thus $f(t,x,y) = y$,
     $g(t,x,y) = t - 3x - t^2y$, $t_0 = 0$, $x_0 = 1$ and $y_0 = 2$. If a
     program has been written for an earlier problem, then its
     best to use that. Otherwise, the first two steps are as
     follows:

$$k_{01} = \begin{pmatrix} 2 \\ -3 \end{pmatrix}$$

$$\mathbf{k}_{02} = \begin{pmatrix} 2+(-.15) \\ .05-3(1.1)-(.05)^2(1.85) \end{pmatrix} = \begin{pmatrix} 1.85 \\ -3.25463 \end{pmatrix}$$

$$\mathbf{k}_{03} = \begin{pmatrix} 2+(-.16273) \\ .05-3(1.0925)-(.05)^2(1.83727) \end{pmatrix} = \begin{pmatrix} 1.83727 \\ -3.23209 \end{pmatrix}$$

$$\mathbf{k}_{04} = \begin{pmatrix} 2+(-.32321) \\ .1-3(1.18373)-(.1)^2(1.67679) \end{pmatrix} = \begin{pmatrix} 1.67679 \\ -3.46796 \end{pmatrix}$$

and thus

$x_1 = 1+(.1/6)[2 + 2(1.85)+2(1.83727)+(1.67679)]=1.18419,$
$y_1 = 2+(.1/6)[-3-2(3.25463)-2(3.23209)-3.46796]=1.67598,$

which are approximations to $x(.1)$ and $y(.1) = x'(.1)$.
In a similar fashion we find

$$\mathbf{k}_{11} = \begin{pmatrix} 1.67598 \\ -3.46933 \end{pmatrix} \qquad \mathbf{k}_{12} = \begin{pmatrix} 1.50251 \\ -3.68777 \end{pmatrix}$$

$$\mathbf{k}_{13} = \begin{pmatrix} 1.49159 \\ -3.66151 \end{pmatrix} \qquad \mathbf{k}_{14} = \begin{pmatrix} 1.30983 \\ -3.85244 \end{pmatrix}$$

and thus

$x_2=x_1+(.1/6)[1.67598+2(1.50251)+2(1.49159)+1.30983]=1.33376$
$y_2=y_1-(.1/6)[3.46933+2(3.68777)+2(3.66151)+3.85244]=1.30897.$

Three more steps must be taken in order to approximate
$x(.5)$ and $y(.5) = x'(.5)$. The intermediate steps yield
$x(.3) \approx 1.44489$, $y(.3) \approx .9093062$ and $x(.4) \approx 1.51499$,
$y(.4) \approx .4908795$.

CHAPTER 9

Section 9.1, Page 494

For Problems 1 through 16, once the eigenvalues have been
found, Table 9.1.1 will, for the most part, quickly yield the
type of critical point and the stability. In all cases it
can be easily verified that **A** is nonsingular.

1a. The eigenvalues are found from the equation det(**A**-r**I**)=0.

Substituting the values for **A** we have $\begin{vmatrix} 3-r & -2 \\ 2 & -2-r \end{vmatrix} =$

$r^2 - r - 2 = 0$ and thus the eigenvalues are $r_1 = -1$ and

$r_2 = 2$. For $r_1 = -1$, we have $\begin{pmatrix} 4 & -2 \\ 2 & -1 \end{pmatrix}\begin{pmatrix} \xi_1 \\ \xi_2 \end{pmatrix} = \begin{pmatrix} 0 \\ 0 \end{pmatrix}$ and thus

$\xi^{(1)} = \begin{pmatrix} 1 \\ 2 \end{pmatrix}$ and for r_2 we have $\begin{pmatrix} 1 & -2 \\ 2 & -4 \end{pmatrix}\begin{pmatrix} \xi_1 \\ \xi_2 \end{pmatrix} = \begin{pmatrix} 0 \\ 0 \end{pmatrix}$ and thus

$\xi^{(2)} = \begin{pmatrix} 2 \\ 1 \end{pmatrix}$.

1b. Since the eigenvalues differ in sign, the critical point
     is a saddle point and is unstable.

1d.

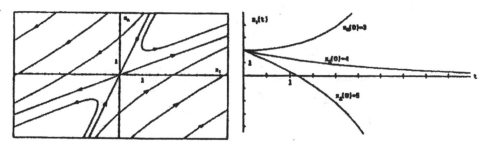

4a. Again the eigenvalues are given by $\begin{vmatrix} 1-r & -4 \\ 4 & -7-r \end{vmatrix} =$

$r^2 + 6r + 9 = 0$ and thus $r_1 = r_2 = -3$. The eigenvectors

are solutions of $\begin{pmatrix} 4 & -4 \\ 4 & -4 \end{pmatrix}\begin{pmatrix} \xi_1 \\ \xi_2 \end{pmatrix} = \begin{pmatrix} 0 \\ 0 \end{pmatrix}$ and hence there is

just one eigenvector $\xi = \begin{pmatrix} 1 \\ 1 \end{pmatrix}$.

4b. Since the eigenvalues are negative, (0,0) is an improper
     node which is asymptotically stable. If we had found
     that there were two independent eigenvectors then (0,0)
     would have been a proper node, as indicated in Case 3a.

4d.

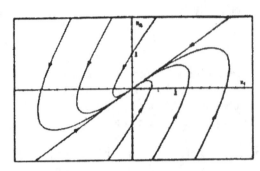 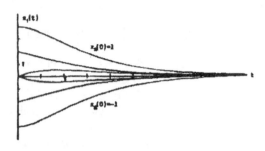

7a. In this case $\det(\mathbf{A} - r\mathbf{I}) = r^2 - 2r + 5$ and thus the eigenvalues are $r_{1,2} = 1 \pm 2i$. For $r_1 = 1 + 2i$ we have

$$\begin{pmatrix} 2-2i & -2 \\ 4 & -2-2i \end{pmatrix} \begin{pmatrix} \xi_1 \\ \xi_2 \end{pmatrix} = \begin{pmatrix} 2-2i & -2 \\ 8-8i & -8 \end{pmatrix} \begin{pmatrix} \xi_1 \\ \xi_2 \end{pmatrix} = \begin{pmatrix} 0 \\ 0 \end{pmatrix} \text{ and thus}$$

$\xi^{(1)} = \begin{pmatrix} 1 \\ 1-i \end{pmatrix}$ and $\xi^{(2)} = \begin{pmatrix} 1 \\ 1+i \end{pmatrix}$, which is the complex conjugate of $\xi^{(1)}$.

7b. Since the eigenvalues are complex with positive real part, we conclude that the critical point is a spiral point and is unstable.

7d.

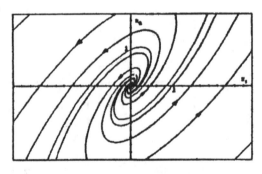 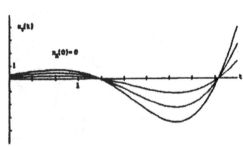

10a. Again, $\det(\mathbf{A}-r\mathbf{I}) = r^2 + 9$ and thus we have $r_{1,2} = \pm 3i$.

For $r_1 = 3i$ we have $\begin{pmatrix} 1-3i & 2 \\ -5 & -1-3i \end{pmatrix} \begin{pmatrix} \xi_1 \\ \xi_2 \end{pmatrix} = \begin{pmatrix} 0 \\ 0 \end{pmatrix}$ and thus

$\xi^{(1)} = \begin{pmatrix} 2 \\ -1+3i \end{pmatrix}$ and $\xi^{(2)} = \begin{pmatrix} 2 \\ -1-3i \end{pmatrix}$, which is the complex conjugate.

10b. Since the eigenvalues are pure imaginary the critical point is a center, which is stable.

10d.

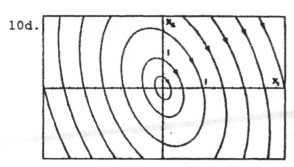 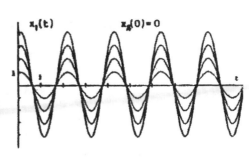

13. If we let $x = x^0 + u$ then $x' = u'$ and thus the system

becomes $u' = \begin{pmatrix} 1 & 1 \\ 1 & -1 \end{pmatrix} x^0 + \begin{pmatrix} 1 & 1 \\ 1 & -1 \end{pmatrix} u - \begin{pmatrix} 2 \\ 0 \end{pmatrix}$ which will be in

the form of Eq.(2) if $\begin{pmatrix} 1 & 1 \\ 1 & -1 \end{pmatrix} x^0 = \begin{pmatrix} 2 \\ 0 \end{pmatrix}$. Using row

operations, this last set of equations is equivalent to
$\begin{pmatrix} 1 & 1 \\ 0 & -2 \end{pmatrix} x^0 = \begin{pmatrix} 2 \\ -2 \end{pmatrix}$ and thus $x_1^0 = 1$ and $x_2^0 = 1$. Since

$u' = \begin{pmatrix} 1 & 1 \\ 1 & -1 \end{pmatrix} u$ has $(0,0)$ as the critical point, we

conclude that $x = x^0 + u = \begin{pmatrix} 1 \\ 1 \end{pmatrix}$ is the critical point of

the original system. As in the earlier problems, the

eigenvalues are given by $\begin{vmatrix} 1-r & 1 \\ 1 & -1-r \end{vmatrix} = r^2 - 2 = 0$ and thus

$r_{1,2} = \pm\sqrt{2}$. Hence the critical point $(1,1)$ is an
unstable saddle point.

17. The equivalent system is $dx/dt = y, dy/dt = -(k/m)x-(c/m)y$
which is written in the form of Eq.(2) as
$\frac{d}{dt}\begin{pmatrix} x \\ y \end{pmatrix} = \begin{pmatrix} 0 & 1 \\ -k/m & -c/m \end{pmatrix}\begin{pmatrix} x \\ y \end{pmatrix}$. The point $(0,0)$ is clearly a
critical point, and since A is nonsingular, it is the
only one. The characteristic equation is $r^2+(c/m)r+k/m=0$
so $r_1, r_2 = [-c \pm (c^2 - 4km)^{1/2}]/2m$. In the underdamped
case $c^2 - 4km < 0$, the characteristic roots are complex
with negative real parts, since $c > 0$, and thus the
critical point $(0,0)$ is an asymptotically stable spiral
point. In the overdamped case $c^2 - 4km > 0$, the
characteristic roots are real, unequal, and negative and
hence the critical point $(0,0)$ is an asymptotically
stable node. In the critically damped case $c^2 - 4km = 0$,
the characteristic roots are equal and negative. As

indicated in the solution to Prob.4, to determine whether
this is an improper or proper node we must determine
whether there are one or two linearly independent
eigenvectors. The eigenvectors satisfy the equation
$\begin{pmatrix} c/2m & 1 \\ -k/m & -c/2m \end{pmatrix} \begin{pmatrix} \xi_1 \\ \xi_2 \end{pmatrix} = \begin{pmatrix} 0 \\ 0 \end{pmatrix}$, which has just one solution if
$c^2 - 4km = 0$. Thus the critical point $(0,0)$ is an
asymptotically stable improper node.

18a. If **A** has one zero eigenvalue then for $r = 0$ we have
$\det(\mathbf{A} - r\mathbf{I}) = \det\mathbf{A} = 0$. Hence **A** is singular which means
$\mathbf{Ax} = \mathbf{0}$ has infinitely many solutions and consequently
there are infinitely many critical points. Since **A** is
2x2 the homogeneous equation $\mathbf{Ax} = \mathbf{0}$ will yield the
solution $x_2 = cx_1$, which indicates that the critical
points lie on a straight line through the origin.

18b. From Chapter 7, the solution is $\mathbf{x}(t) = c_1\xi^{(1)} + c_2\xi^{(2)}e^{r_2t}$,
which can be written in scalar form as
$x_1 = c_1\xi_1^{(1)} + c_2\xi_1^{(2)}e^{r_2t}$ and $x_2 = c_1\xi_2^{(1)} + c_2\xi_2^{(2)}e^{r_2t}$.
Assuming $\xi_1^{(2)} \neq 0$, the first equation can be solved for
$c_2e^{r_2t}$, which is then substituted into the second
equation to yield $x_2 = c_1\xi_2^{(1)} + [\xi_2^{(2)}/\xi_1^{(2)}][x_1 - c_1\xi_1^{(1)}]$.
These are straight lines parallel to the vector $\xi^{(2)}$.
Note that the family of lines is independent of c_2. If
$\xi_1^{(2)} = 0$, then the lines are vertical. If $r_2 > 0$, the
direction of motion will be in the same direction as
indicated for $\xi^{(2)}$. If $r_2 < 0$, then it will be in the
opposite direction.

19a. $\det(\mathbf{A} - r\mathbf{I}) = r^2 - (a_{11} + a_{22})r + a_{11}a_{22} - a_{21}a_{12} = 0$. If
$a_{11} + a_{22} = 0$, then $r^2 = -(a_{11}a_{22} - a_{21}a_{12}) < 0$ if
$a_{11}a_{22} - a_{21}a_{12} > 0$.

19b. Eq.(i) can be written in scalar form as
$dx/dt = a_{11}x + a_{12}y$ and $dy/dt = a_{21}x + a_{22}y$, which then
yields Eq.(iii). Ignoring the middle quotient in
Eq.(iii), we can rewrite that equation as
$(a_{21}x + a_{22}y)dx - (a_{11}x + a_{12}y)dy = 0$, which is exact
since $a_{22} = -a_{11}$ from Eq.(ii)..

19c. Integrating $\phi_x = a_{21}x + a_{22}y$ we obtain
$\phi = a_{21}x^2/2 + a_{22}xy + g(y)$ and thus

$\phi_y = a_{22}x + g' = -a_{11}x - a_{12}y$ or $g' = -a_{12}y$ using Eq.(ii).
Hence $\phi(x,y) = a_{21}x^2/2 + a_{22}xy - a_{12}y^2/2 = k/2$ is the
solution to Eq.(iii). The quadratic equation
$Ax^2 + Bxy + Cy^2 = D$ is an ellipse provided $B^2 - 4AC < 0$.
Hence for our problem if $a_{22}^2 + a_{21}a_{12} < 0$ then Eq.(iv) is
an ellipse. From $a_{11} + a_{22} = 0$ we have $a_{22}^2 = -a_{11}a_{22}$ and
hence the condition becomes $-a_{11}a_{22} + a_{21}a_{12} < 0$ or
$a_{11}a_{22} - a_{21}a_{12} > 0$, which is true by Eqs.(ii). Thus
Eq.(iv) is an ellipse under the conditions of Eqs.(ii).

20. The given system can be written as $\dfrac{d}{dt}\begin{pmatrix} x \\ y \end{pmatrix} = \begin{pmatrix} a_{11} & a_{12} \\ a_{21} & a_{22} \end{pmatrix}\begin{pmatrix} x \\ y \end{pmatrix}$.

    Thus the eigenvalues are given by
    $r^2 - (a_{11}+a_{22})r + a_{11}a_{22} - a_{12}a_{21} = 0$ and using the given
    definitions we rewrite this as $r^2 - pr + q = 0$ and thus
    $r_{1,2} = (p \pm \sqrt{p^2-4q})/2 = (p \pm \sqrt{\Delta})/2$. The results are
    now obtained using Table 9.1.1.

Section 9.2, Page 506

1. Solutions of the D.E. for x
    are y are $x = Ae^{-t}$ and
    $y = Be^{-2t}$ respectively.
    $x(0) = 4$ and $y(0) = 2$ yield
    $A = 4$ and $B = 2$, so $x = 4e^{-t}$
    and $y = 2e^{-2t}$. Solving the
    first equation for e^{-t} and
    then substituting into the

    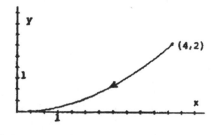

    second yields $y = 2[x/4]^2 = x^2/8$, which is a parabola.
    From the original D.E., or from the parametric solutions,
    we find that $0 < x \le 4$ and $0 < y \le 2$ for $t \ge 0$ and thus
    only the portion of the parabola shown is the trajectory,
    with the direction of motion indicated.

3. Utilizing the approach indicated in Eq.(17), we have
    $dy/dx = -x/y$, which separates into $xdx + ydy = 0$.
    Integration then yields the trajectory (a circle)
    $x^2 + y^2 = c^2$, where $c^2 = 16$ for both sets of I.C. For the
    first I.C. $y'(0) > 0$, so $y(t)$ is increasing and for the
    second I.C. $x'(0) < 0$, so $x(t)$ is decreasing. Hence the
    direction of motion is counterclockwise for both I.C. To
    obtain the parametric equations, we write the system in

the form

$$\frac{d}{dt}\begin{pmatrix} x \\ y \end{pmatrix} = \begin{pmatrix} 0 & -1 \\ 1 & 0 \end{pmatrix}\begin{pmatrix} x \\ y \end{pmatrix},$$ which has the characteristic

equation $\begin{vmatrix} -r & -1 \\ 1 & -r \end{vmatrix} = r^2 + 1 = 0$, or $r = \pm i$. Following

the procedures of Sect.7.6, we find that one solution of

the above system is $\begin{pmatrix} 1 \\ -i \end{pmatrix}e^{it} = \begin{pmatrix} \cos t + i\sin t \\ \sin t - i\cos t \end{pmatrix}$ and thus two

real solutions are $\mathbf{u}(t) = \begin{pmatrix} \cos t \\ \sin t \end{pmatrix}$ and $\mathbf{v}(t) = \begin{pmatrix} \sin t \\ -\cos t \end{pmatrix}$. The

general solution of the system is then

$\begin{pmatrix} x \\ y \end{pmatrix} = c_1\mathbf{u}(t) + c_2\mathbf{v}(t)$ and hence the first I.C. yields

$c_1 = 4$, $c_2 = 0$, or $x = 4\cos t$, $y = 4\sin t$. The second I.C.
yields $c_1 = 0$, $c_2 = -4$, or $x = -4\sin t$, $y = 4\cos t$. Note
that both these parametric representations satisfy the
form of the trajectories found in the first part of this
problem.

5a. The critical points are given by the solutions
   of $x(1-y) = 0$ and $y(1+2x) = 0$. Clearly $(0,0)$ is
   one critcal point and if $x \neq 0$, then $y-1=0$ and
   $2x+1=0$, so $(-\frac{1}{2},1)$ is a second critical point.

5b.

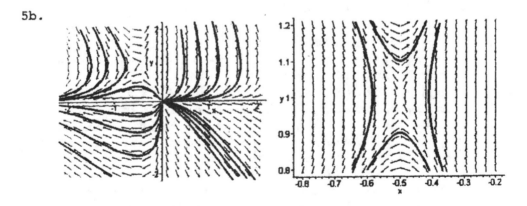

5c. From Part(b), the critical point $(0,0)$ is unstable since
   all trajectories leave this point. The trajectories and
   direction field near $(-\frac{1}{2},1)$ indicate that it is a
   saddle point and thus is unstable.

5d. There is no basin of attraction since all critical points are unstable.

12a. The critical points are given by y = 0 and
x(1 - x^2/6 - y/5) = 0, so (0,0), ($\sqrt{6}$,0) and (-$\sqrt{6}$,0) are the only critical points.

12b.

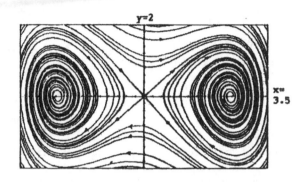

12c. Clearly ($\sqrt{6}$,0) and (-$\sqrt{6}$,0) are spiral points which are asymptotically stable since the trajectories tend to each point, respectively. (0,0) is a saddle point which is unstable, since there are trajectories that leave and trajectories that approach the point.

12d. The basin of attaction is defined by the trajectory starting at (0,0), encircling each of the critical points ($\sqrt{6}$,0) and (-$\sqrt{6}$,0), and ending up at (0,0) again.

17a. $\dfrac{dy}{dx} = \dfrac{dy/dt}{dx/dt} = \dfrac{8x}{2y}$, so 4xdx - ydy = 0 and thus 4x^2 - y^2 = c, which are hyperbolas for c ≠ 0 and straight lines y = ±2x for c = 0.

17b.   21b.

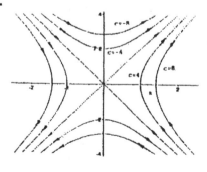 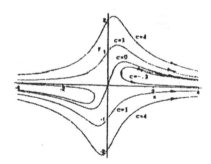

21a. $\dfrac{dy}{dx} = \dfrac{y-2xy}{-x+y+x^2}$, so (y-2xy)dx + (x-y-x^2)dy = 0, which is an
exact D.E. Therefore φ(x,y) = xy - x^{2}y + g(y) and hence

$\dfrac{\partial \phi}{\partial y} = x - x^2 + g'(y) = x - y - x^2$, so $g'(y) = -y$ and

$g(y) = -y^2/2$. Thus $2x^2y - 2xy + y^2 = c$ (after multiplying by -2) is the desired solution.

23a. $\dfrac{dy}{dx} = \dfrac{-\sin x}{y}$, so

$y\,dy + \sin x\,dx = 0$ and
thus $y^2/2 - \cos x = c$.

23b.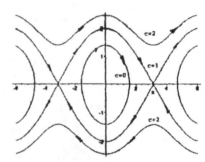

25. We know that $\phi'(t) = F[\phi(t),\psi(t)]$ and
$\psi'(t) = G[\phi(t),\psi(t)]$ for $\alpha < t < \beta$. By direct
substitution we have
$\Phi'(t) = \phi'(t-s) = F[\phi(t-s),\psi(t-s)] = F[\Phi(t),\Psi(t)]$ and
$\Psi'(t) = \psi'(t-s) = G[\phi(t-s),\psi(t-s)] = G[\Phi(t),\Psi(t)]$ for
$\alpha < t - s < \beta$ or $\alpha + s < t < \beta + s$.

26. Suppose that $t_1 > t_0$. Let $s = t_1 - t_0$. Since the system
is autonomous, the result of Prob.25, with s replaced
by $-s$ shows that $x = \phi_1(t+s)$ and $y = \psi_1(t+s)$ generates
the same trajectory (C_1) as $x = \phi_1(t)$ and $y = \psi_1(t)$. But
at $t = t_0$ we have $x = \phi_1(t_0+s) = \phi_1(t_1) = x_0$ and
$y = \psi_1(t_0+s) = \psi_1(t_1) = y_0$. Thus the solution
$x = \phi_1(t+s)$, $y = \psi_1(t+s)$ satisfies <u>exactly</u> the same
initial conditions as the solution $x = \phi_0(t)$, $y = \psi_0(t)$
which generates the trajectory C_0. Hence C_0 and C_1 are
the same.

27. From the existence and uniqueness theorem we know that if
the two solutions $x = \phi(t)$, $y = \psi(t)$ and $x = x_0$, $y = y_0$
satisfy $\phi(a) = x_0$, $\psi(a) = y_0$ and $x = x_0$, $y = y_0$ at $t = a$,
then these solutions are identical. Hence $\phi(t) = x_0$ and
$\psi(t) = y_0$ for all t contradicting the fact that the
trajectory generated by $[\phi(t),\psi(t)]$ started at a
noncritical point.

28. By direct substitution
$\Phi'(t) = \phi'(t+T) = F[\phi(t+T),\psi(t+T)] = F[\Phi(t),\Psi(t)]$ and
$\Psi'(t) = \psi'(t+T) = G[\phi(t+T),\psi(t+T)]$, $G[\Phi(t),\Psi(t)]$.
Furthermore $\Phi(t_0) = x_0$ and $\Psi(t_0) = y_0$. Thus by the
existence and uniqueness theorem $\Phi(t) = \phi(t)$ and
$\Psi(t) = \psi(t)$ for all t.

Section 9.3, Page 516

In Problems 1 through 4, write the system in the form of
Eq.(4). Then if $g(0) = 0$ we may conclude that $(0,0)$ is a
critical point. In addition, if g satisfies Eq.(5) or
Eq.(6), then the system is locally linear. In this case the
linear system, Eq.(1), will determine, in most cases, the
type and stability of the critical point $(0,0)$ of the locally
linear system. These results are summarized in Table 9.3.1.

3. In this case the system can be written as

$$\frac{d}{dt}\begin{pmatrix} x \\ y \end{pmatrix} = \begin{pmatrix} 0 & 0 \\ -1 & 0 \end{pmatrix}\begin{pmatrix} x \\ y \end{pmatrix} + \begin{pmatrix} (1+x)\sin y \\ 1 - \cos y \end{pmatrix}.$$ However, the

coefficient matrix is singular and $g_1(x,y) = (1+x)\sin y$

does not satisfy Eq.(6). However, if we consider the
Taylor series for $\sin y$, we see that
$(1+x)\sin y - y = \sin y - y + x\sin y$

$$= -y^3/3! + y^5/5! + \cdots + x(y - y^3/3! + \cdots),$$

which does satisfy Eq.(6), using $x = r\cos\theta$, $y = r\sin\theta$.
Thus the first equation now becomes

$$\frac{dx}{dt} = y + [(1+x)\sin y - y] \text{ and hence}$$

$$\frac{d}{dt}\begin{pmatrix} x \\ y \end{pmatrix} = \begin{pmatrix} 0 & 1 \\ -1 & 0 \end{pmatrix}\begin{pmatrix} x \\ y \end{pmatrix} + \begin{pmatrix} (1+x)\sin y - y \\ 1-\cos y \end{pmatrix},$$ where the coefficient

matrix is now nonsingular and

$$g(x,y) = \begin{pmatrix} (1+x)\sin y - y \\ 1-\cos y \end{pmatrix} \text{ satisfies Eq.(6). The linear}$$

system, represented by the matrix $A = \begin{pmatrix} 0 & 1 \\ -1 & 0 \end{pmatrix}$, has

eigenvalues $r_{1,2} = \pm i$, and thus the origin is a center
which is stable and the nonlinear system has either a
center or spiral point at the origin and the stability is
indeterminent, from Table 9.3.1.

4. In this case the system can be written as

$$\frac{d}{dt}\begin{pmatrix} x \\ y \end{pmatrix} = \begin{pmatrix} 1 & 0 \\ 1 & 1 \end{pmatrix}\begin{pmatrix} x \\ y \end{pmatrix} + \begin{pmatrix} y^2 \\ 0 \end{pmatrix} \text{ and thus } A = \begin{pmatrix} 1 & 0 \\ 1 & 1 \end{pmatrix} \text{ and}$$

$$g = \begin{pmatrix} y^2 \\ 0 \end{pmatrix}. \text{ Since } g(0) = \begin{pmatrix} 0 \\ 0 \end{pmatrix} \text{ we conclude that } (0,0) \text{ is a}$$

critical point. Following the procedure of Ex.1, we let
$x = r\cos\theta$ and $y = r\sin\theta$ and thus

$g_1(x,y)/r = \dfrac{r^2\sin^2\theta}{r} \to 0$ as $r \to 0$ and thus the system is locally linear. Since $\det(\mathbf{A}-r\mathbf{I}) = (r-1)^2$, we find that the eigenvalues are $r_1 = r_2 = 1$. Since the roots are equal, we must determine whether there are one or two eigenvectors to classify the type of critical point. The eigenvectors are determined by $\begin{pmatrix} 0 & 0 \\ 1 & 0 \end{pmatrix}\begin{pmatrix} \xi_1 \\ \xi_2 \end{pmatrix} = \begin{pmatrix} 0 \\ 0 \end{pmatrix}$ and hence there is only one eigenvector $\xi = \begin{pmatrix} 0 \\ 1 \end{pmatrix}$. Thus the critical point for the linear system is an unstable improper node, from Sect. 9.1 From Table 9.3.1 we then conclude that the given system, which is locally linear, has a critical point near $(0,0)$ which is either a node or spiral point (depending on how the roots bifurcate) which is unstable.

6a. The critical points are the solutions of $x(1-x-y) = 0$ and $y(3-x-2y) = 0$. Solutions are $x = 0$, $y = 0$; $x = 0$, $3 - 2y = 0$ which gives $y = 3/2$; $y = 0$ and $1 - x = 0$ which give $x = 1$; and $1 - x - y = 0$, $3 - x - 2y = 0$ which give $x = -1$, $y = 2$. Thus the critical points are $(0,0),(0,3/2)$, $(1,0)$, and $(-1,2)$.

6b, For the critical point $(0,0)$ the D.E. is of the form

6c. $\mathbf{x}' = \begin{pmatrix} 1 & 0 \\ 0 & 3 \end{pmatrix}\mathbf{x} - \mathbf{g}(x,y)$ so the corresponding linear system is $du/dt = u$, $dv/dt = 3v$ which has the eigenvalues $r_1 = 1$ and $r_2 = 3$. Thus the critical point $(0,0)$ is an unstable node. Each of the other three critical points is dealt with in the same manner; we consider only the critical point $(-1,2)$. In order to translate this critical point to the origin we set $x(t) = -1 + u(t)$, $y(t) = 2 + v(t)$ and substitute in the D.E. to obtain
$u' = -1 + u - (-1+u)^2 - (-1+u)(2+v) = u + v - u^2 - uv$
and
$v' = 3(2+v) - (-1+u)(2+v) - 2(2+v)^2 = -2u - 4v - uv - 2v^2$.
Writing this in the form of Eq.(4) we find that
$\mathbf{A} = \begin{pmatrix} 1 & 1 \\ -2 & -4 \end{pmatrix}$ and $\mathbf{g} = -\begin{pmatrix} u^2 + uv \\ uv + 2v^2 \end{pmatrix}$ which is a locally linear system. The eigenvalues of the corresponding linear system are $r = (-3 \pm \sqrt{17})/2$ and hence the critical point $(-1,2)$, of the original system, is an unstable saddle point.

6d. The phase portrait near (-1,2) is shown.

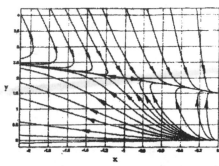

10a. The critical points are solutions of $x + x^2 + y^2 = 0$ and
$y(1-x) = 0$, which yield $(0,0)$ and $(-1,0)$.

10b. For $(0,0)$ the D.E. is already in the form of a locally
linear system and thus $u' = u$ and $v' = v$ is the
corresponding linear system. For $(-1,0)$ we let $u = x+1$,
$v = y$ so that substituting $x = u-1$ and $y = v$ into the
D.E. we obtain $\dfrac{du}{dt} = -u + u^2 + v^2$ and $\dfrac{dv}{dt} = 2v - uv$. Thus
the corresponding linear system is $u' = -u$ and $v' = 2v$.

10c. For $(0,0)$ $\mathbf{A} = \begin{pmatrix} 1 & 0 \\ 0 & 1 \end{pmatrix}$ we have $r_1 = r_2 = 1$, so that $(0,0)$,

for the nonlinear system, will be either a node or spiral
point, depending on how the roots bifurcate. In any
case, since r_1 and r_2 are positive, the system will

be unstable. For $(-1,0)$ $\mathbf{A} = \begin{pmatrix} -1 & 0 \\ 0 & 2 \end{pmatrix}$ and thus

$r_1 = -1$ and $r_2 = 2$, and hence the nonlinear system, from
Table 9.3.1, has an unstable saddle point at $(-1,0)$.

18a. The critical points are solutions of $(1-y)(2x-y) = 0$ and
$(2+x)(x-2y) = 0$. Thus there are four critical points:
$(0,0)$, $(2,1)$, $(-2,-4)$, and $(-2,1)$.

18b. For $(0,0)$ we have $\mathbf{x}' = \begin{pmatrix} 2 & -1 \\ 2 & -4 \end{pmatrix}\mathbf{x} + \mathbf{g}(x,y)$. The eigenvalues

18c. of the matrix are given by $r^2+2r+6 = 0$, or $r = -1\pm\sqrt{7}$.
Thus $(0,0)$ is an unstable saddle point. For $(2,1)$ let

$u = x-2$ and $v = y-1$, which gives $\begin{pmatrix} u' \\ v' \end{pmatrix} = \begin{pmatrix} 0 & -3 \\ 4 & -8 \end{pmatrix}\begin{pmatrix} u \\ v \end{pmatrix} + \mathbf{g}(u,v)$.

The eigenvalues of $\mathbf{A}$ are $r = -2, -6$ and thus $(2,1)$ is an
asymptotically stable node. The results for the other
critical points are found in a similar manner.

20a. The system is $\dfrac{d}{dt}\begin{pmatrix} x \\ y \end{pmatrix} = \begin{pmatrix} 1 & 0 \\ 0 & -2 \end{pmatrix}\begin{pmatrix} x \\ y \end{pmatrix} + \begin{pmatrix} 0 \\ x^3 \end{pmatrix}$ and thus is

locally linear using the procedures outlined in the earlier problems. The corresponding linear system has the eigenvalues $r_1 = 1$, $r_2 = -2$ and thus $(0,0)$ is an unstable saddle point for both the linear and locally linear systems.

20b. The trajectories of the linear system are the solutions of $dx/dt = x$ and $dy/dt = -2y$ and thus $x(t) = c_1 e^t$ and $y(t) = c_2 e^{-2t}$. To sketch these, solve the first equation
for e^t and substitute into
the second to obtain
$y = c_1^2 c_2 / x^2$, $c_1 \neq 0$.
Several trajectories
are shown in the figure.
Since $x(t) = c_1 e^t$, we
must pick $c_1 = 0$ for
$x \to 0$ as $t \to \infty$. Thus
$x = 0$, $y = c_2 e^{-2t}$ (the
vertical axis) is the
only trajectory for
which $x \to 0$, $y \to 0$ as $t \to \infty$.

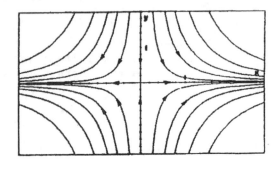

20c. For $x \neq 0$ we have $dy/dx = (dy/dt)/(dx/dt) = (-2y+x^3)/x$.
This is a linear equation, and the general solution is
$y = x^3/5 + k/x^2$, where k is an arbitrary constant. In
addition the system of equations has the solution $x = 0$,
$y = Be^{-2t}$. Any solution with its initial point on the
y-axis (x=0) is given by the latter solution. The
trajectories corresponding to these solutions approach
the origin as $t \to \infty$.
The trajectory that passes
through the origin and
divides the family of
curves is given by $k = 0$,
namely $y = x^3/5$. This
trajectory corresponds to
the trajectory $y = 0$ for
the linear problem.
Several trajectories are
shown in the figure.

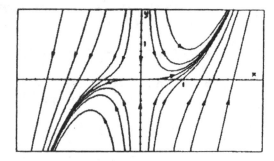

24a.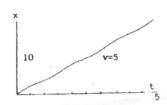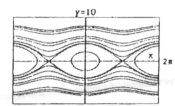

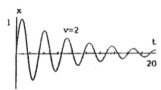 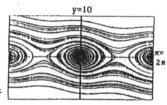

The graph for v = 5 indicates the initial velocity causes
the pendulum to rotate beyond its upper critical point.
Since there is no damping, the x value continues to
increase indefinitely.

24b. From the graphs in Part(a), we see that v_c is between
v = 2 and v = 5. Using several values for v, we estimate
$v_c \cong 4.00$.

25a

For v = 2, the motion is damped oscillatory about x = 0.
For v = 5, the pendulum swings all the way around once
and then is a damped oscillation about x = 2π (after one
full rotation). For Prob. 24, this latter case is not
damped, so x continues to increase, as shown earlier.

29a. Setting c = 0 in Eq.(10) of Sect. 9.2 we obtain
$mL^2 d^2\theta/dt^2 + mgL\sin\theta = 0$. Considering $d\theta/dt$ as a function
of θ and using the chain rule we have

$$\frac{d}{dt}\left(\frac{d\theta}{dt}\right) = \frac{d}{d\theta}\left(\frac{d\theta}{dt}\right)\frac{d\theta}{dt} = \frac{1}{2}\frac{d}{d\theta}\left(\frac{d\theta}{dt}\right)^2 .$$ Thus

$(1/2)mL^2 d[(d\theta/dt)^2]/d\theta = -mgL\sin\theta$. Now integrate both
sides from α to θ where $d\theta/dt = 0$ at θ = α:
$(1/2)mL^2(d\theta/dt)^2 = mgL(\cos\theta - \cos\alpha)$. Thus
$(d\theta/dt)^2 = (2g/L)(\cos\theta - \cos\alpha)$. Since we are releasing
the pendulum with zero velocity from a positive angle α,
the angle θ will initially be decreasing so $d\theta/dt < 0$.
If we restrict attention to the range of θ from θ = α to
θ = 0, we can assert $d\theta/dt = -\sqrt{2g/L}\sqrt{\cos\theta - \cos\alpha}$.
Solving for dt gives $dt = -\sqrt{L/2g}\, d\theta/\sqrt{\cos\theta - \cos\alpha}$.

29b. Since there is no damping, the pendulum will swing from
its initial angle α through 0 to -α, then back through 0
again to the angle α in one period. It follows that
θ(T/4) = 0. Integrating the last equation and noting
that as t goes from 0 to T/4, θ goes from α to 0 yields
$T/4 = -\sqrt{L/2g} \int_\alpha^0 (1/\sqrt{\cos\theta - \cos\alpha})\, d\theta$.

30a. If $\dfrac{dx}{dt} = y$, then $\dfrac{d^2x}{dt^2} = \dfrac{dy}{dt} = -g(x) - c(x)y$.

30b. Under the given assumptions we have

$g(x) = g(0) + g'(0)x + g''(\xi_1)x^2/2$ and

$c(x) = c(0) + c'(\xi_2)x$, where $0 < \xi_1$, $\xi_2 < x$ and $g(0) = 0$.

Hence $\dfrac{dy}{dt} = -g'(0)x - c(0)y - [g''(\xi_1)x^2/2 - c'(\xi_2)xy]$ and

thus the system can be written as

$$\frac{d}{dt}\begin{pmatrix} x \\ y \end{pmatrix} = \begin{pmatrix} 0 & 1 \\ -g'(0) & -c(0) \end{pmatrix}\begin{pmatrix} x \\ y \end{pmatrix} - \begin{pmatrix} 0 \\ -g''(\xi_1)x^2/2 - c'(\xi_2)xy \end{pmatrix},$$

from which the results follow.

Section 9.4, Page 530

3b. $x(1.5 - .5x - y) = 0$ and $y(2 - y - 1.125x) = 0$ yield $(0,0)$, $(0,2)$ and $(3,0)$ very easily. The fourth critical point is the intersection of $.5x + y = 1.5$ and $1.125x + y = 2$, which is $(.8,1.1)$.

3c. From Eq.(5) we get $\dfrac{d}{dt}\begin{pmatrix} u \\ v \end{pmatrix} = \begin{pmatrix} 1.5-x_0-y_0 & -x_0 \\ -1.125y_0 & 2-2y_0-1.125x_0 \end{pmatrix}\begin{pmatrix} u \\ v \end{pmatrix}$. For

$(0,0)$ we get $u' = 1.5u$ and $v' = 2v$, so $r = 3/2$ and $r = 2$, and thus $(0,0)$ is an unstable node. For $(0,2)$ we have $u' = -.5u$ and $v' = -2.25u-2v$, so $r = -.5$, -2 and thus $(0,2)$ is an asymptotically stable node. For $(3,0)$ we get $u' = -1.5u-3v$ and $v' = -1.375v$, so $r = -1.5$, -1.375 and hence $(3,0)$ is an symptotically stable node. For $(.8,1.1)$ we have $u' = -.4u -.8v$ and $v' = -1.2375u - 1.1v$ which give $r = -1.80475$, $.30475$ and thus $(.8,1.1)$ is an unstable saddle point.

3e.

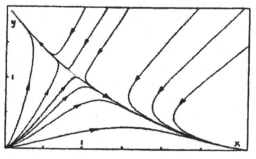

3f. As in Ex.2, one species will die out, depending on the I.C. For an I.C. lying below the separatrix [not shown, but which is a curve starting at $(0,0)$ and passing

through (.8,1.1)], the species denoted by x will survive,
while if the I.C. is above the separatrix the species
denoted by y will survive.

5b.   The critical points are found by setting dx/dt = 0 and
      dy/dt = 0 and thus we need to solve x(1 - x - y) = 0 and
      y(1.5 - y - x) = 0. The first yields x = 0 or y = 1 - x
      and the second yields y = 0 or y = 1.5 - x. Thus (0,0),
      (0,3/2) and (1,0) are the only critical points since the
      two straight lines do not intersect in the first quadrant
      (or anywhere in this case). This is an example of one of
      the cases shown in Figure 9.4.5 a or b.

5e.

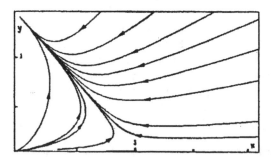

5f.   Only the species denoted by y will survive as t → ∞.

6b.   The critical points are found by setting dx/dt = 0 and
      dy/dt = 0 and thus we need to solve x(1-x + y/2) = 0 and
      y(5/2 - 3y/2 + x/4) = 0. The first yields x = 0 or
      y = 2x - 2 and the second yields y = 0 or y = x/6 + 5/3.
      Thus we find the critical points (0,0), (1,0), (0,5/3)
      and (2,2). The last point is the intersection of the two
      straight lines, which will be used again in Part(d).

6c.   For (0,0) the linearized system is x′ = x and y′ = 5y/2,
      which has the eigenvalues r_1 = 1 and r_2 = 5/2. Thus the
      origin is an unstable node. For (2,2) we let
      x = u + 2 and y = v + 2 in the given system to find
      (since x′ = u′ and y′ = v′) that
      du/dt = (u+2)[1 - (u+2) + (v+2)/2] = (u+2)(-u+v/2) and
      dv/dt = (v+2)[5/2 - 3(v+2)/2 + (u+2)/4] = (v+2)(u/4 - 3v/2).
      Hence, keeping just the linear terms, the linearized

      equations are $\begin{pmatrix} u \\ v \end{pmatrix}' = \begin{pmatrix} -2 & 1 \\ 1/2 & -3 \end{pmatrix} \begin{pmatrix} u \\ v \end{pmatrix}$ which has the eigenvalues

      $r_{1,2}$ = (-5 ± $\sqrt{3}$)/2. Since these are both negative we
      conclude that (2,2) is an asymptotically stable node. In a
      similar fashion for (1,0) we let x = u + 1 and y = v to

obtain the linearized system

$$\begin{pmatrix} u \\ v \end{pmatrix}' = \begin{pmatrix} -1 & 1/2 \\ 0 & 11/4 \end{pmatrix} \begin{pmatrix} u \\ v \end{pmatrix}.$$ This has $r_1 = -1$ and $r_2 = 11/4$ as

eigenvalues and thus $(1,0)$ is an unstable saddle point.
Likewise, for $(0,5/3)$ we let $x = u$, $y = v + 5/3$ to find

$$\begin{pmatrix} u \\ v \end{pmatrix}' = \begin{pmatrix} 11/6 & 0 \\ 5/12 & -5/2 \end{pmatrix} \begin{pmatrix} u \\ v \end{pmatrix}$$ as the corresponding linear system.

Thus $r_1 = 11/6$ and $r_2 = -5/2$ and thus $(0,5/3)$ is an
unstable saddle point.

6d. To sketch the required trajectories, we must find the
eigenvectors for each of the linearized systems and then
analyze the behavior of the linear solution near the
critical point. Using this approach we find that the
solution near $(0,0)$ has the form

$$\begin{pmatrix} x \\ y \end{pmatrix} = c_1 \begin{pmatrix} 1 \\ 0 \end{pmatrix} e^t + c_2 \begin{pmatrix} 0 \\ 1 \end{pmatrix} e^{5t/2}$$ and thus the origin is

approached only for large negative values of t. In this
case e^t dominates $e^{5t/2}$ and hence in the neighborhood of
the origin all trajectories are tangent to the x-axis
except for one pair ($c_1 = 0$) that lies along the y-axis.

For $(2,2)$ we find the eigenvector corresponding to
$r = (-5 + \sqrt{3})/2 = -1.63$ is given by $(1-\sqrt{3})\xi_1/2 + \xi_2 = 0$

and thus $\begin{pmatrix} 1 \\ (\sqrt{3}-1)/2 \end{pmatrix} = \begin{pmatrix} 1 \\ .37 \end{pmatrix}$ is one eigenvector. For

$r = (-5 -\sqrt{3})/2 = -3.37$ we have $(1 +\sqrt{3})\xi_1/2 + \xi_2 = 0$ and

thus $\begin{pmatrix} 1 \\ -(\sqrt{3}+1)/2 \end{pmatrix} = \begin{pmatrix} 1 \\ -1.37 \end{pmatrix}$ is the second eigenvector.

Hence the linearized solution is

$$\begin{pmatrix} u \\ v \end{pmatrix} = c_1 \begin{pmatrix} 1 \\ .37 \end{pmatrix} e^{-1.63t} + c_2 \begin{pmatrix} 1 \\ -1.37 \end{pmatrix} e^{-3.37t}.$$ For large

positive values of t the first term is the dominant one
and thus we conclude that all trajectories but two
approach $(2,2)$ tangent to the straight line with
slope .37. If $c_1 = 0$, we see that there are exactly two
($c_2 > 0$ and $c_2 < 0$) trajectories that lie on the straight
line with slope -1.37. In similar fashion, we find the

linearized solutions near $(1,0)$ and $(0,5/3)$ to be,

respectively,

$$\begin{pmatrix} u \\ v \end{pmatrix} = c_1 \begin{pmatrix} 1 \\ 0 \end{pmatrix} e^{-t} + c_2 \begin{pmatrix} 1 \\ 15/2 \end{pmatrix} e^{11t/4}$$

and

$$\begin{pmatrix} u \\ v \end{pmatrix} = c_1 \begin{pmatrix} 0 \\ 1 \end{pmatrix} e^{-5t/2} + c_2 \begin{pmatrix} 1 \\ 5/52 \end{pmatrix} e^{11t/6},$$

which, along with the above
analysis, yields the sketch
shown.

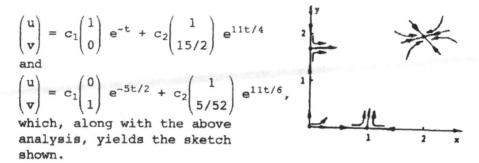

6e. From the above sketch, it appears that $(x,y) \to (2,2)$ as
6f. $t \to \infty$ as long as (x,y) starts in the first quadrant.
To ascertain this, we need to prove that x and y cannot
become unbounded as $t \to \infty$. From the given system, we
can observe that, since $x > 0$ and $y > 0$, that dx/dt and
dy/dt have the same sign as the quantities $1 - x + y/2$
and $5/2 - 3y/2 + x/4$ respectively. If we set these
quantities equal to zero we get the straight lines
$y = 2x - 2$ and $y = x/6 + 5/3$, which divide the first

quadrant into the four
sectors shown. The signs
of x' and y' are indicated,
from which it can be
concluded that x and y
must remain bounded [and
in fact approach $(2,2)$]
as $t \to \infty$. The discussion
leading up to Fig.9.4.4
is also useful here.

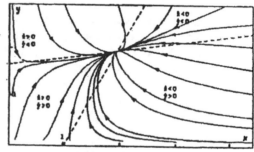

8a. Setting the right sides of the equations equal to zero
gives the critical points $(0,0)$, $(0, \varepsilon_2/\sigma_2)$, $(\varepsilon_1/\sigma_1, 0)$
[which are no fish, no bluegill or no redear respectively],
and possibly
$([\varepsilon_1\sigma_2 - \varepsilon_2\alpha_1]/[\sigma_1\sigma_2 - \alpha_1\alpha_2], [\varepsilon_2\sigma_1 - \varepsilon_1\alpha_2]/[\sigma_1\sigma_2 - \alpha_1\alpha_2])$.
(The last point can be obtained from Eq.(36) also). The
conditions $\varepsilon_2/\alpha_2 > \varepsilon_1/\sigma_1$ and $\varepsilon_2/\sigma_2 > \varepsilon_1/\alpha_1$ imply that
$\varepsilon_2\sigma_1 - \varepsilon_1\alpha_2 > 0$ and $\varepsilon_1\sigma_2 - \varepsilon_2\alpha_1 < 0$. Thus either the x
coordinate or the y coordinate of the last critical point
is negative so both species cannot survive. The
linearized system for $(0,0)$ is $x' = \varepsilon_1 x$ and $y' = \varepsilon_2 y$ and
thus $(0,0)$ is an unstable equilibrium point. Similarly,
it can be shown [by linearizing the given system or by
using Eq.(35)] that $(0, \varepsilon_2/\sigma_2)$ is an asymptotically

stable critical point and that $(\varepsilon_1\sigma_1, 0)$ is an unstable critical point. Thus the fish represented by y(redear) survive.

8b. The conditions $\varepsilon_1/\sigma_1 > \varepsilon_2/\alpha_2$ and $\varepsilon_1/\alpha_1 > \varepsilon_2/\sigma_2$ imply that $\varepsilon_2\sigma_1 - \varepsilon_1\alpha_2 < 0$ and $\varepsilon_1\sigma_2 - \varepsilon_2\alpha_1 > 0$ so again one of the coordinates of the fourth point in Part(a) is negative and hence a mixed state is not possible. An analysis similar to that in Part(a) shows that (0,0) and $(0,\varepsilon_2/\sigma_2)$ are unstable while $(\varepsilon_1/\sigma_1,0)$ is stable. Hence the bluegill (represented by x) survive in this case.

9a. $x' = \varepsilon_1 x(1 - \dfrac{\sigma_1}{\varepsilon_1}x - \dfrac{\alpha_1}{\varepsilon_1}y) = \varepsilon_1 x(1 - \dfrac{1}{B}x - \dfrac{\gamma_1}{B}y)$

$y' = \varepsilon_2 y(1 - \dfrac{\sigma_2}{\varepsilon_2}y - \dfrac{\alpha_2}{\varepsilon_2}x) = \varepsilon_2 y(1 - \dfrac{1}{R}y - \dfrac{\gamma_2}{R}x)$. The coexistence

equilibrium point is given by $\dfrac{1}{B}x + \dfrac{\gamma_1}{B}y = 1$ and $\dfrac{\gamma_2}{R}x + \dfrac{1}{R}y = 1$.
Solving these yields $X = (B - \gamma_1 R)/(1 - \gamma_1\gamma_2)$ and
$Y = (R - \gamma_2 B)/(1 - \gamma_1\gamma_2)$.

9b. If B is reduced, it is clear from the answer to Part(a) that X is reduced and Y is increased. To determine whether the bluegill will die out, we give an intuitive argument which can be confirmed by doing the analysis. Note that $B/\gamma_1 = \varepsilon_1/\alpha_1 > \varepsilon_2/\sigma_2 = R$ and $R/\gamma_2 = \varepsilon_2/\alpha_2 > \varepsilon_1/\sigma_1 = B$ so that the graph of the lines $1 - x/B - \gamma_1 y/B = 0$ and $1 - y/R - \gamma_2 x/R = 0$ must appear as indicated in the figure, where critical points are inidcated by heavy dots.
As B is decreased,
X decreases, Y increases
(as indicated above) and
the point of intersection
moves closer to (0,R). If
$B/\gamma_1 < R$ coexistence is not
possible, and the only
critical points are (0,0),(0,R)
and (B,0).

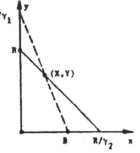

It can be shown that (0,0) and (B,0) are unstable and (0,R) is asymptotically stable. Hence we conlcude, when coexistence is no longer possible, that $x \to 0$ and $y \to R$ and thus the bluegill population will die out.

13a. The nullclines are given by $x' = 0$ and $y' = 0$, or $y = 4x - x^2$

and $y = \dfrac{3\alpha}{2}$. Thus there are two critical points, where the horizontal line $y = \dfrac{3\alpha}{2}$, for $\alpha < \dfrac{8}{3}$, intersects the parabola $y = 4x - x^2$. As α increases the y value of the critical points increases until $\alpha = \dfrac{8}{3}$.

13b. The critical points are determined by $4x - x^2 = \dfrac{3\alpha}{2}$ or $x = 2 \pm \sqrt{4 - 3\alpha/2}$. Thus the critical points are $(2 \pm \sqrt{4 - 3\alpha/2}, 3\alpha/2)$.

13c. For $\alpha = 2$ the critical points are $(1,3)$ and $(3,3)$. The corresponding linear system has the coefficient matrix $\begin{pmatrix} -4+2x_0 & 1 \\ 0 & -1 \end{pmatrix}$ and thus for $(1,3)$ the eigenvalues are $r_{1,2} = -1,-2$ which indicate $(1,3)$ is an asymptotically stable node. For $(3,3)$, the eigenvalues are $r_{1,2} = -1,2$, so $(3,3)$ is an unstable saddle point.

13c. 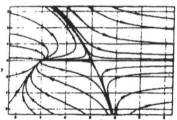   13d.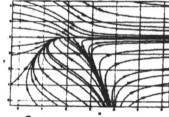

13d. The bifurcation point occurs when $\alpha = \dfrac{8}{3}$ since the horizontal line $y = \dfrac{3\alpha}{2}$ is tangent to $y = 4x - x^2$ and there is only one critical point $(2,4)$. Using the linear matrix of Part(c) we find that $r_{1,2} = 0,-1$ are the eigenvalues. The phase portrait is shown above.

13e. If $\alpha = 10/3$ we have $x' = -4x + y + x^2$ and $y' = 5 - y$, which has no critical points, consistent with the phase portrait shown.

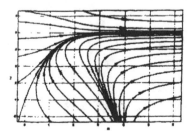

17a. The nullclines are given by $y = 1-x$ and $y = \alpha - x/2$. For
     $\alpha > 1$ there are no points of intersection (in the 1st
     quadrant) and for $\alpha \le 1$ the nullclines will resemble those
     in Fig.9.4.5 (c) and (d).

17b. By solving $x(1-x-y) = 0$ and $y(\alpha-y-x/2) = 0$ we easily
     obtain $P_1(0,0), P_2(1,0)$ and $P_3(0,\alpha)$. The fourth point is
     the intersection of the nullclines and is $P_4(2-2\alpha,-1+2\alpha)$,
     which which lies in the first quadrant for $.5 \le \alpha \le 1$.

17c. When $\alpha = 0$ P_3 coincides with P_1, when $\alpha = .5$ P_4 coincides
     with P_2 and when $\alpha = 1$ P_4 coincides with P_3. These are
     the bifurcation points , since in each case two of the
     four points have the same coordinates.

17d. With $F = x(1-x-y)$ and $G = y(\alpha-y-x/2)$ the Jacobian matrix
     (Eq.(13), Sect.9.3) is $J = \begin{pmatrix} 1-2x-y & -x \\ -y/2 & \alpha-2y-x/2 \end{pmatrix}$.

17e. For P_1 $J = \begin{pmatrix} 1 & 0 \\ 0 & \alpha \end{pmatrix}$ which has the eigenvalues $r = 1, \alpha$. Thus

     P_1 is an unstable node for $\alpha > 0$. For P_2 $J = \begin{pmatrix} -1 & -1 \\ 0 & \alpha-.5 \end{pmatrix}$ so

     $r = -1$, $r = .5 - \alpha$, so P_2 is an asymptotically stable node
     for $0 < \alpha < .5$ and a saddle point for $\alpha > .5$. Similar
     calculations will yield the results for P_3 and P_4.

17f.

$\alpha = 0.25$ $\alpha = 0.75$ $\alpha = 1.25$

Section 9.5, Page 540

3b. We have $x = 0$ or $(1 - .5x - .5y) = 0$ and $y = 0$ or
    $(-.25 + .5x) = 0$ and thus we have three critical points:
    $(0,0)$, $(2,0)$ and $(1/2,3/2)$.

3c. For (0,0) the linear system is dx/dt = x and

dy/dt = -.25y and hence A = $\begin{pmatrix} 1 & 0 \\ 0 & -1/4 \end{pmatrix}$ which has

eigenvalues r_1 = 1 and r_2 = -1/4 and corresponding

eigenvectors $\begin{pmatrix} 1 \\ 0 \end{pmatrix}$ and $\begin{pmatrix} 0 \\ 1 \end{pmatrix}$. Thus (0,0) is an unstable

saddle point.

For (2,0), we let x = 2 + u and y = v in the given

equations and obtain $\dfrac{du}{dt}$ = -(u+v) - $\dfrac{1}{2}$u(u+v) and

$\dfrac{dv}{dt}$ = $\dfrac{3}{4}$v + $\dfrac{1}{2}$uv. The linear portion of this has matrix

A = $\begin{pmatrix} -1 & -1 \\ 0 & 3/4 \end{pmatrix}$, which has the eigenvalues r_1 = -1,

r_2 = 3/4 and corresponding eigenvectors $\begin{pmatrix} 1 \\ 0 \end{pmatrix}$ and $\begin{pmatrix} -4 \\ 7 \end{pmatrix}$.

Thus (2,0) is also an unstable saddle point.

For $\left(\dfrac{1}{2}, \dfrac{3}{2} \right)$ we let x = 1/2 + u and y = 3/2 + v in the

given equations, which yields $\dfrac{du}{dt}$ = $-\dfrac{1}{4}$u - $\dfrac{1}{4}$v, $\dfrac{dv}{dt}$ = $\dfrac{3}{4}$u

as the linear portion. Thus A = $\begin{pmatrix} -\dfrac{1}{4} & -\dfrac{1}{4} \\ \dfrac{3}{4} & 0 \end{pmatrix}$, which has

eigenvalues $r_{1,2}$ = $(-1 \pm \sqrt{11}\, i)/8$. Thus $\left(\dfrac{1}{2}, \dfrac{3}{2} \right)$ is an

asymptotically stable spiral point since the eigenvalues
are complex with negative real part. Using

r_1 = $(-1 + \sqrt{11}\, i)/8$ we find that one eigenvector is

$\begin{pmatrix} -2 \\ 1 + \sqrt{11}\, i \end{pmatrix}$ and by Sect. 7.6 the second eigenvector is

the complex conjugate $\begin{pmatrix} -2 \\ 1 - \sqrt{11}\, i \end{pmatrix}$.

3e.

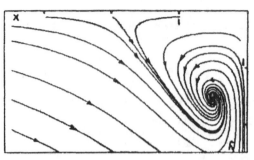

3f. For (x,y) above the line $x + y = 2$ we see that $x' < 0$ and
thus x must remain bounded. For (x,y) to the right of
$x = 1/2$, $y' > 0$ so it appears that y could grow large
asymptotic to $x = $ constant. However, this implies a
contradiction ($x = $ constant implies $x' = 0$, but as y gets
larger, x' gets increasingly negative) and hence we
conclude y must remain bounded and hence
$(x,y) \to (1/2, 3/2)$ as $t \to \infty$, again assuming they start
in the first quadrant.

7a. The amplitude ratio is $(cK/\gamma)/(\sqrt{ac}\, K/\alpha) = \alpha\sqrt{c}/\gamma\sqrt{a}$.

7b. From Eq.(2) $\alpha = .5$, $a = 1$, $\gamma = .25$ and $c = .75$, so the
ratio is $.5\sqrt{.75}/.25\sqrt{1} = 2\sqrt{.75} = \sqrt{3}$.

7c. A rough measurement of the amplitudes is $(6.2 - 1.2)/2 = 2.50$
and $(3.8 - .9)/2 = 1.45$ and thus the ratio is approximately
1.72. In this case the linear approximation is a good
predictor.

11a. Setting $x' = 0$ and $y' = 0$ we find the three critical
points $P_1(0,0)$, $P_2(\frac{1}{\sigma}, 0)$ and $P_3(3, 2-6\sigma)$. As σ increases

from 0 to $\frac{1}{3}$ the x coordinate of P_2 decreases and thus P_2

moves to the left. Likewise, the y coordinate of P_3
decreases and thus P_3 moves down and coincides with P_2

when $\sigma = \frac{1}{3}$.

11b. The Jacoban for this system is $\begin{pmatrix} 1-2\sigma x - .5y & -.5x \\ .25y & -.75+.25x \end{pmatrix}$. For

P_1 we have $\mathbf{A} = \begin{pmatrix} 1 & 0 \\ 0 & -.75 \end{pmatrix}$ so $r = 1, -.75$ and thus P_1 is a saddle point. For P_2 we have $\mathbf{A} = \begin{pmatrix} -1 & \dfrac{-1}{2\sigma} \\ 0 & -.75+\dfrac{.25}{\sigma} \end{pmatrix}$ so

$r = -1$ and $-.75 + \dfrac{.25}{\sigma}$ so P_2 is a saddle point for $\sigma < 1/3$ and an asymptotically stable node for $\sigma > 1/3$. For P_3

$J = \begin{pmatrix} -3\sigma & -1.5 \\ .5-1.5\sigma & 0 \end{pmatrix}$ which gives $r = -1.5(\sigma \pm \sqrt{\sigma^2+\sigma-1/3})$. By completing the square under the radical we find that for $\sigma < (\sqrt{7/3}-1)/2 = \sigma_1$ the radicand is negative, so the roots are complex and P_3 is an asymptotically stable spiral point. For $\sigma_1 < \sigma < 1/3$ the roots are real and negative so P_3 is an asymptotically stable node. For $\sigma > 1/3$ the roots differ in sign, so P_3 is a saddle point in the 4th quadrant.

11c. $\sigma = 0.2$ $\sigma = 0.3$

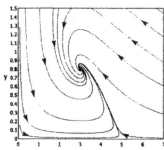
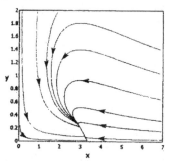

11d. For $0 < \sigma < \sigma_1$ the phase protrait is a spiral and thus the populations will oscillate about the equilibrium point with dacaying amplitudes. For $\sigma_1 < \sigma < 1/3$ the populations will no longer oscillate but will still move to the equilibrium point as t increases.

14. The new equalibrium solution is given by $a - \alpha y - E_1 = 0$ and $-c + \gamma x - E_2 = 0$ which yield $(\dfrac{c+E_2}{\gamma}, \dfrac{a-E_1}{\alpha})$. The results for (b),(c) and (d) follow directly from this.

Section 9.6, Page 551

1.   Assuming that $V(x,y) = ax^2 + cy^2$ we find $V_x(x,y) = 2ax$, $V_y = 2cy$ and thus Eq.(7) yields

$\dot{V}(x,y) = 2ax(-x^3 + xy^2) + 2cy(-2x^2y - y^3)$
$= -[2ax^4 + 2(2c-a)x^2y^2 + 2cy^4]$.   If we choose a

and c to be any positive real numbers with $2c > a$, then $\dot{V}$ is a negative definite. By definition V is positive definite, so by Theorem 9.6.1 the origin is an asymptotically stable critical point.

3. Assuming the same form for $V(x,y)$ as in Prob. 1, we have
$$\dot{V}(x,y) = 2ax(-x^3 + 2y^3) + 2cy(-2xy^2) = -2ax^4 + 4(a-c)xy^3.$$
    If we choose $a = c > 0$, then $\dot{V}(x,y) = -2ax^4 \le 0$ in any

    neighborhood containing the origin and thus $\dot{V}$ is negative semidefinite and V is positive definite. Theorem 9.6.1 then concludes that the origin is a stable critical point. However, the origin may still be asymptotically stable since the $V(x,y)$ used here is not sufficient to prove that.

6a. The correct system is $dx/dt = y$ and $dy/dt = -g(x)$. Since $g(0) = 0$, we conclude that $(0,0)$ is a critical point.

6b. From the given conditions, the graph of g must be positive for $0 < x < k$ and negative for $-k < x < 0$. Thus
    if $0 < x < k$ then $\int_0^x g(s)ds > 0$,

    if $-k < x < 0$ then $\int_0^x g(s)ds = -\int_x^0 g(s)ds > 0$.

    Since $V(0,0) = 0$ it follows that $V(x,y) = y^2/2 + \int_0^x g(s)ds$
    is positive definite for $-k < x < k$, $-\infty < y < \infty$. Next,
    we have $\dot{V}(x,y) = V_x\dfrac{dx}{dt} + V_y\dfrac{dy}{dt} = g(x)y + y[-g(x)] = 0$.

    Since $\dot{V}(x,y)$ is never positive, we may conclude that it is negative semidefinite and hence by Theorem 9.6.1 $(0,0)$ is at least a stable critical point.

7b. V is positive definite by Theorem 9.6.4. Since
    $V_x(x,y) = 2x$, $V_y(x,y) = 2y$, we obtain
    $$\dot{V}(x,y) = 2xy - 2y^2 - 2y\sin x = 2y[-y + (x - \sin x)].$$ Since

    $x - \sin x < 0$ for $x < 0$ we have $\dot{V}(x,y) < 0$ for all $y > 0$.
    If $x > 0$, choose y so that $0 < y < x - \sin x$.
    Then $\dot{V}(x,y) > 0$. Hence V is not a Liapunov function.

7c. Since $V(0,0) = 0$, $1 - \cos x > 0$ for $0 < |x| < 2\pi$ and $y^2 > 0$
    for $y \ne 0$, it follows that $V(x,y)$ is positive definite
    in a neighborhood of the origin. Next $V_x(x,y) = \sin x$,

$V_y(x,y) = y$, so $\dot{V}(x,y) = (\sin x)(y) + y(-y - \sin x) = -y^2$.

Hence $\dot{V}$ is negative semidefinite and $(0,0)$ is a stable critical point by Theorem 9.6.1.

7d. $V(x,y) = (x+y)^2/2 + x^2 + y^2/2 = 3x^2/2 + xy + y^2$ is positive definite by Theorem 9.6.4. Next $V_x(x,y) = 3x + y$, $V_y(x,y) = x + 2y$ so

$$\begin{aligned}
\dot{V}(x,y) &= (3x+y)y - (x+2y)(y+\sin x)\\
&= 2xy - y^2 - (x+2y)\sin x\\
&= 2xy - y^2 - (x+2y)(x - \alpha x^3/6) \quad [\text{from the hint}]\\
&= -x^2 - y^2 + \alpha(x+2y)x^3/6\\[6pt]
&= -r^2 + \alpha r^4(\cos\theta + 2\sin\theta)(\cos^3\theta)/6
\end{aligned}$$

$< -r^2 + r^4/2 = -r^2(1-r^2/2)$. Thus $\dot{V}$ is negative definite for $r < \sqrt{2}$. From Theorem 9.6.1 it follows that the origin is an asymptotically stable critical point.

8.   Let $x = u$ and $y = du/dt$ to obtain the system $dx/dt = y$ and $dy/dt = -c(x)y - g(x)$. Now consider

$$V(x,y) = y^2/2 + \int_0^x g(s)\,ds, \quad \text{which yields}$$

$\dot{V} = g(x)y + y[-c(x)y - g(x)] = -y^2 c(x)$, which is negative semidefinite by Theorem 9.6.1.

10b. Since $V_x(x,y) = 2Ax + By$, $V_y(x,y) = Bx + 2Cy$, we have

$$\begin{aligned}
\dot{V}(x,y) &= (2Ax + By)(a_{11}x + a_{12}y) + (Bx + 2Cy)(a_{21}x + a_{22}y)\\
&= (2Aa_{11} + Ba_{21})x^2 + [2(Aa_{12} + Ca_{21}) + B(a_{11}+a_{22})]xy\\
&\qquad\qquad + (2Ca_{22} + Ba_{12})y^2.
\end{aligned}$$

We choose A, B, and C so that $2Aa_{11} + Ba_{21} = -1$, $2(Aa_{12} + Ca_{21}) + B(a_{11}+a_{22}) = 0$, and $2Ca_{22} + Ba_{12} = -1$. The first and third equations give us A and C in terms of B, respectively. We substitute in the second equation to find B and then calculate A and C. The result is given in the text.

10c. Since $a_{11}a_{22} - a_{12}a_{21} > 0$ and $a_{11} + a_{22} < 0$, we see that $\Delta < 0$ and so $A > 0$. Using the expressions for A, B, and C found in Part(b) we obtain

$$\begin{aligned}
(4AC-B^2)\Delta^2 &= [a_{21}^2+a_{22}^2 + (a_{11}a_{22}-a_{12}a_{21})][a_{11}^2+a_{12}^2 + (a_{11}a_{22}-a_{12}a_{21})]\\
&\qquad\qquad - (a_{12}a_{22}+a_{11}a_{21})^2\\
&= (a_{11}^2+a_{12}^2+a_{21}^2+a_{22}^2)(a_{11}a_{22}-a_{12}a_{21}) + (a_{11}^2+a_{12}^2)(a_{21}^2+a_{22}^2)\\
&\qquad\qquad + (a_{11}a_{22}-a_{12}a_{21})^2 - (a_{12}a_{22}+a_{11}a_{21})^2\\
&= (a_{11}^2+a_{12}^2+a_{21}^2+a_{22}^2)(a_{11}a_{22}-a_{12}a_{21}) + 2(a_{11}a_{22}-a_{12}a_{21})^2.
\end{aligned}$$

Since $a_{11}a_{22} - a_{12}a_{21} > 0$ it follows that $4AC - B^2 > 0$.

11a. For $V(x,y) = Ax^2 + Bxy + Cy^2$ we have

$$\dot{V} = (2Ax + By)(a_{11}x + a_{12}y + F_1(x,y)) + (Bx + 2Cy)(a_{21}x + a_{22}y + G_1(x,y))$$

$$= (2Ax + By)(a_{11}x + a_{12}y) + (Bx + 2Cy)(a_{21}x + a_{22}y)$$
$$+ (2Ax + By)F_1(x,y) + (Bx + 2Cy)G_1(x,y)$$

$$= -x^2 - y^2 + (2Ax + By)F_1(x,y) + (Bx + 2Cy)G_1(x,y), \text{ if } A, B \text{ and}$$
C are chosen as in Prob. 10.

11b. Substituting $x = r\cos\theta$, $y = r\sin\theta$ we find that

$$\dot{V}[x(r,\theta), y(r,\theta)] = -r^2 + r(2A\cos\theta + B\sin\theta)F_1[x(r,\theta), y(r,\theta)]$$
$$+ r(B\cos\theta + 2C\sin\theta)G_1[x(r,\theta), y(r,\theta)].$$

Now we make use of the facts that: (1) there exists an M
such that $|2A| \leq M$, $|B| \leq M$, and $|2C| \leq M$; and (2) given
any $\varepsilon > 0$ there exists a circle $r = R$ such that
$|F_1(x,y)| < \varepsilon r$ and $|G_1(x,y)| < \varepsilon r$ for $0 \leq r < R$. We have
$|2A\cos\theta + B\sin\theta| \leq 2M$ and $|B\cos\theta + 2C\sin\theta| \leq 2M$. Hence

$$\dot{V}[x(r,\theta), y(r,\theta)] \leq -r^2 + 2Mr(\varepsilon r) + 2Mr(\varepsilon r) = -r^2(1 - 4M\varepsilon).$$

If we choose $\varepsilon = M/8$ we obtain $\dot{V}[x(r,\theta), y(r,\theta)] \leq -r^2/2$

for $0 \leq r < R$. Hence $\dot{V}$ is negative definite in $0 \leq r < R$
and from Prob.10c V is positive definite and thus V is a
Liapunov function for the almost linear system.

Section 9.7, Page 560

1. Note that $r = 1$, $\theta = t + t_0$ satisfy the two equations for
all t and is thus a periodic solution. If $r < 1$, then
$dr/dt > 0$, and the direction of motion on a trajectory is
outward. If $r > 1$, then the direction of motion is
inward. It follows that the periodic solution $r = 1$,
$\theta = t + t_0$ is an asymptotically stable limit cycle.

2. $r = 1$, $\theta = -t + t_0$ is a periodic solution. If $r < 1$,
then $dr/dt > 0$, and the direction of motion on a
trajectory is outward. If $r > 1$, the $dr/dt > 0$, and the
direction of motion is still outward. It follows that
the solution $r = 1$, $\theta = -t + t_0$ is a semistable limit cycle.

4. $r = 1$, $\theta = -t + t_0$ and $r = 2$, $\theta = -t + t_0$ are periodic
solutions. If $r < 1$, then $dr/dt < 0$, and the direction
of motion on a trajectory is inward. If $1 < r < 2$, then
$dr/dt > 0$, and the direction of motion is outward.
Similarly, if $r > 2$, the direction of motion is inward.
It follows that the periodic solution $r = 1$,
$\theta = -t + t_0$ is unstable and the periodic solution $r = 2$,

$\theta = -t + t_0$ is an asymptotically stable limit cycle.

7.   Differentiating x and y with respect to t we find that
$dx/dt = (dr/dt)\cos\theta - (r\sin\theta)d\theta/dt$ and
$dy/dt = (dr/dt)\sin\theta + (r\cos\theta)d\theta/dt$.   Hence
$ydx/dt - xdy/dt = (r\sin\theta\cos\theta)dr/dt - (r^2\sin^2\theta)d\theta/dt$
$$- (r\cos\theta\sin\theta)dr/dt - (r^2\cos\theta)d\theta/dt$$
$$= - r^2 d\theta/dt.$$

8a. Multiplying the first equation by x and the second by y
and adding yields $xdx/dt + ydy/dt = (x^2+y^2)f(r)/r$, or
$rdr/dt = rf(r)$, as in the derivation of Eq.(8), and thus
$dr/dt = f(r)$.   To obtain an equation for θ multiply the
first equation by y, the second by x and substract to
obtain $ydx/dt - xdy/dt = -x^2-y^2$, or $-r^2 d\theta/dt = -r^2$, using
the results of Prob. 7. Thus $d\theta/dt = 1$.   It follows that
periodic solutions are given by $r = c$, $\theta = t + t_0$ where
$f(c) = 0$.   Since $\theta = t + t_0$, the motion is
counterclockwise.

8b. First note that $f(r) = r(r-2)^2(r-3)(r-1)$.   Thus $r = 1$,
$\theta = t + t_0$; $r = 2$, $\theta = t + t_0$; and $r = 3$, $\theta = t + t_0$ are
periodic solutions.   If $r < 1$, then $dr/dt > 0$, and the
direction of motion on a trajectory is outward.   If
$1 < r < 2$, then $dr/dt < 0$ and the direction of motion is
inward.   Thus the periodic solution $r = 1$, $\theta = t + t_0$ is
an asymptotically stable limit cycle.   If $2 < r < 3$, then
$dr/dt < 0$, and the direction of motion is inward.   Thus
the periodic solution $r = 2$, $\theta = t + t_0$ is a semistable
limit cycle.   If $r > 3$, then $dr/dt > 0$, and the direction
of motion is outward.   Thus the periodic solution $r = 3$,
$\theta = t + t_0$ is unstable.

9.   Setting $x = r\cos\theta$, $y = r\sin\theta$ and using the techniques of
Prob. 8 the equations transform to $dr/dt = r^2 - 2$,
$d\theta/dt = -1$.   This system has a periodic solution $r = \sqrt{2}$,
$\theta = -t + t_0$.   If $r < \sqrt{2}$, then $dr/dt < 0$, and the
direction of motion along a trajectory is inward.
f $r > \sqrt{2}$, then $dr/dt > 0$, and the direction of motion is
outward.   Thus the periodic solution $r = \sqrt{2}$, $\theta = -t + t_0$
is unstable.

11. If $F(x,y) = x+y+x^3-y^2$, $G(x,y) = -x+2y+x^2y+y^3/3$, then
$F_x(x,y) + G_y(x,y) = 1+3x^2+2+x^2+y^2 = 3+4x^2+y^2$.   Since the
conditions of Theorem 9.7.2 are satisfied for all x and
y, and since $F_x + G_y > 0$ for all x and y, it follows that
the system has no periodic nonconstant solution.

13. Since $x = \phi(t)$, $y = \psi(t)$ is a solution of Eqs.(15), we
    have $d\phi/dt = F[\phi(t),\psi(t)]$, $d\psi/dt = G[\phi(t),\psi(t)]$. Hence
    on the curve C,
    $F(x,y)dy - G(x,y)dx = \phi'(t)\psi'(t)dt - \psi'(t)\phi'(t)( )= 0$. It
    follows that the line integral around C is zero.
    However, if $F_x + G_y$ has the same sign throughout D, then
    the double integral cannot be zero. This gives a
    contradiction. Thus either the solution of Eqs.(15) is
    not periodic or if it is, it cannot lie entirely in D.

18a. Setting x=0 and y=0 in x'=0and y'=0 yields (0,0) and (2,0)
     as critical points. The third critical point comes from
     setting the terms in parenthises equal to zero and
     solving to get (2,8).

18b. The Jacobian for this system is given by

$$J = \begin{pmatrix} 2.4-.4x-\dfrac{12y}{(x+6)^2} & \dfrac{-2x}{x+6} \\ \dfrac{6y}{(x+6)^2} & -.25+\dfrac{x}{x+6} \end{pmatrix}. \text{ For (2,8) } J = \begin{pmatrix} .1 & -.5 \\ .75 & 0 \end{pmatrix} \text{ so}$$

$r = (.1 \pm \sqrt{.01 - 1.5})/2$ and thus (2,8) is an unstable

spiral point. Likewise for (0,0) $J = \begin{pmatrix} 2.4 & 0 \\ 0 & -.25 \end{pmatrix}$ and for

(12,0) $J = \begin{pmatrix} -2.4 & -4/3 \\ 0 & -.25+2/3 \end{pmatrix}$, which both yield saddle points.

18c.

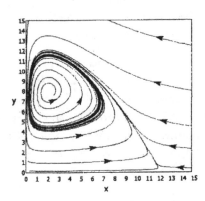

21a. Setting x' = 0 and solving for y yields $y = x^3/3 - x + k$.
     Substituting this into y'= 0 then gives $W(x) = x$
     $+.8(x^3/3 - x + k) -.7 = .8x^3/3 + .2x + (.8k -.7) = 0$. Since
     $W'(x) = .8x^2 + .2$ is never zero, we conclude that W
     always has a positive slope and thus the cubic equation

crosses the x axis only once for all values of k.

21b. Using an equation solver on the expression in Part(a) we
obtain x = 1.1994, y = −.62426 for k = 0 and x = .80485,
y = −.13106 for k = .5.   To determine the type of critical
points these are, we use Eq.(13) of Section 9.3 to find the

linear coefficient matrix to be $A = \begin{pmatrix} 3(1-x_c^2) & 3 \\ -1/3 & -.8/3 \end{pmatrix}$, where x_c

is the critical point. For $x_c = 1.1994$ we obtain complex
conjugate eigenvalues with a negative real part, and
therefore k = 0 yields an asymptotically stable spiral
point. For $x_c = .80485$ the eigenvalues are also complex
conjugates, but with positive real parts, so k = .5 yields
an unstable spiral point.

21c. Letting k = .1, .2, .3, .4 in the cubic equation of Part(a)
and finding the corresponding eigenvalues from the matrix in
Part(b), we find that the real part of the eigenvalues
changes sign between k = .3 and k = .4. Continuing to
iterate in this fashion we find that for k = .3464 that the
real part of the eigenvalue is −.0002 while for k = .3465
the real part is .00005, which indicates $k_0 = .3465$ is the
critical point for which the system changes from stable to
unstable.

21e. Again, iterating as in Part(c), we find for k = 1.403 that
$x_c = −.9541$ and for k = 1.404 that $x_c = −.9549$. Substituting
these values into the coefficient matrix of Part(b) and
finding the eigenvalues we find that the real part changes
sign beween k = 1.403 and k = 1.404. Thus the critical point
again becomes asymptotically stable.

Section 9.8, Page 573

1a. From Eq.(9), $\lambda_1 = -8/3$ is clearly one eigenvalue and the
other two may be found from $\lambda^2 + 11\lambda - 10(r-1) = 0$ using
the quadratic formula.

1b. For $\lambda = \lambda_1$ we have
$$\begin{pmatrix} -10+8/3 & 10 & 0 \\ r & -1+8/3 & 0 \\ 0 & 0 & 0 \end{pmatrix}\begin{pmatrix} \xi_1 \\ \xi_2 \\ \xi_3 \end{pmatrix} = \begin{pmatrix} 0 \\ 0 \\ 0 \end{pmatrix},$$ which requires $\xi_1 = \xi_2 = 0$

and ξ_3 arbitrary and thus $\xi^{(1)} = (0,0,1)^T$.
For $\lambda = \lambda_3 = (-11 + \alpha)/2$, where $\alpha = \sqrt{81+40r}$, we have

$$\begin{pmatrix} -10+(11-\alpha)/2 & 10 & 0 \\ r & -1+(11-\alpha)/2 & 0 \\ 0 & 0 & -8/3+(11-\alpha)/2 \end{pmatrix} \begin{pmatrix} \xi_1 \\ \xi_2 \\ \xi_3 \end{pmatrix} = \begin{pmatrix} 0 \\ 0 \\ 0 \end{pmatrix}.$$

The last line implies $\xi_3 = 0$ and multiplying the first line by

$(-9+\alpha)/2$ we obtain $\begin{pmatrix} (81-\alpha^2)/4 & 10(-9+\alpha)/2 \\ r & (9-\alpha)/2 \end{pmatrix} \begin{pmatrix} \xi_1 \\ \xi_2 \end{pmatrix} = \begin{pmatrix} 0 \\ 0 \end{pmatrix}.$

Substituting $\alpha^2 = 81+40r$ we have

$$\begin{pmatrix} -10r & -10(9-\alpha)/2 \\ r & (9-\alpha)/2 \end{pmatrix} \begin{pmatrix} \xi_1 \\ \xi_2 \end{pmatrix} = \begin{pmatrix} 0 \\ 0 \end{pmatrix}. \quad \text{Thus } \xi^{(3)} = \begin{pmatrix} 9 - \sqrt{81+40r} \\ -2r \\ 0 \end{pmatrix},$$

which is proportional to the answer given in the text. Similar calculations give $\xi^{(2)}$.

1c. Simply substitute $r = 28$ into the answers in Parts(a) and (b).

2a. Using Eq.(3), the linear system is given by

$$\begin{pmatrix} u \\ v \\ w \end{pmatrix}' = \begin{pmatrix} F_x & F_y & F_z \\ G_x & G_y & G_z \\ H_x & H_y & H_z \end{pmatrix}_{(x_0,y_0,z_0)} \begin{pmatrix} u \\ v \\ w \end{pmatrix}, \text{ where } F = -10x + 10y,$$

$G = rx - y - xz$ and $H = -8z/3 + xy$ and thus

$$\begin{pmatrix} u \\ v \\ w \end{pmatrix}' = \begin{pmatrix} -10 & 10 & 0 \\ r-z_0 & -1 & -x_0 \\ y_0 & x_0 & -8/3 \end{pmatrix} \begin{pmatrix} u \\ v \\ w \end{pmatrix}, \text{ which is Eq.(11) for } P_2$$

since $x_0 = y_0 = \sqrt{8(r-1)/3}$, $z_0 = r-1$.

2b. Eq.(12) is found by evaluating $\begin{vmatrix} -10-\lambda & 10 & 0 \\ 1 & -1-\lambda & -\beta \\ \beta & \beta & -8/3-\lambda \end{vmatrix} = 0.$

where $\beta = \sqrt{8(r-1)/3}$.

2c. If $r = 28$, then Eq.(12) is $3\lambda^3 + 41\lambda^2 + 304\lambda + 4320 = 0$, which has the roots -13.8546 and $.093956 \pm 10.1945i$.

3b. If r_1, r_2, r_3 are the three roots of a cubic polynomial, then the polynomial can be factored as $(x-r_1)(x-r_2)(x-r_3)$. Expanding this and equating to the given polynomial we have $A = -(r_1+r_2+r_3)$,

$B = r_1r_2 + r_1r_3 + r_2r_3$ and $C = -r_1r_2r_3$. We are interested in the case when the real part of the complex conjugate roots changes sign. Thus let $r_2 = \alpha+i\beta$ and $r_3 = \alpha-i\beta$, which yields
$A = -(r_1+2\alpha)$, $B = 2\alpha r_1 + \alpha^2 + \beta^2$ and $C = -r_1(\alpha^2+\beta^2)$.
Hence, if $AB = C$, we have
$-(r_1+2\alpha)(2\alpha r_1+\alpha^2+\beta^2) = -r_1(\alpha^2+\beta^2)$ or
$-2\alpha[r_1^2 + 2\alpha r_1 + (\alpha^2+\beta^2)] = 0$ or $-2\alpha[(r_1+\alpha)^2+\beta^2] = 0$.
Since the square bracket term is positive, we conclude that if $AB = C$, then $\alpha = 0$. That is, the conjugate complex roots are pure imaginary. Note that the converse is also true. That is, if the conjugate complex roots are pure imaginary then $AB = C$.

3c. Comparing Eq.(12) to that of Part(b), we have $A = 41/3$, $B = 8(r+10)/3$ and $C = 160(r-1)/3$. Thus $AB = C$ yields $r = 470/19$.

4. We have $\dot{V} = 2x[\sigma(-x+y)] + 2\sigma y[rx-y-xz] + 2\sigma z[-bz+xy]$
$= -2\sigma x^2 + 2\sigma xy + 2\sigma rxy - 2\sigma y^2 - 2\sigma bz^2$
$= 2\sigma\{-[x^2-(r+1)xy+y^2]-bz^2\}$. For $r < 1$, the term in the square brackets remains positive for all values of

x and y, by Theorem 9.6.4, and thus $\dot{V}$ is negative definite. Thus, by the extension of Theorem 9.6.1 to three equations, we conclude that the origin is an asymptotically stable critical point.

5a. $V = rx^2 + \sigma y^2 + \sigma(z-2r)^2 = c > 0$ yields
$\dfrac{dv}{dt} = 2rx[\sigma(-x+y)] + 2\sigma y(rx-y-xz) + 2\sigma(z-2r)(-bz+xy)$. Thus

$\dot{V} = -2\sigma[rx^2+y^2 + b(z^2 - 2rz)] = -2\sigma[rx^2 + y^2 + b(z-r)^2 - br^2]$.

5b. From the proof of Theorem 9.6.1, we find that we need to show that $\dot{V}$, as found in Part(a), is always negative as it crosses $V(x,y,z) = c$. (Actually, we need to use the extension of Theorem 9.6.1 to three equations, but the proof is very similar using the vector calculus approach.) From Part(a) we see that

$\dot{V} < 0$ if $rx^2 + y^2 + b(z-r)^2 > br^2$, which holds if (x,y,z) lies outside the ellipsoid $\dfrac{x^2}{br} + \dfrac{y^2}{br^2} + \dfrac{(z-r)^2}{r^2} = 1$, Eq.(i).
Thus we need to choose c such that $V = c$ lies outside Eq.(i). Writing $V = c$ in the form of Eq.(i) we obtain
the ellipsoid $\dfrac{x^2}{c/r} + \dfrac{y^2}{c/\sigma} + \dfrac{(z-2r)^2}{c/\sigma} = 1$, Eq.(ii). Now let

$M = \max(\sqrt{br}, r\sqrt{b}, r)$, then the ellipsoid (i) is
contained inside the sphere S1: $\dfrac{x^2}{M^2} + \dfrac{y^2}{M^2} + \dfrac{(z-r)^2}{M^2} = 1$.

Let S2 be a sphere centered at $(0,0,2r)$ with radius

$M+r$: $\dfrac{x^2}{(M+r)^2} + \dfrac{y^2}{(M+r)^2} + \dfrac{(z-2r)}{(M+r^2)} = 1$, then S1 is contained

in S2. Thus, if we choose c, in Eq.(ii), such that
$\dfrac{c}{r} > (M+r)^2$ and $\dfrac{c}{\sigma} > (M+r)^2$, then $\dot{V} < 0$ as the trajectory

crosses $V(x,y,z) = c$. Note that this is a sufficient
condition and there may be many other "better" choices
using different techniques.

8b. Several cases are shown. Results may vary, particularly
for $r = 24$, due to the closeness of r to $r_3 \cong 24.06$.

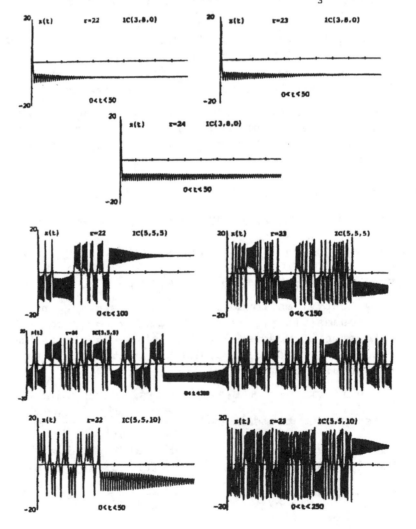

11a. let a = .25 and b = .5 in Eq.(i) to get x' = -y-z,
     y' = x+.25y and z' = .5+z(x-c). Setting these equal to zero
     yields x = .25z and y = -z from the first two. Substitution
     into the third gives .25z² - cz + .5 = 0, and thus
     z = 2(c±√(c²-.5)), which gives the desired results.

11b. If we let F = -y-z, G = x+ay and H = b+z(x-c) then, by

     Eq.(3), the Jacobian matrix is given by $J = \begin{pmatrix} 0 & -1 & -1 \\ 1 & a & 0 \\ z & 0 & x-c \end{pmatrix}$. If

     we set a = .25, c = √.5, x = √2/4, and z = √2 the
     eigenvalues for this critical point are given by

$$\begin{vmatrix} -\lambda & -1 & -1 \\ 1 & .25-\lambda & 0 \\ \sqrt{2} & 0 & -.25\sqrt{2}-\lambda \end{vmatrix} = -\lambda[\lambda^2-.25(1-\sqrt{2})\lambda+(.9375\sqrt{2}+1)] = 0,$$

     which gives the desired eigenvalues.

11c.   *Initial condition = (0,0,0)*

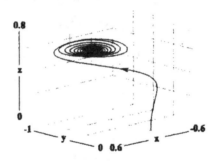

11d.

                        Initial condition = (1,1,1)

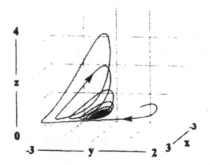

CHAPTER 10

Section 10.1 Page 611

2. $y(x) = c_1\cos\sqrt{2}\,x + c_2\sin\sqrt{2}\,x$ is the general solution of the
    D.E. Thus $y'(x) = -\sqrt{2}\,c_1\sin\sqrt{2}\,x + \sqrt{2}\,c_2\cos\sqrt{2}\,x$ and hence
    $y'(0) = \sqrt{2}\,c_2 = 1$, which gives $c_2 = 1/\sqrt{2}$. Now,
    $y'(\pi) = -\sqrt{2}\,c_1\sin\sqrt{2}\,\pi + \cos\sqrt{2}\,\pi = 0$ then yields
    $c_1 = \dfrac{\cos\sqrt{2}\,\pi}{\sqrt{2}\,\sin\sqrt{2}\,\pi} = \cot\sqrt{2}\,\pi/\sqrt{2}$. Thus the desired solution
    is $y = \left(\cot\sqrt{2}\,\pi\cos\sqrt{2}\,x + \sin\sqrt{2}\,x\right)/\sqrt{2}$.

3. We have $y(x) = c_1\cos x + c_2\sin x$ as the general solution and
    hence $y(0) = c_1 = 0$ and $y(L) = c_2\sin L = 0$. If $\sin L \neq 0$, then
    $c_2 = 0$ and $y(x) = 0$ is the only solution. If $\sin L = 0$, then
    $y(x) = c_2\sin x$ is a solution for arbitrary c_2.

7. $y(x) = c_1\cos 2x + c_2\sin 2x$ is the solution of the related
    homogeneous equation and $y_p(x) = \dfrac{1}{3}\cos x$ is a particular
    solution (using undetermined coefficients), yielding
    $y(x) = c_1\cos 2x + c_2\sin 2x + \dfrac{1}{3}\cos x$ as the general solution of
    the D.E. Thus $y(0) = c_1 + \dfrac{1}{3} = 0$ and $y(\pi) = c_1 - \dfrac{1}{3} = 0$ and
    hence there is no solution since there is no value of c_1 that
    will satisfy both boundary conditions.

12. This is an Euler Equation (Sect.5.4), so solutions of the
    homogeneous equation are of the form $y = x^{\alpha}$. Substituting
    this into the D.E. yields $(\alpha+1)^2 = 0$ and thus
    $y_c = (c_1 + c_2\ln x)x^{-1}$. For the particular solution we assume
    $y_p = Ax^2$, which upon substitution in the D.E. yields
    $A = 1/9$. Hence the general solution is
    $y(x) = (c_1 + c_2\ln x)x^{-1} + x^2/9$, so
    $y(1) = c_1 + 1/9 = 0$ and $y(e) = (c_1 + c_2)e^{-1} + e^2/9 = 0$.
    Solving for c_1 and c_2 yields the desired solution.

14. If $\lambda < 0$ set $\lambda = -\mu^2$, then the general solution of the D.E. is
    $y = c_1\sinh\mu x + c_2\cosh\mu x$. The two B.C. require that $c_2 = 0$ and
    $c_1 = 0$, so $\lambda < 0$ is not an eigenvalue. If $\lambda = 0$, the general
    solution of the D.E. is $y = c_1 + c_2 x$. The B.C. require that

$c_1 = 0$, $c_2 = 0$, so again $\lambda = 0$ is not an eigenvalue.
If $\lambda > 0$, the general solution of the D.E. is
$y = c_1\sin\sqrt{\lambda}\,x + c_2\cos\sqrt{\lambda}\,x$. The B.C. require that $c_2 = 0$
and $\sqrt{\lambda}\,c_1\cos\sqrt{\lambda}\,\pi = 0$. The second condition is satisfied
for $\lambda \neq 0$ and $c_1 \neq 0$ if $\sqrt{\lambda}\,\pi = (2n-1)\pi/2$, $n = 1,2,\ldots$.
Thus the eigenvalues are $\lambda_n = (2n-1)^2/4$, $n = 1,2,3\ldots$
with the corresponding eigenfunctions (with $c_1 = 1$)
$y_n(x) = \sin[(2n-1)x/2]$, $n = 1,2,3\ldots$.

18. For $\lambda < 0$ there are no eigenvalues, using similar steps as in
    Prob. 14. For $\lambda = 0$ we have $y(x) = c_1 + c_2x$, so $y'(0) = c_2 = 0$
    and $y'(L) = c_2 = 0$, and thus $\lambda = 0$ is an eigenvalue, with
    $y_0(x) = 1$ as the eigenfunction. For $\lambda > 0$ we again have
    $y(x) = c_1\sin\sqrt{\lambda}\,x + c_2\cos\sqrt{\lambda}\,x$, so $y'(0) = \sqrt{\lambda}\,c_1 = 0$ (so $c_1 = 0$)
    and $y'(L) = -c_2\sqrt{\lambda}\,\sin\sqrt{\lambda}\,L = 0$. We know $\lambda > 0$, in this case,
    so the eigenvalues are given by $\sin\sqrt{\lambda}\,L = 0$ or $\sqrt{\lambda}\,L = n\pi$.
    Thus $\lambda_n = (n\pi/L)^2$ and $y_n(x) = \cos(n\pi x/L)$ for $n = 1,2,3\ldots$.

21a. Multiplying D.E. by r^2 we obtain $r^2w'' + rw' = -\dfrac{Gr^2}{\mu}$, which is

    an Euler Eq. that has repeated roots given by $\alpha^2 = 0$. Thus,
    as in Prob. 12, we have $w(r) = (c_1 + c_2\ln r) - Gr^2/4\mu$, where
    the particular solution $(- Gr^2/4\mu)$ is found by assuming
    $x_p = Ar^2$ and solving for A. We are given that $w(r)$ is bounded
    for $0 < r < R$ which implies $c_2 = 0$ and $w(R) = 0$ implies
    $c_1 = GR^2/4\mu$, so $w(r) = G(R^2 - r^2)/4\mu$.

21b. Intergrating over a cross section, in polar coordinates, gives
    the integral $\int_0^{2\pi}\int_0^R w(r)r\,dr\,d\theta$, which when evaluated yields
    $Q = \pi R^4 G/8\mu$.

21c. Setting $R = .75$ we find $(.75)^4 = .3164$, which is the desired
    reduction since all other values are constant.

Section 10.2, Page

3. We look for values of T for which $\sinh 2(x+T) = \sinh 2x$ for
    all x. Expanding the left side of this equation gives
    $\sinh 2x\cosh 2T + \cosh 2x\sinh 2T = \sinh 2x$, which will be
    satisfied for all x if we can choose T so that $\cosh 2T = 1$
    and $\sinh 2T = 0$. The only value of T satisfying these
    two constraints is $T = 0$. Since T is not positive we
    conclude that the function $\sinh 2x$ is not periodic.

5. We look for values of T for which $\tan\pi(x+T) = \tan\pi x$.
    Expanding the left side gives
    $\tan\pi(x+T) = (\tan\pi x + \tan\pi T)/(1-\tan\pi x\tan\pi T)$ which is equal
    to $\tan\pi x$ only for $\tan\pi T = 0$. The only positive solutions
    of this last equation are $T = 1,2,3...$ and hence $\tan\pi x$ is
    periodic with fundamental period $T = 1$.

7. To start, let $n = 0$, then $f(x) = \begin{cases} 0 & -1 \leq x < 0 \\ 1 & 0 \leq x < 1 \end{cases}$; for $n = 1$,

    $f(x) = \begin{cases} 0 & 1 \leq x < 2 \\ 1 & 2 \leq x < 3 \end{cases}$; and for $n = 2$, $f(x) = \begin{cases} 0 & 3 \leq x < 4 \\ 1 & 4 \leq x < 5 \end{cases}$. By

    continuing in this fashion, and drawing a graph, it can be
    seen that $f(x)$ is periodic with fundamental period $T = 2$.

10. The graph of $f(x)$ is:
    We note that $f(x)$ is
    a straight line with
    a slope of one in any
    interval. Thus $f(x)$ has the form $x+b$ in any interval for
    the correct value of b. Since $f(x+2) = f(x)$, we may set
    $x = -1/2$ to obtain $f(3/2) = f(-1/2)$. Noting that 3/2 is
    on the interval $1 < x < 2[f(3/2) = 3/2 + b]$ and that $-1/2$
    is on the interval $-1 < x < 0[f(-1/2) = -1/2 + 1]$, we
    conclude that $3/2 + b = -1/2 + 1$, or $b = -1$ for the
    interval $1 < x < 2$. For the interval $8 < x < 9$ we have
    $f(x+8) = f(x+6) = ... = f(x)$ by successive applications
    of the periodicity condition. Thus for $x = 1/2$ we have
    $f(17/2) = f(1/2)$ or $17/2 + b = 1/2$ so $b = -8$ on the
    interval $8 < x < 9$.

In Problems 13 through 18 it is often necessary to use
integration by parts to evaluate the coefficients, although
all the details will not be shown here.

13a. The function represents
     a sawtooth wave. It is
     periodic with period 2L.

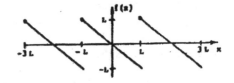

13b. The Fourier series is of the form

$$f(x) = a_0/2 + \sum_{m=1}^{\infty} [a_m\cos(m\pi x/L) + b_m\sin(m\pi x/L)], \text{ where the}$$

coefficients are computed from Eqs. (13) and (14).
Substituting for $f(x)$ in these equations yields

$a_0 = (1/L)\int_{-L}^{L}(-x)dx = 0$ and $a_m = (1/L)\int_{-L}^{L}(-x)\cos(m\pi x/L)dx = 0$,

$m = 1,2\ldots$ (these can be shown by direct integration, or using the fact that $\int_{-a}^{a}g(x)dx = 0$ when $g(x)$ is an odd function). Finally,

$b_m = (1/L)\int_{-L}^{L}(-x)\sin(m\pi x/L)dx$

$= (x/m\pi)\cos(m\pi x/L)\Big|_{-L}^{L} - (1/m\pi)\int_{-L}^{L}\cos(m\pi x/L)dx$

$= (2L\cos m\pi)/m\pi - (L/m^2\pi^2)\sin(m\pi x/L)\Big|_{-L}^{L} = 2L(-1)^m/m\pi$,

since $\cos m\pi = (-1)^m$. Substituting these terms in the above Fourier series for $f(x)$ yields the desired answer.

15a. See below.

15b. In this case $f(x)$ is periodic of period 2π and thus $L = \pi$ in Eqs. (9), (13,) and (14). The constant a_0 is found to be $a_0 = (1/\pi)\int_{-\pi}^{0}xdx = -\pi/2$ since $f(x)$ is zero on the interval $[0,\pi]$. Likewise $a_n = (1/\pi)\int_{-\pi}^{0}x\cos nxdx = [1 - (-1)^n]/n^2\pi$, using integration by parts and recalling that $\cos n\pi = (-1)^n$. Thus $a_n = 0$ for n even and $a_n = 2/n^2\pi$ for n odd, which may be written as $a_{2n-1} = 2/(2n-1)^2\pi$ since 2n-1 is always an odd number. In a similar fashion $b_n = (1/\pi)\int_{-\pi}^{0}x\sin nxdx = (-1)^{n+1}/n$ and thus the desired solution is obtained. Notice that in this case both cosine and sine terms appear in the Fourier series for the given $f(x)$.

15a.   21a.

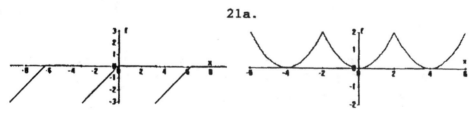

21b. $a_0 = \frac{1}{2}\int_{-2}^{2}\frac{x^2}{2}dx = \frac{1}{12}x^3\Big|_{-2}^{2} = \frac{4}{3}$, so $\frac{a_0}{2} = \frac{2}{3}$ and

$a_n = \frac{1}{2}\int_{-2}^{2}\frac{x^2}{2}\cos\frac{n\pi x}{2}dx$

$= \frac{1}{4}[\frac{2x^2}{n\pi}\sin\frac{n\pi x}{2} + \frac{8x}{n^2\pi^2}\cos\frac{n\pi x}{2} - \frac{16}{n^3\pi^3}\sin\frac{n\pi x}{2}]\Big|_{-2}^{2}$

$= (8/n^2\pi^2)\cos(n\pi) = (-1)^n 8/n^2\pi^2$

where the second line for a_n is found by integration by parts or a computer algebra system. Similarly,

$b_n = \frac{1}{2}\int_{-2}^{2}\frac{x^2}{2}\sin\frac{n\pi x}{2}dx = 0$, since $x^2\sin\frac{n\pi x}{2}$ is an odd

function. Thus $f(x) = \frac{2}{3} + \frac{8}{\pi^2}\sum_{n=1}^{\infty}\frac{(-1)^n}{n^2}\cos\frac{n\pi x}{2}$.

21c. As in Eq. (27), we have $s_m(x) = \frac{2}{3} + \frac{8}{\pi^2}\sum_{n=1}^{m}\frac{(-1)^n}{n^2}\cos\frac{n\pi x}{2}$

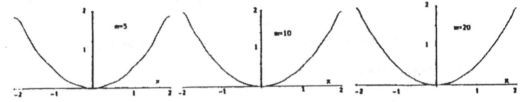

21d. Observing the graphs we see that the Fourier series converges quite rapidly, except, at $x = -2$ and $x = 2$, where there is a sharp "point" in the periodic function.

25a.

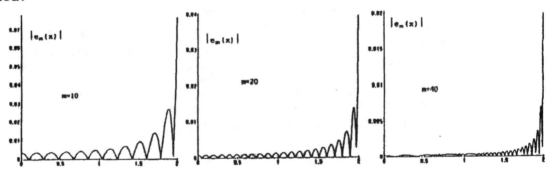

25b. Clearly the maximum error occurs at $x = 2$, so use a computer algebra system to evaluate $e_m(2)$ to show that at least 81 terms are needed for $|e_m(2)| \le .01$.

27a. First we have $\int_{T}^{a+T}g(x)dx = \int_{0}^{a}g(s)ds$ by letting $x = s + T$ in the left integral. Now, if $0 \le a \le T$, then from elementary calculus we know that

$\int_{a}^{a+T}g(x)dx = \int_{a}^{T}g(x)dx + \int_{T}^{a+T}g(x)dx = \int_{a}^{T}g(x)dx + \int_{0}^{a}g(x)dx$

using the equality derived above. This last sum is $\int_{0}^{T}g(x)dx$ and thus we have the desired result.

Section 10.3, Page 629

2a. Substituting for f(x) in Eqs.(2) and (3) with L = π
yields $a_0 = (1/\pi)\int_0^\pi x dx = \pi/2$;

$a_m = (1/\pi)\int_0^\pi x\cos mx dx = (\cos m\pi - 1)/\pi m^2 = 0$ for m even and

$= -2/\pi m^2$ for m odd; and

$b_m = (1/\pi)\int_0^\pi x\sin mx dx = -(\pi\cos m\pi)/m\pi = (-1)^{m+1}/m$,

m = 1,2... . Substituting these values into Eq.(1) with
L = π yields the desired solution.

2b. The function to which the series converges is indicated
in the figure and is periodic with period 2π. Note that

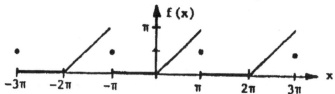

the Fourier series converges to π/2 at x = -π, π, etc.,
even though the function is defined to be zero there.
This value (π/2) represents the mean value of the left
and right hand limits at those points. In (-π, 0),
f(x) = 0 and f'(x) = 0 so both f and f' are continuous and
have finite limits as x → -π from the right and as
x → 0 from the left. In (0, π), f(x) = x and f'(x) = 1
and again both f and f' are continuous and have limits as
x → 0 from the right and as x → π from the left. Since
f and f' are piecewise continuous on [-π, π] the
conditions of the Fourier theorem are satisfied.

4a. Substituting for f(x) in Eqs.(2) and (3), with L = 1
yields $a_0 = \int_{-1}^1 (1-x^2)dx = 4/3$;

$a_n = \int_{-1}^1 (1-x^2)\cos n\pi x dx = (2/n\pi)\int_{-1}^1 x\sin n\pi x dx$

$= (-2/n^2\pi^2)[x\cos n\pi x\Big|_{-1}^1 - \int_{-1}^1 \cos n\pi x dx]$

$= 4(-1)^{n+1}/n^2\pi^2$; and

$b_n = \int_{-1}^1 (1-x^2)\sin n\pi x dx = 0$. Substituting these values

into Eq.(1) gives the desired series.

4b. The function to which the series converges is shown in
the figure and is periodic of fundamental period 2. In
[-1,1] f(x) and f'(x) = -2x are both continuous and have
finite limits as the endpoints of the interval are

approached from within the interval.

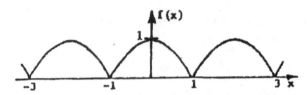

7a. As in Prob. 15, Sect. 10.2, we have

$$f(x) = -\frac{\pi}{4} + \sum_{n=1}^{\infty} [\frac{2\cos(2n-1)x}{\pi(2n-1)^2} + \frac{(-1)^{n+1}\sin nx}{n}].$$

7b. $e_n(x) = f(x) + \frac{\pi}{4} - \sum_{k=1}^{n} [\frac{2\cos(2k-1)x}{\pi(2k-1)^2} + \frac{(-1)^{k+1}\sin kx}{k}].$

Using a computer algebra system, we find that for
n = 5, 10 and 20 the maximum error occurs at $x = -\pi$ in
each case and is 1.6025, 1.5867 and 1.5787 respectively.
Note that the author's n values are 10, 20 and 40, since
he has included the zero cosine coefficient terms and the
sine terms are all zero at $x = -\pi$.

7c. It's not possible in this case, due to Gibb's phenomenon,
to satisfy $|e_n(x)| \le 0.01$ for all x.

12a. $a_0 = \int_{-1}^{1}(x-x^3)dx = 0$ and $a_n = \int_{-1}^{1}(x-x^3)\cos n\pi x dx = 0$ since
$(x-x^3)$ and $(x-x^3)\cos n\pi x$ are odd functions.
$b_n = \int_{-1}^{1}(x-x^3)\sin n\pi x dx$

$$= [\frac{x^3}{n\pi}\cos n\pi x - \frac{3x^2}{n^2\pi^2}\sin n\pi x - \frac{(n^2\pi^2+6)}{n^3\pi^3}x\cos n\pi x + \frac{(n^2\pi^2+6)}{n^4\pi^4}\sin n\pi x]_{-1}^{1}$$

$$= \frac{-12}{n^3\pi^3}\cos n\pi, \text{ so } f(x) = -\frac{12}{\pi^3}\sum_{n=1}^{\infty}\frac{(-1)^n}{n^3}\sin n\pi x.$$

12b. $e_n(x) = f(x) + \frac{12}{\pi^3}\sum_{k=1}^{n}\frac{(-1)^k}{k^3}\sin k\pi x.$ These errors will be

much smaller than in the earlier problems due to the n^3
factor in the denominator. Convergence is much more
rapid in this case.

14. The solution to the corresponding homogeneous equation is
found by methods presented in Sect. 3.3 and is
$y(t) = c_1\cos\omega t + c_2\sin\omega t.$ For the nonhomogeneous terms
we use the method of superposition and consider the
sequence of equations $y_n'' + \omega^2 y_n = b_n\sin nt$ for
n = 1,2,3... . If $\omega > 0$ is not equal to an integer,

then the particular solution to this last equation has the form $Y_n = a_n \cos nt + d_n \sin nt$, as previously discussed in Sect. 3.5. Substituting this form for Y_n into the equation and solving, we find $a_n = 0$ and $d_n = b_n/(\omega^2 - n^2)$. Thus the formal general solution of the original nonhomogeneous D.E. is

$$y(t) = c_1 \cos \omega t + c_2 \sin \omega t + \sum_{n=1}^{\infty} b_n (\sin nt)/(\omega^2 - n^2), \text{ where we}$$

have superimposed all the Y_n terms to obtain the infinite sum. To evalute c_1 and c_2 we set $t = 0$ to obtain

$$y(0) = c_1 = 0 \text{ and } y'(0) = \omega c_2 + \sum_{n=1}^{\infty} nb_n/(\omega^2 - n^2) = 0 \text{ where}$$

we have formally differentiated the infinite series term by term and evaluated it at $t = 0$. (Differentiation of a Fourier Series has not been justified yet and thus we can only consider this method a formal solution at this

point). Thus $c_2 = -(1/\omega) \sum_{n=1}^{\infty} nb_n/(\omega^2 - n^2)$, which when

substituted into the above series yields the desired solution.

If $\omega = m$, a positive integer, then the particular solution of $y_m'' + m^2 y_m = b_m \sin mt$ has the form $Y_m = t(a_m \cos mt + d_m \sin mt)$ since $\sin mt$ is a solution of the related homogeneous D.E.[all other particular solutions are the same as previously, with $\omega = m$]. Substituting Y_m into the D.E. yields $a_m = -b_m/2m$ and $d_m = 0$ and thus the general solution of the D.E. (when $\omega = m$) is now

$$y(t) = c_1 \cos mt + c_2 \sin mt - b_m t(\cos mt)/2m + \sum_{n=1, n \neq m}^{\infty} b_n (\sin nt)/(m^2 - n^2).$$

To evaluate c_1 and c_2 we set $y(0) = 0 = c_1$ and

$$y'(0) = c_2 m - b_m/2m + \sum_{n=1, n \neq m}^{\infty} b_n n/(m^2 - n^2) = 0. \text{ Thus}$$

$$c_2 = b_m/2m^2 - \sum_{n=1, n \neq m}^{\infty} b_n n/m(m^2 - n^2), \text{ which when substituted}$$

into the equation for $y(t)$ yields the desired solution.

15. In order to use the results of Prob. 14, we must first find the Fourier series for the given $f(t)$. From Eqs.(2) and (3) with $L = \pi$, we find that [Using the result of

Prob. 27, Sect 10.2] $a_0 = (1/\pi)\int_0^\pi dx - (1/\pi)\int_\pi^{2\pi}dx = 0$;

$a_n = (1/\pi)\int_0^\pi \cos nx\, dx - (1/\pi)\int_\pi^{2\pi}\cos nx\, dx = 0$; and

$b_n = (1/\pi)\int_0^\pi \sin nx\, dx - (1/\pi)\int_\pi^{2\pi}\sin nx\, dx = 0$ for n even and
$= 4/n\pi$ for n odd. Note that these values for a_n and
b_n are the same as those in Prob.1. Thus

$$f(t) = (4/\pi)\sum_{n=1}^{\infty}\sin(2n-1)t/(2n-1).$$ Comparing this to the

forcing function of Prob. 14 we see that b_n of Prob. 14
has the specific values $b_{2n} = 0$ and $b_{2n-1} = (4/\pi)/(2n-1)$
in this example. Substituting these into the answer to
Prob. 14 yields the desired solution. Note that we have
asumed ω is not a positive integer. Note also, that if
the solution to Prob. 14 is not available, the procedure
for solving this problem would be exactly the same as
shown in Prob. 14.

16. From Prob. 8, the Fourier series for f(t) is given by

$$f(t) = 1/2 + (4/\pi^2)\sum_{n=1}^{\infty}\cos(2n-1)\pi t/(2n-1)^2$$ and thus we may

not use the form of the answer in Prob. 14. The procedure
outlined for solving Prob.14, however, is applicable and
will yield the desired solution for this problem.

18a. We will assume f(x) is continuous for this part. For the
case where f(x) has jump discontinuities, a more detailed
proof can be developed, as shown in Part(b). From Eq.(3)
we have $b_n = \dfrac{1}{L}\int_{-L}^{L}f(x)\sin\dfrac{n\pi x}{L}dx$. If we let u = f(x) and dv

$= \sin\dfrac{n\pi x}{L}dx$, then du = f'(x)dx and v $= \dfrac{-L}{n\pi}\cos\dfrac{n\pi x}{L}$. Thus

$b_n = \dfrac{1}{L}[\dfrac{-L}{n\pi}f(x)\cos\dfrac{n\pi x}{L}\Big|_{-L}^{L} + \dfrac{L}{n\pi}\int_{-L}^{L}f'(x)\cos\dfrac{n\pi x}{L}dx]$

$\quad = -\dfrac{1}{n\pi}[f(L)\cos n\pi - f(-L)\cos(-n\pi)] + \dfrac{1}{n\pi}\int_{-L}^{L}f'(x)\cos\dfrac{n\pi x}{L}dx$

$\quad = \dfrac{1}{n\pi}\int_{-L}^{L}f'(x)\cos\dfrac{n\pi x}{L}dx$, since f(L) = f(-L) and

$\cos(-n\pi) = \cos n\pi$. Hence $nb_n = \dfrac{1}{\pi}\int_{-L}^{L}f'(x)\cos\dfrac{n\pi x}{L}dx$, which

exists for all n since f'(x) is piecewise continuous.
Thus nb_n is bounded as $n \to \infty$. Likewise, for a_n, we

obtain $na_n = -\dfrac{1}{\pi}\displaystyle\int_{-L}^{L} f'(x)\sin\dfrac{n\pi x}{L}dx$ and hence na_n is also bounded as $n \to \infty$.

18b. Note that f and f' are continuous at all points where f'' is continuous. Let $x_1, \ldots, x_m$ be the points in $(-L,L)$ where f'' is not continuous. By splitting up the interval of integration at these points, and integrating Eq.(3) by parts twice, we obtain

$$n^2 b_n = \frac{n}{\pi}\sum_{i=1}^{m}[f(x_i+)-f(x_i-)]\cos\frac{n\pi x_i}{L} - \frac{n}{\pi}[f(L-)-f(-L+)]\cos n\pi$$

$$-\frac{L}{\pi^2}\sum_{i=1}^{m}[f'(x_i+)-f'(x_1-)]\sin\frac{n\pi x_i}{L} - \frac{L}{\pi^2}\int_{-L}^{L}f''(x)\sin\frac{n\pi x}{L}dx, \text{ where}$$

we have used the fact that cosine is continuous. We want the first two terms on the right side to be zero, for otherwise they grow in magnitude with n. Hence f must be continuous throughout the closed interval $[-L,L]$. The last two terms are bounded, by the hypotheses on f' and f''. Hence $n^2 b_n$ is bounded as $n\to\infty$; similarly $n^2 a_n$ is bounded as $n\to\infty$.

If $f(x)$ is continuous on $[-L,L]$ and $f(-L) = f(L)$ [since f is periodic of period $2L$], then the periodic extension of $f(x)$ is continuous for all x. Thus there will be no jump discontinuities, as might be the case in Part(a).

18c. If $n^2 a_n$ is bounded as as $n\to\infty$ then $|a_n| \le \dfrac{M}{n^2}$ as $n\to\infty$ and

hence $\displaystyle\sum_{n=1}^{\infty}|a_n|$ converges by the comparison test using $\displaystyle\sum_{n=1}^{\infty}\dfrac{1}{n^2}$.

18d. Since $|a_n\cos(n\pi x/L) + b_n\sin(n\pi x/L)| \le |a_n| + |b_n|$ and

since $\displaystyle\sum_{n=1}^{\infty}|a_n|$ and $\displaystyle\sum_{n=1}^{\infty}|b_n|$ converge from Part(c), we conclude that the Fourier Series (4) converges absolutely.

Section 10.4, Page 637

3. Let $f(x) = \tan 2x$, then
$$f(-x) = \tan(-2x) = \frac{\sin(-2x)}{\cos(-2x)} = \frac{-\sin(2x)}{\cos(2x)} = -\tan 2x = -f(x)$$
and thus $\tan 2x$ is an odd function.

6. Let $f(x) = e^{-x}$, then $f(-x) = e^x$ so that $f(-x) \neq f(x)$ and
    $f(-x) \neq -f(x)$ and thus e^{-x} is neither even nor odd.

7.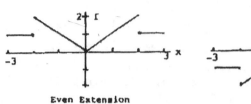

    Even Extension   Odd Extension

10.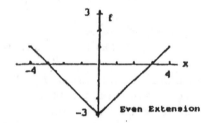

    Even Extension   Odd Extension

13. By the hint $f(-x) = g(-x) + h(-x) = g(x) - h(x)$, since g
    is an even function and f is an odd function. Thus
    $f(x) + f(-x) = 2g(x)$ and hence $g(x) = [f(x) + f(-x)]/2$
    defines $g(x)$. Likewise
    $f(x)-f(-x) = g(x)-g(-x) + h(x)-h(-x) = 2h(x)$ and thus
    $h(x) = [f(x) - f(-x)]/2$.

All functions and their derivatives in Problems 14 through 30
are piecewise continuous on the given intervals and their
extensions. Thus the Fourier Theorem applies in all cases.

14. For the cosine series we use the even extension of the
    function given in Eq.(13) and hence
    $$f(x) = \begin{cases} 0 & -2 \leq x < -1 \\ 1+x & -1 \leq x < 0 \end{cases} \text{ on the interval } -2 \leq x < 0.$$
    However, we don't really need this, as the coefficients
    in this case are given by Eqs.(7), which just use the
    original values for $f(x)$ on $0 < x \leq 2$. Applying Eqs.(7)
    we have $L = 2$ and thus
    $a_0 = (2/2)\int_0^1 (1-x)dx + (2/2)\int_1^2 0dx = 1/2$. Similarly,
    $a_n = (2/2)\int_0^1 (1-x)\cos(n\pi x/2)dx = 4[1-\cos(n\pi/2)]/n^2\pi^2$ and
    $b_n = 0$. Substituting these values in the Fourier series
    yields the desired results.
    For the sine series, we use Eqs.(8) with $L = 2$. Thus
    $a_n = 0$ and
    $b_n = (2/2)\int_0^1 (1-x)\sin(n\pi x/2)dx = 4[n\pi/2 - \sin(n\pi/2)]/n^2\pi^2$.

15a. Using Eqs.(7) with L = 2 we have

$a_0 = \int_0^1 dx = 1$ and $a_n = \int_0^1 \cos(n\pi x/2)dx = 2\sin(n\pi/2)/n\pi$.

Thus $a_n = 0$ for n even, $a_n = 2/n\pi$ for n = 1,5,9,...and $a_n = -2/n\pi$ for n = 3,7,11,... . Hence we may write $a_{2n} = 0$ and $a_{2n-1} = 2(-1)^{n+1}/(2n-1)\pi$, which when substituted into the series gives the desired answer. This is the same series as in Prob. 5 of Sect. 10.3.

15b.

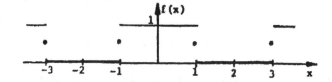

18a. Since we want a sine series, we use Eqs.(8) to find, with $L = \pi$, that $b_n = (2/\pi)\int_0^\pi \sin nx\,dx = 2[1-(-1)^n]/n\pi$ and thus $b_n = 0$ for n even and $b_n = 4/n\pi$ for n odd.

18b.

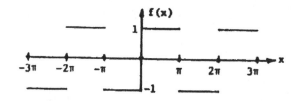

20a. We note that f(x) is specified over its entire fundamental period (T = 1) and hence we cannot extend f to make it either an odd or an even function. Using Eqs.(2) and (3) from Sect. 10.3 we have (L = 1/2)

$a_0 = 2\int_0^1 x\,dx = 1$, $a_n = 2\int_0^1 x\cos(2n\pi x)dx = 0$ and

$b_n = 2\int_0^1 x\sin(2n\pi x)dx = -1/n\pi$. [Note: We have used the results of Prob. 27 of Sect. 10.2 in writing these integrals. That is, if f(x) is periodic of period T, then every integral of f over an interval of length T has the same value. Thus we integrate from 0 to 1 here, rather than -1/2 to 1/2.] Substituting the above values into Eq.(1) of Sect. 10.3 yields the desired solution. It can also be observed from the above graph that g(x) = f(x) - 1/2 is an odd function. If Eqs.(8) are used with g(x), then it is found that

$$g(x) = (-1/\pi) \sum_{n=1}^{\infty} \sin(2n\pi x)/n$$ and thus we obtain the same

series for f(x) as found above.

20b.

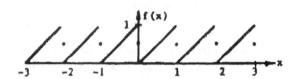

25a. $b_n = \dfrac{2}{2}\displaystyle\int_0^2 (2-x^2)\sin\dfrac{n\pi x}{2}dx$

$= [\dfrac{-2}{n\pi}(2-x^2)\cos\dfrac{n\pi x}{2} - \dfrac{8x}{n^2\pi^2}\sin\dfrac{n\pi x}{2} - \dfrac{16}{n^3\pi^3}\cos\dfrac{n\pi x}{2}]\Big|_0^2$

$= \dfrac{4}{n\pi}(1+\cos n\pi) + \dfrac{16}{n^3\pi^3}(1-\cos n\pi)$ and thus

$f(x) = \displaystyle\sum_{n=1}^{\infty}(\dfrac{4n^2\pi^2(1+\cos n\pi) + 16(1-\cos n\pi)}{n^3\pi^3})\sin\dfrac{n\pi x}{2}$

25b.

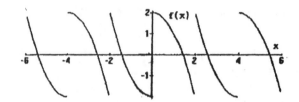

25c.

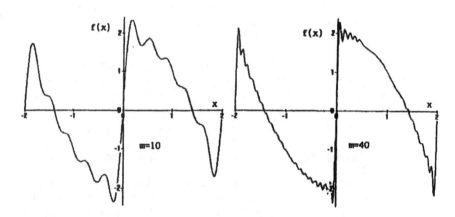

m=10

m=40

28b. For the cosine series (even extension) we have

$a_0 = \dfrac{2}{2}\displaystyle\int_0^1 xdx = \dfrac{1}{2}$

$a_n = \dfrac{2}{2}\displaystyle\int_0^1 x\cos\dfrac{n\pi x}{2}dx = [\dfrac{2x}{n\pi}\sin\dfrac{n\pi x}{2} + \dfrac{4}{n^2\pi^2}\cos\dfrac{n\pi x}{2}]\Big|_0^1$

$= \dfrac{2}{n\pi}\sin\dfrac{n\pi}{2} + \dfrac{4}{n^2\pi^2}\cos\dfrac{n\pi}{2} - \dfrac{4}{n^2\pi^2}$, so

$$g(x) = \frac{1}{4} + \sum_{n=1}^{\infty} \frac{4\cos(n\pi/2) + 2n\pi\sin(n\pi/2) - 4}{n^2\pi^2} \cos\frac{n\pi x}{2}.$$

For the sine series (odd extension) we have

$$b_n = \frac{2}{2}\int_0^1 x\sin\frac{n\pi x}{2}dx = [\frac{-2x}{n\pi}\cos\frac{n\pi x}{2} + \frac{4}{n^2\pi^2}\sin\frac{n\pi x}{2}]\Big|_0^1$$

$$= \frac{-2}{n\pi}\cos\frac{n\pi}{2} + \frac{4}{n^2\pi^2}\sin\frac{n\pi}{2}, \text{ so}$$

$$h(x) = \sum_{n=1}^{\infty} \frac{4\sin(n\pi/2) - 2n\pi\cos(n\pi/2)}{n^2\pi^2}\sin\frac{n\pi x}{2}.$$

28c.

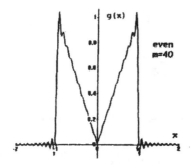 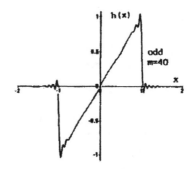

28d. The maximum error does not approach zero in either case, due to Gibb's phenomenon. Note that the coefficients in both series behave like 1/n as $n \to \infty$ since there is an n in the numerator.

31. We have $\int_{-L}^{L} f(x)dx = \int_{-L}^{0} f(x)dx + \int_0^L f(x)dx$. Now, if we let x = -y in the first integral on the right, then

$$\int_{-L}^0 f(x)dx = \int_L^0 f(-y)(-dy) = \int_0^L f(-y)dy = -\int_0^L f(y)dy. \text{ Thus}$$

$$\int_{-L}^L f(x)dx = -\int_0^L f(y)dy + \int_0^L f(x)dx = 0.$$

32. To prove property 2 let f_1 and f_2 be odd functions and let $f(x) = f_1(x) \pm f_2(x)$. Then $f(-x) = f_1(-x) \pm f_2(-x) = -f_1(x) \pm [-f_2(x)] = -[f_1(x) \pm f_2(x)] = -f(x)$, so $f(x)$ is odd. Now let $g(x) = f_1(x)f_2(x)$, then $g(-x) = f_1(-x)f_2(-x) = [-f_1(x)][-f_2(x)] = f_1(x)f_2(x) = g(x)$ and thus $g(x)$ is even. Finally, let $h(x) = f_1(x)/f_2(x)$ and hence $h(-x) = f_1(-x)/f_2(-x) = [-f_1(x)]/[-f_2(x)] = f_1(x)/f_2(x) = h(x)$, which says $h(x)$ is also even. Property 3 is proven in a similar manner.

34. Since $F(x) = \int_0^x f(t)dt$ we have

$$F(-x) = \int_0^{-x} f(t)dt = -\int_0^x f(-s)ds \text{ by letting } t = -s. \text{ If } f$$

is an even function, $f(-s) = f(s)$, we then have
$F(-x) = -\int_0^x f(s)ds = -F(x)$ from the original definition of
F. Thus $F(x)$ is an odd function. The argument is
similar if f is odd.

35. Set $x = L/2$ in Eq.(6) of Sect. 10.3. Since we know f is
continuous at $L/2$, we may conclude, by the Fourier
theorem, that the series will converge to
$f(L/2) = L$ at this point. Thus we have

$L = L/2 + (2L/\pi) \sum_{n=1}^{\infty} (-1)^{n+1}/(2n-1)$, since

$\sin[(2n-1)\pi/2] = (-1)^{n+1}$. Dividing by L and simplifying

yields $\frac{\pi}{4} = \sum_{n=1}^{\infty} \frac{(-1)^{n+1}}{(2n-1)} = \sum_{n=0}^{\infty} \frac{(-1)^n}{2n+1}$, by changing the

summation index.

37a. Multiplying both sides of the equation by $f(x)$ and
integrating from 0 to L gives

$$\int_0^L [f(x)]^2 dx = \int_0^L [f(x) \sum_{n=1}^{\infty} b_n \sin(n\pi x/L)]dx$$

$$= \sum_{n=1}^{\infty} b_n \int_0^L f(x)\sin(n\pi x/L)dx = (L/2) \sum_{n=1}^{\infty} b_n^2, \text{ by Eq.(8).}$$

This result is identical to that of Prob. 17 of Sect. 10.3
if we set $a_n = 0$, $n = 0,1,2,\ldots$, since
$\frac{1}{L}\int_{-L}^{L} [f(x)]^2 dx = \frac{2}{L}\int_0^L [f(x)]^2 dx$. In a similar manner, it can

be shown that $(2/L)\int_0^L [f(x)]^2 dx = a_0^2/2 + \sum_{n=1}^{\infty} a_n^2$.

37b. From Eq.(9) we have $b_n = 2L(-1)^{n+1}/n\pi$, so

$\sum_{n=1}^{\infty} b_n^2 = \sum_{n=1}^{\infty} 4L^2/n^2\pi^2 = 4L^2/\pi^2 \sum_{n=1}^{\infty} (1/n^2)$. Also $f(x) = x$, so

$(2/L)\int_0^L [f(x)]^2 dx = (2/L)\int_0^L x^2 dx = 2L^2/3$. From Part(a)

then, $2L^2/3 = 4L^2/\pi^2 \sum_{n=1}^{\infty} (1/n^2)$, or $\pi^2/6 = \sum_{n=1}^{\infty} (1/n^2)$.

38. We assume that the extensions of f and f' are piecewise
continuous on $[-2L,2L]$. Since f is an odd periodic

function of fundamental period 4L it follows from
properties 2 and 3 that $f(x)\cos(n\pi x/2L)$ is odd and
$f(x)\sin(n\pi x/2L)$ is even. Thus the Fourier coefficients
of f are given by Eqs.(8) with L replaced by 2L; that is
$a_n = 0$, $n = 0,1,2,\ldots$ and

$b_n = (2/2L)\int_0^{2L} f(x)\sin(n\pi x/2L)dx$, $n = 1,2,\ldots$. The

Fourier sine series for f is $f(x) = \displaystyle\sum_{n=1}^{\infty} b_n\sin(n\pi x/2L)$.

39. From Prob. 38 we have $b_n = (1/L)\int_0^{2L} f(x)\sin(n\pi x/2L)dx$

$= (1/L)\int_0^{L} f(x)\sin(n\pi x/2L)dx + (1/L)\int_L^{2L} f(2L-x)\sin(n\pi x/2L)dx$

$= (1/L)\int_0^{L} f(x)\sin(n\pi x/2L)dx - (1/L)\int_L^{0} f(s)\sin[n\pi(2L-s)/2L]ds$

$= (1/L)\int_0^{L} f(x)\sin(n\pi x/2L)dx - (1/L)\int_0^{L} f(s)\cos(n\pi)\sin(n\pi s/2L)ds$

and thus $b_n = 0$ for n even and

$b_n = (2/L)\int_0^{L} f(x)\sin(n\pi x/2L)dx$ for n odd. The Fourier

series for f is given in Prob. 38, where the b_n are given
above.

Section 10.5, Page 647

3. We seek solutions of the form $u(x,t) = X(x)T(t)$.
    Substituting into the P.D.E. yields $X''T + X'T' + XT' =$
    $X''T + (X' + X)T' = 0$. Formally dividing by the quantity
    $(X' + X)T$ gives the equation $X''/(X' + X) = -T'/T$ in which
    the variables are separated. Using the same concepts as
    those leading to Eq.(8) we conclude that in order for
    this equation to be valid on the domain of u it is
    necessary that both sides be equal to the same constant
    λ. Hence $X''/(X' + X) = -T'/T = \lambda$ or equivalently,
    $X'' - \lambda(X' + X) = 0$ and $T' + \lambda T = 0$.

5. We seek solutions of the form $u(x,y) = X(x)Y(y)$.
    Substituting into the P.D.E. yields $X''Y + (x+y)XY'' = 0$.
    Formally dividing by XY yields $X''/X + (x + y)Y''/Y = 0$.
    From this equation we see that the presence of the
    independent variable x multiplying the term u_{yy} in the
    original equation leads to the term xY''/Y when we attempt
    to separate the variables. It follows that the argument
    for a separation constant does not carry through and we
    cannot replace the P.D.E. by two O.D.E.

8.   Following the procedures of Eqs.(5) through (8), we set
      $u(x,t) = X(x)T(t)$ in the P.D.E. to obtain $X''T = 4XT'$, or
      $X''X = 4T'/T$, which must be a constant. As stated in the text
      this separation constant must be $-\lambda^2$ (we choose $-\lambda^2$ so that
      when a square root is used later, the symbols are simpler)
      and thus $X'' + \lambda^2X = 0$ and $T' + (\lambda^2/4)T = 0$. Now
      $u(0,t) = X(0)T(t) = 0$, for all $t > 0$, yields $X(0) = 0$, as
      discussed after Eq.(11) and similarly $u(2,t) = X(2)T(t) = 0$,
      for all $t > 0$, implies $X(2) = 0$. The D.E. for X has the
      solution $X(x) = C_1\cos\lambda x + C_2\sin\lambda x$ and $X(0) = 0$ yields $C_1 = 0$.
      Setting $x = 2$ in the remaining form of X yields
      $X(2) = C_2\sin2\lambda = 0$, which has the solutions $2\lambda = n\pi$ or
      $\lambda = n\pi/2$, $n = 1,2,\ldots$. Note that we exclude $n = 0$ since
      then $\lambda = 0$ would yield $X(x) = 0$, which gives the trivial
      solution for $u(x,t)$. Hence $X(x) = \sin(n\pi x/2)$, $n = 1,2,\ldots$.
      Finally, the solution of the D.E. for T yields
      $T(t) = \exp(-\lambda^2 t/4) = \exp(-n^2\pi^2 t/16)$. Thus we have found
      $u_n(x,t) = \exp(-n^2\pi^2 t/16)\sin(n\pi x/2)$, which is Eq.(17) for this
      problem. Setting $t = 0$ in this last expression indicates
      that $u_n(x,0)$ has, for $n = 1,2$ and 4, the same form as the
      terms in $u(x,0)$, the initial condition. Using the principle
      of superposition we know that
      $u(x,t) = c_1u_1(x,t) + c_2u_2(x,t) + c_4u_4(x,t)$ satisfies the
      P.D.E. and the B.C. and hence we let $t = 0$ to obtain
      $u(x,0) = c_1u_1(x,0) + c_2u_2(x,0) + c_4u_4(x,0)$
              $= c_1\sin\pi x/2 + c_2\sin\pi x + c_4\sin2\pi x$. If we choose
      $c_1 = 2$, $c_2 = -1$ and $c_4 = 4$ then $u(x,0)$ here will match
      the given initial condition, and hence substituting these
      values in $u(x,t)$ above then gives the desired solution.

10.   Since the B.C. for this heat conduction problem are
      $u(0,t) = u(40,t) = 0$, $t > 0$, the solution $u(x,t)$ is given
      by Eq.(19) with $\alpha^2 = 1$ cm^2/sec, $L = 40$ cm, and the
      coefficients c_n determined by Eq.(21) with the I.C.

$$u(x,0) = f(x) = \begin{cases} x & 0 \le x < 20 \\ 40 - x & 20 \le x \le 40 \end{cases} . \text{ Thus}$$

$$c_n = \frac{1}{20}\left[\int_0^{20} x\sin\frac{n\pi x}{40}dx + \int_{20}^{40}(40-x)\sin\frac{n\pi x}{40}dx\right] = \frac{160}{n^2\pi^2}\sin\frac{n\pi}{2}.$$

It follows that $u(x,t) = \dfrac{160}{\pi^2}\displaystyle\sum_{n=1}^{\infty}\dfrac{\sin(n\pi/2)}{n^2}e^{-n^2\pi^2 t/1600}\sin\dfrac{n\pi x}{40}.$

15a.

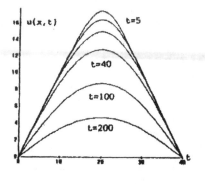

15b.

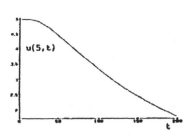

15c.

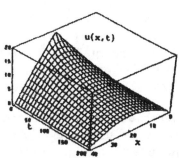

15d. As in Ex. 1, the maximum temperature will be at the midpoint, x = 20. We use just the first term (n=1) of the series found in Prob. 10, since t will be large, so additional terms will be neligible. Thus

$$u(20,t) = 1 = \frac{160}{\pi^2}\sin(\pi/2)e^{-\pi^2 t/1600}\sin(20\pi/40).$$ Solving

for t, we obtain $e^{-\pi^2 t/1600} = \pi^2/160$, or

$$t = \frac{1600}{\pi^2}\ln\frac{160}{\pi^2} = 451.60 \text{ sec. If t is not so large in}$$

other problems, then two or more terms must be used, which would require an equation solver or numerical solution to find t.

18. From the problem statement we conclude that the I.C. is u(x,0) = 100°C and the B.C. are u(0,t) = u(20,t) = 0,

t > 0. The solution $u(x,t)$ is given by Eq.(19) with
L = 20 cm, and the coefficients c_n determined by Eq.(21)
with $f(x) = 100$. Thus
$c_n = (1/10)\int_0^{20}(100)\sin(n\pi x/20)dx = -200[(-1)^n - 1]/n\pi$ and hence
$c_{2n} = 0$ and $c_{2n-1} = 400/(2n-1)\pi$. Substituting these values
into Eq.(19) yields

$$u(x,t) = \frac{400}{\pi}\sum_{n=1}^{\infty}\frac{e^{-(2n-1)^2\pi^2\alpha^2 t/400}}{2n-1}\sin\frac{(2n-1)\pi x}{20}, \text{ which is the}$$

same as the answer in the text since the zero terms have
not been included here.

18b. For aluminum, we have $\alpha^2 = .86$ cm^2/sec (Table 10.5.1) and
thus the first two terms give

$$u(10,30) = \frac{400}{\pi}\left\{e^{-\pi^2(.86)30/400} - \frac{1}{3}e^{-9\pi^2(.86)30/400}\right\}$$

$$= 67.228°C.$$

If an additional term is used, the temperature is
increased by $\dfrac{80}{\pi}e^{-25\pi^2(.86)30/400} = 3\times10^{-6}$ degrees C.

19b. Using only one term in the series for $u(x,t)$ of Prob. 18,
we must solve the equation $(400/\pi)\exp[-\pi^2(.86)t/400] = 5$
for t. Taking the logarithm of both sides and solving
for t yields $t \approx 400\ln(80/\pi)/\pi^2(.86) = 152.56$ sec.

20a. Applying the chain rule to partial differentiation of u
with respect to x we see that $u_x = u_\xi\xi_x = u_\xi(1/L)$ and
$u_{xx} = u_{\xi\xi}(1/L)^2$. Substituting $u_{\xi\xi}/L^2$ for u_{xx} in the heat
equation gives $\alpha^2 u_{\xi\xi}/L^2 = u_t$ or $u_{\xi\xi} = (L^2/\alpha^2)u_t$.

20b. In a similar manner, $u_t = u_\tau\tau_t = u_\tau(\alpha^2/L^2)$ and hence $\dfrac{L^2}{\alpha^2}u_t$

$= u_\tau$ and thus $u_{\xi\xi} = u_\tau$.

22. Substituting $u(x,y,t) = X(x)Y(y)T(t)$ in the P.D.E. yields
$\alpha^2(X''YT + XY''T) = XYT'$. Dividing by $\alpha^2 XYT$ we obtain
$\dfrac{X''}{X} + \dfrac{Y''}{Y} = \dfrac{T'}{\alpha^2 T}$. By keeping the independent variables x
and y fixed and varying t we see that $T'/\alpha^2 T$ must equal
some constant σ_1 since the left side of the equation is
fixed. Hence, $X''/X + Y''/Y = T'/\alpha^2 T = \sigma_1$, or
$X''/X = \sigma_1 - Y''/Y$ and $T' - \sigma_1\alpha^2 T = 0$. By keeping x fixed
and varying y in the equation involving X and Y we see
that $\sigma_1 - Y''/Y$ must equal some constant σ_2 since the left

side of the equation is fixed. Hence,
$X''/X = \sigma_1 - Y''/Y = \sigma_2$ so $X'' - \sigma_2 X = 0$ and $Y'' - (\sigma_1 - \sigma_2)Y = 0$.
For $T' - \sigma_1 \alpha^2 T = 0$ to have solutions that remain bounded as
$t \to \infty$ we must have $\sigma_1 < 0$. Thus, setting $\sigma_1 = -\lambda^2$, we have
$T' + \alpha^2 \lambda^2 T = 0$. For $X'' - \sigma_2 X = 0$ and homogeneous B.C., we
conclude, as in Sect. 10.1, that $\sigma_2 < 0$ and, if we let
$\sigma_2 = -\mu^2$, then $X'' + \mu^2 X = 0$. With these choices for σ_1 and σ_2
we then have $Y'' + (\lambda^2 - \mu^2)Y = 0$.

Section 10.6, Page 657

3. The steady-state temperature distribution $v(x)$ must
    satisfy Eq.(9) and also satsify the B.C. $v_x(0) = 0$,
    $v(L) = 0$. The general solution of $v'' = 0$ is
    $v(x) = Ax + B$. The B.C. $v_x(0) = 0$ implies $A = 0$ and then
    $v(L) = 0$ implies $B = 0$, so the steady state solution is
    $v(x) = 0$.

7. Again, $v(x)$ must satisfy $v'' = 0$, $v'(0) - v(0) = 0$ and $v(L) = T$.
    The general solution of $v'' = 0$ is $v(x) = ax + b$, so $v(0) = b$,
    $v'(0) = a$ and $v(L) = T$. Thus $a - b = 0$ and $aL + b = T$,
    which give $a = b = T/(L+1)$. Hence $v(x) = T(x+1)/(L+1)$.

9a. Since the B.C. are not homogeneous, we must first find
    the steady state solution. Using Eqs.(9) and (10) we have
    $v'' = 0$ with $v(0) = 0$ and $v(20) = 60$, which has the
    solution $v(x) = 3x$. Thus the transient solution $w(x,t)$
    satisfies the equations $\alpha^2 w_{xx} = w_t$, $w(0,t) = 0$,
    $w(20,t) = 0$ and $w(x,0) = 25 - 3x$, which are obtained from
    Eqs.(13) to (15). The solution of this problem is given
    by Eq.(4) with the c_n given by Eq.(6):
    $$c_n = \frac{1}{10}\int_0^{20}(25-3x)\sin\frac{n\pi x}{20}dx = (70\cos n\pi + 50)/n\pi, \text{ and thus}$$

    $$u(x,t) = 3x + \sum_{n=1}^{\infty}\frac{70\cos n\pi + 50}{n\pi}e^{-0.86n^2\pi^2 t/400}\sin\frac{n\pi x}{20} \text{ since}$$

    $\alpha^2 = .86$ for aluminum.

9b.   9c.

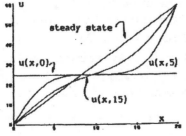
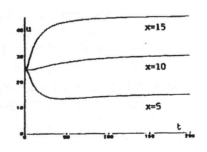

9d. Using just the first term of the sum, we have

$$u(5,t) = 15 - \frac{20}{\pi}e^{-0.86\pi^2 t/400}\sin\frac{\pi}{4} = 15 - .15. \quad \text{Thus}$$

$\frac{20}{\pi}e^{-0.86\pi^2 t/400}\sin\frac{\pi}{4} = .15$, which yields $t = 160.30$ sec.

To obtain the answer in the text, the first two terms of the sum must be used, which requires an equation solver to solve for t. Note that this reduces t by only .01 seconds.

12a. Since the B.C. are $u_x(0,t) = u_x(L,t) = 0$, $t > 0$, the solution $u(x,t)$ is given by Eq.(35) with the coefficients c_n determined by Eq.(37). Substituting the I.C. $u(x,0) = f(x) = \sin(\pi x/L)$ into Eq.(37) yields

$$c_0 = (2/L)\int_0^L \sin(\pi x/L)dx = 4/\pi,$$

$$c_1 = (2/L)\int_0^L \sin(\pi x/L)\cos(\pi x/L)dx = (1/\pi)(\sin(\pi x/L)^2\big|_0^L = 0,$$

$$c_n = (2/L)\int_0^L \sin(\pi x/L)\cos(n\pi x/L)dx$$

$$= (1/L)\int_0^L \{\sin[(n+1)\pi x/L] - \sin[(n-1)\pi x/L]\}dx$$

$$= (1/\pi)\{[1 - \cos(n+1)\pi]/(n+1) - [1 - \cos(n-1)\pi]/(n-1)\},$$

for $n \neq 1$. Thus $c_n = \begin{cases} 0 & n \text{ odd} \\ -4/(n^2-1)\pi & n \text{ even} \end{cases}$. Hence

$$u(x,t) = 2/\pi-(4/\pi)\sum_{n=1}^{\infty}\exp[-4n^2\pi^2\alpha^2 t/L^2]\cos(2n\pi x/L)/(4n^2-1)$$

where we are now summing over even terms by setting $n = 2n$.

12b. As $t \to \infty$ we see that all terms in the series decay to zero except the constant term, $2/\pi$. Hence $\lim\limits_{t\to\infty} u(x,t) = 2/\pi$.

12c.

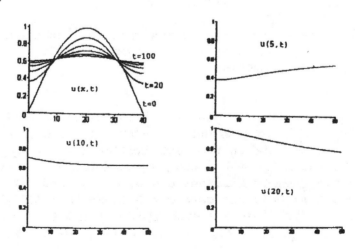

12d. The original heat in the rod is redistributed to give the final temperature distribution, since no heat is lost.

14a. Since the ends are insulated, the solution to this problem is given by Eq.(35) and Eq.(37), with $\alpha^2 = 1$ and $L = 30$. Thus

$$u(x,t) = \frac{c_0}{2} + \sum_{n=1}^{\infty} c_n \exp(-n^2\pi^2 t/900)\cos(n\pi x/30), \text{ where}$$

$$c_0 = \frac{2}{30}\int_0^{30} f(x)dx = \frac{1}{15}\int_5^{10} 25dx = \frac{25}{3} \text{ and}$$

$$c_n = \frac{2}{30}\int_0^{30} f(x)\cos\frac{n\pi x}{30}dx = \frac{1}{15}\int_5^{10} 25\cos\frac{n\pi x}{30}dx = \frac{50}{n\pi}[\sin\frac{n\pi}{3} - \sin\frac{n\pi}{6}].$$

14b.

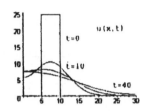
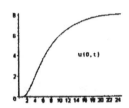
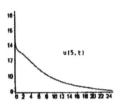
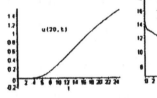
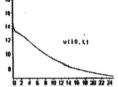

14c.

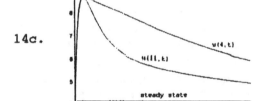

Although $x = 4$ and $x = 11$ are symmetrical to the initial temperature pulse, they are not symmetrical to the insulated end points.

15a. Substituting $u(x,t) = X(x)T(t)$ into Eq.(1) leads to the two O.D.E. $X'' - \sigma X = 0$ and $T' - \alpha^2 \sigma T = 0$. An argument similar to the one in the text implies that we must have $X(0) = 0$ and $X'(L) = 0$. Also, by assuming σ is real and considering the three cases $\sigma < 0$, $\sigma = 0$, and $\sigma > 0$ we can show that only the case $\sigma < 0$ leads to nontrivial solutions of $X'' - \sigma X = 0$ with $X(0) = 0$ and $X'(L) = 0$.

Setting $\sigma = -\lambda^2$, we obtain $X(x) = k_1\sin\lambda x + k_2\cos\lambda x$.
Now, $X(0) = 0 \Rightarrow k_2 = 0$ and thus $X(x) = k_1\sin\lambda x$.
Differentiating and setting $x = L$ yields $\lambda k_1\cos\lambda L = 0$.
Since $\lambda = 0$ and $k_1 = 0$ lead to $u(x,t) \equiv 0$, we must choose
λ so that $\cos\lambda L = 0$, or $\lambda = (2n-1)\pi/2L$, $n = 1,2,3,\ldots$ and
thus $X_n = \sin[(2n-1)\pi x/2L]$. The values for λ imply that
$\sigma = -(2n-1)^2\pi^2/4L^2$ so the solutions $T(t)$ of $T' - \alpha^2\sigma T = 0$
are proportional to $\exp[-(2n-1)^2\pi^2\alpha^2 t/4L^2]$. Combining
the above results leads to the desired set of fundamental
solutions.

15b. In order to satisfy the I.C. $u(x,0) = f(x)$ we assume that
$u(x,t)$ has the form
$$u(x,t) = \sum_{n=1}^{\infty} c_n\exp[-(2n-1)^2\pi^2\alpha^2 t/4L^2]\sin[(2n-1)\pi x/2L].$$ The
coefficients c_n are determined by the requirement that
$$u(x,0) = \sum_{n=1}^{\infty} c_n\sin[(2n-1)\pi x/2L] = f(x).$$ Referring to
Prob. 39 of Sect. 10.4 reveals that such a representation
for $f(x)$ is possible if we choose the coefficients
$c_n = (2/L)\int_0^L f(x)\sin[(2n-1)\pi x/2L]dx$.

19. We must solve $v_1''(x) = 0$, $0 \le x \le a$ and $v_2''(x) = 0$,
$a \le x \le L$ subject to the B.C. $v_1(0) = 0$, $v_2(L) = T$ and
the continuity conditions at $x = a$. For the temperature
to be continuous at $x = a$ we must have $v_1(a) = v_2(a)$ and
for the rate of heat flow to be continuous we must have
$-\kappa_1 A_1 v_1'(a) = -\kappa_2 A_2 v_2'(a)$, from Eq.(2) of Appendix A. The
general solutions to the two O.D.E. are $v_1(x) = C_1 x + D_1$
and $v_2(x) = C_2 x + D_2$. By applying the boundary and
continuity conditions we may solve for C_1, D_1, and C_2 and
D_2 to obtain the desired solution.

21. For the steady state solution we set $u(x,t) = v(x)$ which
gives $u_{xx} = v''$, $u_t = 0$, $u(0,t) = v(0)$, and $u(L,t) = v(L)$.
Substituting these into the original problem gives
$0 = \alpha^2 v'' + s(x)$, $v(0) = T_1$ and $v(L) = T_2$. For the
transient solution we set $u(x,t) = v(x) + w(x,t)$ which
gives $u_{xx} = v'' + w_{xx}$, $u_t = w_t$, $u(0,t) = T_1 + w(0,t) = T_1$,
$u(L,t) = T_2 + w(L,t) = T_2$, and $u(x,0) = v(x) + w(x,0) = f(x)$.
Using these expressions for u in the original equations
yields $w_t = \alpha^2(v'' + w_{xx}) + s(x) = \alpha^2 w_{xx}$, $w(0,t) = 0$,
$w(L,t) = 0$ and $w(x,0) = f(x) - v(x)$.

22a. From Prob. 21 $v'' = -k$ and thus $v(x) = c_1 + c_2 x - kx^2/2$.
$V(0) = T_1 \Rightarrow c_1 = T_1$ and $v(L) = T_2 \Rightarrow c_2 = (T_2 - T_1)/L + kL/2$
and hence $v(x) = T_1 + (T_2 - T_1)x/L + kLx/2 - kx^2/2$.

22b. From Prob. 21 $w(x,t)$ satisfies $w_t = w_{xx}$, $w(0,t) = 0$, $w(20,t) = 0$
and $w(x,0) = -v(x) = (x^2 - 20x)/10$, using the givn values

for T_1, T_2, L and k. Thus $w(x,t) = \sum_{n=1}^{\infty} c_n e^{-n^2\pi^2 t/400} \sin\left(\frac{n\pi x}{20}\right)$,

where $c_n = \dfrac{2}{20}\displaystyle\int_0^{20} \dfrac{x^2-20x}{10}\sin\left(\frac{n\pi x}{20}\right)dx$

$$= \frac{1}{100}\int_0^{20}(x^2-20x)\sin\left(\frac{n\pi x}{20}\right)dx = \begin{cases} 0 & n \text{ even} \\ \dfrac{-320}{n^3\pi^3} & n \text{ odd} \end{cases}.$$

Thus $u(x,t) = v(x) + w(x,t)$

$$= \frac{20x - x^2}{10} - \frac{320}{n^3\pi^3}\sum_{n=1}^{\infty}\frac{e^{-(2n-1)^2\pi^2 t/400}}{(2n-1)^3}\sin\frac{(2n-1)\pi x}{20}.$$

Graphs of $u(x,t)$ are shown for $t = 10,40,100,150$. In all
cases six terms of the series were used.

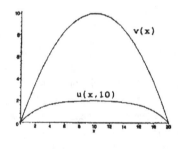

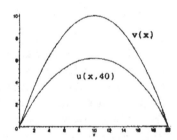

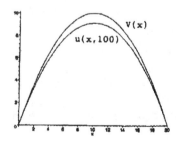

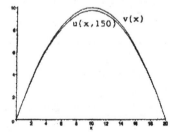

Section 10.7, Page 670

1a. Since the initial velocity is zero, the solution is given
by Eq.(20) with the coefficents c_n given by Eq.(22).
Substituting $f(x)$ into Eq.(22) yields

$$c_n = \frac{2}{L}\left[\int_0^{L/2}\frac{2x}{L}\sin\frac{n\pi x}{L}dx + \int_{L/2}^{L}\frac{2(L-x)}{L}\sin\frac{n\pi x}{L}dx\right]$$

$$= \frac{8}{n^2\pi^2}\sin\frac{n\pi}{2}.\quad\text{Thus Eq. (20) becomes}$$

$$u(x,t) = \frac{8}{\pi^2}\sum_{n=1}^{\infty}\frac{1}{n^2}\sin\frac{n\pi}{2}\sin\frac{n\pi x}{L}\cos\frac{n\pi a t}{L}.$$

1b.

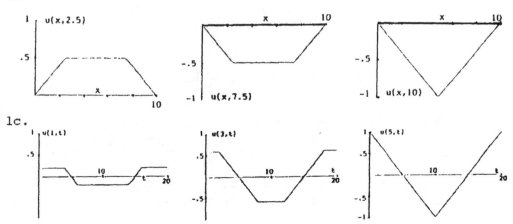

1c.

1e. The graphs in Part(b) can best be understood using
    Eq.(28) (or the results of Probs 13 and 14). The
    original triangular shape is composed of two similar
    triangles of 1/2 the height, one of which moves to the
    right, h(x-at), and the other to the left, h(x+at).
    Recalling that h(x) extends f(x) periodically for all x
    and that h(x+at) is a wave moving left and h(x-at) is a
    wave moving right then gives the results shown. The
    graphs in Part(c) can then be visualized from those in
    Part(b).

6a. The motion is governed by Eqs.(1), (3) and (31), and thus
    the solution is given by Eq.(34) where the k_n are given
    by Eq.(36):

$$k_n = \frac{2}{n\pi a}\left[\int_0^{L/4}\frac{4x}{L}\sin\frac{n\pi x}{L}dx + \int_{L/4}^{3L/4}\sin\frac{n\pi x}{L}dx + \int_{3L}^{L}\frac{4(L-x)}{L}\sin\frac{n\pi x}{L}dx\right]$$

$$= \frac{8L}{n^3\pi^3 a}\left(\sin\frac{n\pi}{4} + \sin\frac{3n\pi}{4}\right).\quad\text{Substituting this in Eq.(34)}$$

$$\text{yields } u(x,t) = \frac{8L}{a\pi^3}\sum_{n=1}^{\infty}\frac{\sin\frac{n\pi}{4} + \sin\frac{3n\pi}{4}}{n^3}\sin\frac{n\pi x}{L}\sin\frac{n\pi a t}{L}.$$

6b.

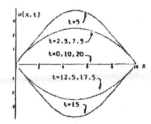

6c.

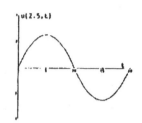 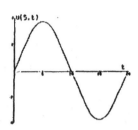

6e. Note that again this is a vibrating system. Of particular
importance, however, is that the solution does not have
any 'corners', as does the initial velocity. This is due
to the integration of the initial velocity, Eq.(36),
which improves the convergence of the series and
consequently gives a smoothing effect.

9. Assuming that $u(x,t) = X(x)T(t)$ and substituting for u in
Eq.(1) leads to the pair of O.D.E. $X'' + \sigma X = 0$,
$T'' + a^2\sigma T = 0$. Applying the B.C. $u(0,t) = 0$ and
$u_x(L,t) = 0$ to $u(x,t)$ we see that we must have $X(0) = 0$ and
$X'(L) = 0$. By considering the three cases $\sigma < 0$, $\sigma = 0$, and
$\sigma > 0$ it can be shown that nontrivial solutions of the
problem $X'' + \sigma X = 0$, $X(0) = 0$, $X'(L) = 0$ are possible if and
only if $\sigma = (2n-1)^2\pi^2/4L^2$, $n = 1,2,\dots$ and the corresponding
solutions for $X(x)$ are proportional to $\sin[(2n-1)\pi x/2L]$.
Using these values for σ we find that $T(t)$ is a linear
combination of $\sin[(2n-1)\pi at/2L]$ and $\cos[(2n-1)\pi at/2L]$. Now,
the I.C. $u_t(x,0)$ implies that $T'(0) = 0$ and thus functions of
the form
$u_n(x,t) = \sin[(2n-1)\pi x/2L]\cos[(2n-1)\pi at/2L]$, $n = 1,2,\dots$
satsify the P.D.E. (1), the B.C. $u(0,t) = 0$, $u_x(L,t) = 0$,
and the I.C. $u_t(x,0) = 0$. We now seek a superposition of
these fundamental solutions u_n that also satisfies the
I.C. $u(x,0) = f(x)$. Thus we assume that
$$u(x,t) = \sum_{n=1}^{\infty} c_n\sin[(2n-1)\pi x/2L]\cos[(2n-1)\pi at/2L]. \quad \text{The}$$
I.C. now implies that we must have

$$f(x) = \sum_{n=1}^{\infty} c_n \sin[(2n-1)\pi x/2L].$$ From Prob. 39 of Sect.10.4

we see that $f(x)$ can be represented by such a series and that $c_n = (2/L)\int_0^L f(x)\sin[(2n-1)\pi x/2L]dx$, $n = 1,2,\ldots$. Substituting these values into the above series for $u(x,t)$ yields the desired solution.

10a. From Prob. 9 we have

$$c_n = \frac{2}{L}\int_{(L-2)/2}^{(L+2)/2} \sin\frac{(2n-1)\pi x}{2L}dx$$

$$= \frac{-4}{(2n-1)\pi}\left[\cos(\frac{(2n-1)\pi(L+2)}{4L}) - \cos(\frac{(2n-1)\pi(L-2)}{4L})\right]$$

$$= \frac{8}{(2n-1)\pi}\sin\frac{(2n-1)\pi}{4}\sin\frac{(2n-1)\pi}{2L} \text{ using the}$$

trigonometric relations for $\cos(A \pm B)$. Substituting this value of c_n into $u(x,t)$ in Prob. 9 yields the desired solution.

10b.

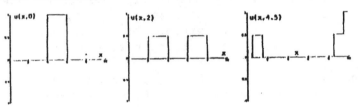

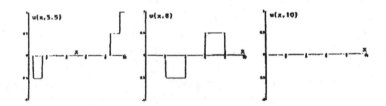

10c.

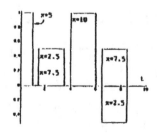

10e. Again visualize the original displacement $f(x)$ as an odd function on $-L < x < 0$ and then periodically for all x. The resulting motion is the result of waves moving left and right from this extension. Note that the original 'corners' persist for all time.

13a. Using the chain rule we obtain $u_x = u_\xi \xi_x + u_\eta \eta_x =$
$u_\xi + u_\eta$ since $\xi_x = \eta_x = 1$. Differentiating a second time
gives $u_{xx} = u_{\xi\xi} + 2u_{\xi\eta} + u_{\eta\eta}$. In a similar way we obtain
$u_t = u_\xi \xi_t + u_\eta \eta_t = -au_\xi + au_\eta$, since $\xi_t = -a$, $\eta_t = a$. Thus
$u_{tt} = a^2(u_{\xi\xi} - 2u_{\xi\eta} + u_{\eta\eta})$. Substituting for u_{xx} and u_{tt}
in the wave equation, we obtain $u_{\xi\eta} = 0$.

13b. Integrating both sides of $u_{\xi\eta} = 0$ with respect to η
yields $u_\xi(\xi,\eta) = \gamma(\xi)$ where γ is an arbitrary function of
ξ. Integrating both sides of $u_\xi(\xi,\eta) = \gamma(\xi)$ with respect
to ξ yields $u(\xi,\eta) = \int\gamma(\xi)d\xi + \psi(\eta) = \phi(\xi) + \psi(\eta)$ where
$\psi(\eta)$ is an arbitrary function of η and $\int\gamma(\xi)d\xi$ is another
arbitrary function of ξ denoted by $\phi(\xi)$. Thus
$u(x,t) = u(\xi(x,t),\eta(x,t)) = \phi(x - at) + \psi(x + at)$.

14a. The graph of $y = \sin(x-at)$ for the various values of t is
indicated in the figure. Note that the graph of $y = \sin x$
is displaced to the right by the distance "at" for each
value of t

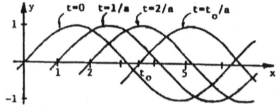

14b. Similarly, the graph of $y = \phi(x + at)$ would be displaced
to the left by a distance "at" for each t. Thus
$\phi(x + at)$ represents a wave moving to the left.

16a. From Prob. 13 we have $u(x,t) = \phi(x-at) + \psi(x+at)$ and
thus $u_t(x,t) = -a\phi'(x-at) + a\psi'(x+at)$. Hence
$u(x,0) = \phi(x) + \psi(x) = f(x)$ and
$u_t(x,0) = -a\phi'(x) + a\psi'(x) = 0$. Dividing the last
equation by a yields the desired result.

16b. Using the hint and the first equation obtained in Part(a)
leads to $\phi(x) + \psi(x) = 2\phi(x) + c = f(x)$ so
$\phi(x) = (1/2)f(x) - c/2$ and $\psi(x) = (1/2)f(x) + c/2$. Hence
$u(x,t) = \phi(x - at) + \psi(x + at)$
$= (1/2)[f(x - at) - c] + (1/2)[f(x + at) + c]$
$= (1/2)[f(x - at) + f(x + at)]$.

16c. For the second part, substitute $x + at$ for x in $f(x)$ to
obtain $f(x + at) = \begin{cases} 2 & -1 < x + at < 1 \\ 0 & \text{otherwise} \end{cases}$.
Subtracting "at" from both sides of the inequality then

$$\text{yields } f(x + at) = \begin{cases} 2 & -1-at < x < 1+at \\ 0 & \text{otherwise} \end{cases}.$$

17a. As in Prob. 16a, we have $u(x,0) = \phi(x) + \psi(x) = 0$ and $u_t(x,0) = -a\phi'(x) + a\psi'(x) = g(x)$.

17b. The first equation gives $\psi(x) = -\phi(x)$ and the second equation gives $-2a\phi'(x) = g(x)$. Integrating we have

$$\phi(x) - \phi(x_0) = \frac{-1}{2a}\int_{x_0}^{x} g(\xi)d\xi \text{ which then gives}$$

$$\psi(x) = \frac{1}{2a}\int_{x_0}^{x} g(\xi)d\xi - \phi(x_0).$$

17c. $u(x,t) = \phi(x-at) + \psi(x+at)$

$$= -(1/2a)\int_{x_0}^{x-at} g(\xi)d\xi + \phi(x_0) + (1/2a)\int_{x_0}^{x+at} g(\xi)d\xi - \phi(x_0)$$

$$= (1/2a)[\int_{x_0}^{x+at} g(\xi)d\xi - \int_{x_0}^{x-at} g(\xi)d\xi]$$

$$= (1/2a)[\int_{x_0}^{x+at} g(\xi)d\xi + \int_{x-at}^{x_0} g(\xi)d\xi]$$

$$= (1/2a\int_{x-at}^{x+at} g(\xi)d\xi.$$

21. Substituting $u(r,\theta,t) = R(r)\Theta(\theta)T(t)$ into the P.D.E. yields $R''\Theta T + R'\Theta T/r + R\Theta''T/r^2 = R\Theta T''/a^2$ or equivalently $R''/R + R'/rR + \Theta''/\Theta r^2 = T''/a^2 T$. In order for this equation to be valid for $0 < r < r_0$, $0 \le \theta \le 2\pi$, $t > 0$, it is necessary that both sides of the equation be equal to the same constant $-\sigma$. Otherwise, by keeping r and θ fixed and varying t, one side would remain unchanged while the other side varied. Thus we arrive at the two equations $T'' + \sigma a^2 T = 0$ and $r^2 R''/R + rR'/R + \sigma r^2 = -\Theta''/\Theta$, where we have multiplied both sides by r^2. By an argument similar to the one above we conclude that both sides of the last equation must be equal to the same constant δ. This leads to the two equations $r^2 R'' + rR' + (\sigma r^2 - \delta)R = 0$ and $\Theta'' + \delta\Theta = 0$. Since the circular membrane is continuous, we must have $\Theta(2\pi) = \Theta(0)$, which requires $\delta = \mu^2$, μ a non-negative integer. The condition $\Theta(2\pi) = \Theta(0)$ is also known as the periodicity condition. Since we also desire solutions which vary periodically in time, it is clear that the separation constant σ should be positive, $\sigma = \lambda^2$. Thus we arrive at the three equations $r^2 R'' + rR' + (\lambda^2 r^2 - \mu^2)R = 0$, $\Theta'' + \mu^2\Theta = 0$, and $T'' + \lambda^2 a^2 T = 0$.

23a. Substituting $u(x,t) = X(x)T(t)$ into the PDE and
separating variables we obtain $\dfrac{T''}{a^2 T} + \gamma^2 = \dfrac{X''}{X} = \sigma$. The
separation constant $\sigma = -\lambda^2$, using the same arguements as
seen earlier. Thus $X'' + \lambda^2 X = 0$, $X(0) = X(L) = 0$ and
$T'' + (\gamma^2 + \lambda^2)a^2 T = 0$, $T'(0) = 0$. The BVP has the
solution $\lambda = n\pi/L$ and $X(x) = \sin(n\pi x/L)$, while the IVP has
the solution $\cos\beta t$, where $\beta = \sqrt{\gamma^2 + \lambda^2}\, a = \sqrt{\gamma^2 + n^2\pi^2/L^2}\, a$.
Thus $u(x,t)$ has the form shown. Satisfying the remaining
IC yields the shown values for c_n.

23b. Use $\sin A \cos B = [\sin(A+B) + \sin(A-B)]/2$ with $A = n\pi x/L$ and
$B = (n\pi/L)\sqrt{1 + \dfrac{\gamma^2 L^2}{n^2\pi^2}}$ at $= (n\pi/L)a_n t$, where

$a_n = \sqrt{1 + \dfrac{\gamma^2 L^2}{n^2\pi^2}}\, a$ is the speed of the wave propagation.

23c. From Part(b), a_n is independent of n when $\gamma = 0$.

Section 10.8, Page 683

1a. Assuming that $u(x,y) = X(x)Y(y)$ leads to the two O.D.E.
$X'' - \sigma X = 0$, $Y'' + \sigma Y = 0$. The B.C. $u(0,y) = 0$,
$u(a,y) = 0$ imply that $X(0) = 0$ and $X(a) = 0$. Thus
nontrivial solutions to $X'' - \sigma X = 0$ which satisfy these
boundary conditions are possible only if $\sigma = -(n\pi/a)^2$,
$n = 1,2...$; the corresponding solutions for $X(x)$ are
proportional to $\sin(n\pi x/a)$. The B.C. $u(x,0) = 0$ implies
that $Y(0) = 0$. Solving $Y'' - (n\pi/a)^2 Y = 0$ subject to this
condition we find that $Y(y)$ must be proportional to
$\sinh(n\pi y/a)$. The fundamental solutions are then
$u_n(x,y) = \sin(n\pi x/a)\sinh(n\pi y/a)$, $n = 1,2,\ldots$, which
satisfy Laplace's equation and the homogeneous B.C. We
assume that $u(x,y) = \displaystyle\sum_{n=1}^{\infty} c_n \sin(n\pi x/a)\sinh(n\pi y/a)$, where
the coefficients c_n are determined from the B.C.
$u(x,b) = \displaystyle\sum_{n=1}^{\infty} c_n \sin(n\pi x/a)\sinh(n\pi b/a) = g(x)$. Recognizing
this as a sine series it follows that
$c_n \sinh(n\pi b/a) = (2/a)\displaystyle\int_0^a g(x)\sin(n\pi x/a)dx$, $n = 1,2,\ldots$.

1b. Substituting for g(x) in the equation for c_n we have

$$c_n\sinh(n\pi b/a) = (2/a)\left\{\int_0^{a/2} x\sin(n\pi x/a)dx + \int_{a/2}^{a}(a-x)\sin(n\pi x/a)dx\right\}$$

$$= [4a\sin(n\pi/2)]/n^2\pi^2, \quad n = 1,2,\ldots, \text{ so}$$

$c_n = (4a/\pi^2)[\sin(n\pi/2)]/[n^2\sinh(n\pi b/a)]$. Substituting these values for c_n in the above series yields the desired solution.

1c.

1d.

2. In solving the D.E. $Y'' - \lambda^2 Y = 0$, one normally writes $Y(y) = c_1\sinh\lambda y + c_2\cosh\lambda y$. However, since we have $Y(b) = 0$, it is advantageous to rewrite Y as $Y(y) = d_1\sinh\lambda(b-y) + d_2\cosh\lambda(b-y)$, where d_1, d_2 are also arbitrary constants and can be related to c_1, c_2 using the appropriate hyperbolic trigonometric identities. The important thing, however, is that the second form also satisfies the D.E. and thus $Y(y) = d_1\sinh\lambda(b-y)$ satisfies the D.E. and the homogeneous B.C. $Y(b) = 0$. The rest of the problem follows the pattern of Prob. 1.

3a. Let $v(x,y)$ satisfy Laplace's equation with the B.C. of Eq.(4) and let $w(x,y)$ satisfy Laplace's equation with the B.C. of Prob. 2. Then using superposition will give $u(x,y) = v(x,y) + w(x,y)$ as the solution of the given problem.

4. Following the pattern of Prob. 3, one could consider
    adding the solutions of four problems, each with only one
    non-homogeneous B.C. It is also possible to consider
    adding the solutions of only two problems, each with only
    two non-homogeneous B.C., as long as they involve the
    same variable. For instance, one such problem would be
    $u_{xx} + u_{yy} = 0$, $u(x,0) = 0$, $u(x,b) = 0$, $u(0,y) = k(y)$,
    $u(a,y) = f(y)$, which has the fundamental solutions
    $u_n(x,y) = [c_n\sinh(n\pi x/b) + d_n\cosh(n\pi x/b)]\sin(n\pi y/b)$.

    Assuming $u(x,y) = \displaystyle\sum_{n=1}^{\infty} u_n(x,y)$ and using the B.C.

    $u(0,y) = k(y)$ we obtain $d_n = (2/b)\displaystyle\int_0^b k(y)\sin(n\pi y/b)dy$. Using

    the B.C. $u(a,y) = f(y)$ we obtain
    $c_n\sinh(n\pi a/b) + d_n\cosh(n\pi a/b) = (2/b)\displaystyle\int_0^b f(y)\sin(n\pi y/b)dy$, which

    can be solved for c_n, since d_n is already known. The second
    problem, in this approach, would be $u_{xx} + u_{yy} = 0$,
    $u(x,0) = h(x)$, $u(x,b) = g(x)$, $u(0,y) = 0$ and $u(a,y) = 0$. This
    has the fundamental solutions
    $u_n(x,y) = [a_n\sinh(n\pi y/a) + b_n\cosh(n\pi y/a)]\sin(n\pi x/a$, so that

    $u(x,y) = \displaystyle\sum_{n=1}^{\infty} u_n(x,y)$. Thus $u(x,0) = h(x)$ gives

    $b_n = (2/a)\displaystyle\int_0^a h(x)\sin(n\pi x/a)dx$ and $u(x,b) = g(x)$ gives

    $a_n\sinh(n\pi b/a) + b_n\cosh(n\pi b/a) = (2/a)\displaystyle\int_0^a g(x)\sin(n\pi x/a)dx$,

    which can be solved for a_n since b_n is known.

5. Using Eq.(20) and following the same arguments as
    presented in the text, we find that $R(r) = k_1 r^n + k_2 r^{-n}$
    and $\Theta(\theta) = c_1\sin n\theta + c_2\cos n\theta$, for n a positive integer,
    and $u_0(r,\theta) = 1$ for n = 0. Since we require that $u(r,\theta)$
    be bounded as $r \to \infty$, we conclude that $k_1 = 0$. The
    fundamental solutions are therefore
    $u_n(r,\theta) = r^{-n}\cos n\theta$, $v_n(r,\theta) = r^{-n}\sin n\theta$, n = 1,2,...
    together with $u_0(r,\theta) = 1$. Assuming that u can be
    expressed as a linear combination of the fundamental
    solutions we have

    $u(r,\theta) = c_0/2 + \displaystyle\sum_{n=1}^{\infty} r^{-n}(c_n\cos n\theta + k_n\sin n\theta)$.   The B.C.

    requires that

    $u(a,\theta) = c_0/2 + \displaystyle\sum_{n=1}^{\infty} a^{-n}(c_n\cos n\theta + k_n\sin n\theta) = f(\theta)$ for

$0 \le \theta < 2\pi$. This is precisely the Fourier series representation for $f(\theta)$ of period 2π and thus

$$a^{-n}c_n = (1/\pi)\int_0^{2\pi}f(\theta)\cos n\theta d\theta, \quad n = 0,1,2,\ldots \text{ and}$$

$$a^{-n}k_n = (1/\pi)\int_0^{2\pi}f(\theta)\sin n\theta d\theta, \quad n = 1,2\ldots .$$

7. Again we let $u(r,\theta) = R(r)\Theta(\theta)$ and thus we have $r^2R'' + rR' - \sigma R = 0$ and $\Theta'' + \sigma\Theta = 0$, with $R(0)$ bounded and the B.C. $\Theta(0) = \Theta(\alpha) = 0$. For $\sigma \le 0$ we find that $\Theta(\theta) \equiv 0$, so we let $\sigma = \lambda^2$ (λ^2 real) and thus $\Theta(\theta) = c_1\cos\lambda\theta + c_2\sin\lambda\theta$. The B.C. $\Theta(0) = 0 \Rightarrow c_1 = 0$ and the B.C. $\Theta(\alpha) = 0 \Rightarrow \lambda = n\pi/\alpha$, $n = 1,2,\ldots$. Substituting these values into Eq.(31) we obtain $R(r) = k_1 r^{n\pi/\alpha} + k_2 r^{-n\pi/\alpha}$. However $k_2 = 0$ since $R(0)$ must be bounded, and thus the fundamental solutions are $u_n(r,\theta) = r^{n\pi/\alpha}\sin(n\pi\theta/\alpha)$. The desired solution may now be formed using previously discussed procedures.

8a. Separating variables, we obtain $X'' + \lambda^2X = 0$, $X(0) = 0$, $X(a) = 0$ and $Y'' - \lambda^2Y = 0$, $Y(y)$ bounded as $y \to \infty$. Thus $X(x) = \sin(n\pi x/a)$, and $\lambda^2 = (n\pi/a)^2$. However, since neither $\sinh y$ nor $\cosh y$ are bounded as $y \to \infty$, we must write the solution to $Y'' - (n\pi/a)^2Y = 0$ as $Y(y) = c_1\exp[n\pi y/a] + c_2\exp[-n\pi y/a]$. Thus we must choose $c_1 = 0$ so that $u(x,y) = X(x)Y(y) \to 0$ as $y \to \infty$. The fundamental solutions are then $u_n(x,t) = e^{-n\pi y/a}\sin(n\pi x/a)$.

$$u(x,y) = \sum_{n=1}^{\infty}c_n u_n(x,y) \text{ then gives}$$

$$u(x,0) = \sum_{n=1}^{\infty}c_n\sin(n\pi x/a) = f(x) \text{ so that } c_n = \frac{2}{a}\int_0^a f(x)\sin\frac{n\pi x}{a}dx.$$

8b. $c_n = \dfrac{2}{a}\displaystyle\int_0^a x(a-x)\sin\dfrac{n\pi x}{a}dx = \dfrac{4a^2}{n^3\pi^3}(1-\cos n\pi)$

8c. Using just the first term ($n=1$) and letting $a = 5$, we have $u(x,y) = \dfrac{200}{\pi^3}e^{-\pi y/5}\sin\dfrac{\pi x}{5}$, which, for a fixed y, has a maximum at $x = 5/2$ and thus we need to find y such that

$u(5/2,y) = \dfrac{200}{\pi^3}e^{-\pi y/5} = .1$. Taking the logarithm of both

sides and solving for y yields $y_0 = 6.6315$. With an equation solver, more terms can be used. However, to four decimal places, three terms yield the same result as above.

13a. Assuming that $u(x,y) = X(x)Y(y)$ and substituting into Eq.(1) leads to the two O.D.E. $X'' - \sigma X = 0$, $Y'' + \sigma Y = 0$. The B.C. $u(x,0) = 0$, $u_y(x,b) = 0$ imply that $Y(0) = 0$ and $Y'(b) = 0$. For nontrivial solutions to exist for $Y'' + \sigma Y = 0$ with these B.C. we find that σ must take the values $(2n-1)^2\pi^2/4b^2$, $n = 1,2,\ldots$; the corresponding solutions for $Y(y)$ are proportional to $\sin[(2n-1)\pi y/2b]$. Solutions to $X'' - [(2n-1)^2\pi^2/4b^2]X = 0$ are of the form $X(x) = A\sinh[(2n-1)\pi x/2b] + B\cosh[(2n-1)\pi x/2b]$. However, the boundary condition $u(0,y) = 0$ implies that $X(0) = B = 0$. It follows that the fundamental solutions are $u_n(x,y) = \sinh[(2n-1)\pi x/2b]\sin[(2n-1)\pi y/2b]$, $n = 1,2,\ldots$. To satisfy the remaining B.C. at $x = a$ we assume that we can represent the solution $u(x,y)$ in the

form $u(x,y) = \sum_{n=1}^{\infty} c_n\sinh[(2n-1)\pi x/2b]\sin[(2n-1)\pi y/2b]$.

The coefficients c_n are determined by the B.C.

$$u(a,y) = \sum_{n=1}^{\infty} c_n\sinh[(2n-1)\pi a/2b]\sin[(2n-1)\pi y/2b] = f(y).$$

By properly extending f as a periodic function of period 4b as in Prob. 39, Sect. 10.4, we find that the coefficients c_n are given by

$$c_n\sinh[(2n-1)\pi a/2b] = (2/b)\int_0^b f(y)\sin[(2n-1)\pi y/2b]dy,$$

$n = 1,2,\ldots$.

16a. If $u(x,z) = X(x)Z(z)$ we obtain $X'' + \lambda^2 X = 0$, $X'(0) = 0$, $X'(a) = 0$ and $Z'' - \lambda^2 Z = 0$, $Z'(0) = 0$. The first B.V.P. yields $\lambda^2 = n^2\pi^2/a^2$ and $X(x) = \cos(n\pi x/a)$, $n = 0,1,2\cdots$, as in Ex.(2) of Sect. 10.6. Solving the second B.V.P we find $Z(z) = \cosh(n\pi z/a)$, so that

$$u(x,z) = c_0/2 + \sum_{n=1}^{\infty} c_n\cosh(n\pi z/a)\cos(n\pi x/a).$$ Setting $z = b$

we get $u(x,b) = c_0/2 + \sum_{n=1}^{\infty} c_n\cosh(n\pi b/a)\cos(n\pi x/a) = b + \alpha x$,

where $c_0 = (2/a)\int_0^a (b + \alpha x)dx = 2b + \alpha a$ and

$$c_n\cosh(n\pi b/a) = (2/a)\int_0^a (b + \alpha x)\cos(n\pi x/a)dx$$

$$= \begin{cases} 0 & n \text{ even} \\ -4\alpha a/n^2\pi^2 & n \text{ odd} \end{cases}.$$ Solving for c_n and

substituting into $u(x,z)$ yields the desired solution.

CHAPTER 11

Section 11.1, Page 700

2.   Since the B.C. at $x = 1$ is nonhomogeneous, the B.V.P. is
     nonhomogeneous.

4.   The D.E. may be written $y'' + (\lambda - x^2)y = 0$ and is thus
     homogeneous, as are both B.C.

5.   Since the D.E. contains the nonhomogeneous term 1, the
     B.V.P. is nonhomogeneous.

9a. If $\lambda > 0$, the general solution of the D.E. is
     $y = c_1 \sin\sqrt{\lambda}\, x + c_2 \cos\sqrt{\lambda}\, x$.   The B.C. require that
     $-\sqrt{\lambda}\, c_1 + c_2 = 0$, and
     $(\sin\sqrt{\lambda} + \sqrt{\lambda}\cos\sqrt{\lambda})c_1 + (\cos\sqrt{\lambda} - \sqrt{\lambda}\sin\sqrt{\lambda})c_2 = 0$.
     Taking the determinant of the coefficients of c_1 and c_2
     we see that in order to have nontrivial solutions λ must
     satisfy $(\lambda - 1)\sin\sqrt{\lambda} - 2\sqrt{\lambda}\cos\sqrt{\lambda} = 0$.   In this case
     $c_2 = \sqrt{\lambda}\, c_1$ and thus $\phi_n = \sin\sqrt{\lambda_n}\, x + \sqrt{\lambda_n}\cos\sqrt{\lambda_n}\, x$.

     For $\lambda < 0$, the discussion follows the pattern of Ex. 1
     yielding $y(x) = c_1\sinh\sqrt{\mu}\, x + c_2\cosh\sqrt{\mu}\, x$.   The B.C. then yield
     $-\sqrt{\mu}\, c_1 + c_2 = 0$ and
     $(\sinh\sqrt{\mu} + \sqrt{\mu}\cosh\sqrt{\mu})c_1 + (\cosh\sqrt{\mu} + \sqrt{\mu}\sinh\sqrt{\mu})c_2 = 0$, which
     have a non-zero solution if and only if
     $(\mu + 1)\sinh\sqrt{\mu} + 2\sqrt{\mu}\cosh\sqrt{\mu} = 0$.   By plotting $y = \tanh\sqrt{\mu}$ and
     $y = -2\sqrt{\mu}/(\mu + 1)$ we see that they intersect only at $\mu = 0$, and
     thus there are no negative eigenvalues.

9b. If $\lambda = 0$, then $y(x) = c_1 x + c_2$ and thus $y(0) = c_2$,
     $y(1) = c_1 + c_2$, $y'(0) = c_1$ and $y'(1) = c_1$.   Hence the B.C.
     yield the two equations $c_2 - c_1 = 0$ and $c_1 + c_2 + c_1 = 0$
     which give $c_1 = c_2 = 0$ and thus $\lambda = 0$ is not an
     eigenvalue.

9c. If $\lambda \neq 1$, the eigenvalue equation is equivalent to
     $\tan\sqrt{\lambda} = 2\sqrt{\lambda}/(\lambda - 1)$ and thus by graphing $f(\sqrt{\lambda}) = \tan\sqrt{\lambda}$
     and $g(\sqrt{\lambda}) = 2\sqrt{\lambda}/(\lambda - 1)$ we can estimate the eigenvalues.
     Since $g(\sqrt{\lambda})$ has a vertical asymptote at $\lambda = 1$ and
     $f(\sqrt{\lambda})$ has a vertical asymptote at $\sqrt{\lambda} = \pi/2$, we see
     that $1 < \sqrt{\lambda_1} < \pi/2$.   By interating numerically, we find

$\sqrt{\lambda_1} \cong 1.30655$ and thus $\lambda_1 \cong 1.7071$. The second eigenvalue will lie to the right of π, the second zero of $\tan\sqrt{\lambda}$. Again iterating numerically, we find $\sqrt{\lambda_2} \cong 3.6732$ and thus $\lambda_2 \cong 13.4924$.

9d. For large values of n, we see from the graph of Part(c) that $\sqrt{\lambda_n} \cong (n-1)\pi$, which are the zeros of $\tan\sqrt{\lambda}$. Thus $\lambda_n \cong (n-1)^2\pi^2$ for large n.

10. If $\lambda < 0$, set $-\lambda = \mu^2$ to obtain $y = c_1\cos\mu x + c_2\sin\mu x$. In this case the B.C. require $c_1 + \mu c_2 = 0$ and $c_1\cos\mu + c_2\sin\mu = 0$ which yields nontrivial solutions for c_1 and c_2 (i.e., $c_1 = -\mu c_2$) if and only if $\tan\mu = \mu$ and thus $\phi_n = \sin\mu_n x - \mu_n\cos\mu_n x$.

If $\lambda > 0$, the general solution of the D.E. is $y(x) = c_1\cosh\sqrt{\lambda}\,x + c_2\sinh\sqrt{\lambda}\,x$. The B.C. respectively require that $c_1 + \sqrt{\lambda}\,c_2 = 0$ and $c_1\cosh\sqrt{\lambda} + c_2\sinh\sqrt{\lambda} = 0$ and thus λ must satisfy $\tanh\sqrt{\lambda} = \sqrt{\lambda}$ in order to have nontrivial solutions. The only solution of this equation is $\lambda = 0$ and thus there are no positive eigenvalues.

10b. If $\lambda = 0$, the general solution of the D.E. is $y = c_1 + c_2 x$. The B.C. $y(0) + y'(0) = 0$ requires $c_1 + c_2 = 0$ and the B.C. $y(1) = 0$ requires $c_1 + c_2 = 0$ and thus $\lambda = 0$ is an eigenvalue with corresponding eigenfunction $\phi_0(x) = 1-x$.

10c. By plotting on the same graph $f(\mu) = \mu$ and $g(\mu) = \tan\mu$, we see that they intersect at $\mu_0 = 0$ ($\mu = 0 \Rightarrow \lambda = 0$, which has already been discussed), $\mu_1 \cong 4.49341$ (which is just to the left of the vertical asymptote of $\tan\mu$ at $\mu = 3\pi/2$, $\mu_2 \cong 7.72525$ (which is just to the left of the vertical asymptote of $\tan\mu$ at $\mu = 5\pi/2$). Since $\lambda_n = -\mu_n^2$, we have $\lambda_1 \cong -20.1907$ and $\lambda_2 \cong -59.6795$,

10d. For larger n values $\mu_n \cong (2n+1)\pi/2$ and $\lambda_n \cong -(2n+1)^2\pi^2/4$.

11a. From Eq.(i) the coefficient of y' is μQ and from Eq.(ii) the coefficient of y' is $(\mu P)'$. Thus $(\mu P)' = \mu Q$, which gives Eq.(iii).

11b. Eq.(iii) is both linear and separable. Using the latter approach we have $d\mu/\mu = [(Q-P')/P]dx$ and thus $\ln\mu = \int_{x_0}^{x}[Q(s)/P(s)]ds - [\ln P(x) - \ln P(x_0)]$. Taking the exponential of both sides yields Eq.(iv). The choice of x_0 simply alters the constant of integration, which is immaterial here.

13. Since $P(x) = x^2$ and $Q(x) = x$, we find that $\mu(x) = (1/x^2)\exp[\int_{x_0}^{x}(s/s^2)ds] = k/x$, where k is an arbitrary constant which may be set equal to 1. It follows that Bessel's equation takes the form $(xy')' + (x - \upsilon^2/x)y = 0$.

18a. Assuming $y = s(x)u$, we have $y' = s'u + su'$ and $y'' = s''u + 2s'u' + su''$ and thus the D.E. becomes $su'' + (2s'+4s)u' + [s'' + 4s' + (4+9\lambda)s]u = 0$. Setting $2s' + 4s = 0$ we find $s(x) = e^{-2x}$ and the D.E. becomes $u'' + 9\lambda u = 0$. The B.C. $y(0) = 0$ yields $s(0)u(0) = 0$, or $u(0) = 0$ since $s(0) \neq 0$. The B.C. at L is $y'(L) = s'(L)u(L) + s(L)u'(L) = e^{-2L}(-2u(L) + u'(L)) = 0$ and thus $u'(L) - 2u(L) = 0$. Thus the B.V.P. satisfied by $u(x)$ is $u'' + 9\lambda u = 0$, $u(0) = 0$, $u'(L) - 2u(L) = 0$.

If $\lambda < 0$, the general solution of the D.E. $u'' + 9\lambda u = 0$ is $u = c_1\sinh 3\mu x + c_2\cosh 3\mu x$ where $-\lambda = \mu^2$. The B.C. require that $c_2 = 0$, $c_1(3\mu\cosh 3\mu L - 2\sinh 3\mu L) = 0$. In order to have nontrivial solutions μ must satisfy the equation $3\mu/2 = \tanh 3\mu L$. A graphical analysis reveals that for $L \le 1/2$ this equation has no solutions for $\mu \neq 0$ so there are no negative eigenvalues for $L \le 1/2$. If $L > 1/2$ there is one solution and hence one negative eigenvalue with eigenfuction $\phi_{-1}(x) = e^{-2x}\sinh 3\mu x$.

If $\lambda = 0$, the general solution of the D.E. $u'' + 9\lambda u = 0$ is $u = c_1 + c_2x$. The B.C. require that $c_1 = 0$, $c_2(1-2L) = 0$ so nontrivial solutions are possible only if $L = 1/2$. In this case the eigenfuction is $\phi_0(x) = xe^{-2x}$.

If $\lambda > 0$, the general solution of the D.E. $u'' + 9\lambda u = 0$ is $u = c_1\sin 3\sqrt{\lambda} x + c_2\cos 3\sqrt{\lambda} x$. The B.C. require that $c_2 = 0$, $c_1(3\sqrt{\lambda}\cos 3\sqrt{\lambda} L - 2\sin 3\sqrt{\lambda} L) = 0$. In order to have nontrivial solutions λ must satisfy the equation $\sqrt{\lambda} = (2/3)\tan 3\sqrt{\lambda} L$. A graphical analysis reveals that there is an infinite number of solutions to this

eigenvalue equation. Thus the eigenfunctions are
$\phi_n(x) = e^{-2x}\sin 3\sqrt{\lambda_n}\,x$ where the eigenvalues λ_n satisfy
$\sqrt{\lambda_n} = (2/3)\tan 3\sqrt{\lambda_n}\,L$.

18b. The roots of the characteristic equation are $r = -2 \pm \sqrt{-9\lambda}$,
so $y(x) = e^{-2x}(c_1\cos 3\sqrt{\lambda}\,x + c_2\sin 3\sqrt{\lambda}\,x)$.

20. This is an Euler equation whose characteristic equation
has roots $r_1 = \lambda$ and $r_2 = 1$. If $\lambda = 1$ the general
solution of the D.E. is $y = c_1 x + c_2 x\ln x$ and the B.C.
require that $c_1 = c_2 = 0$ and thus $\lambda = 1$ is not an
eigenvalue. If $\lambda \neq 1$, $y = c_1 x + c_2 x^\lambda$ is the general
solution and the B.C. require that $c_1 + c_2 = 0$ and
$2c_1 + c_2 2^\lambda - (c_1 + \lambda c_2 2^{\lambda-1}) = 0$. Thus nontrivial solutions
exist if and only if $\lambda = 2(1-2^{-\lambda})$. The graphs of
$f(\lambda) = \lambda$ and $g(\lambda) = 2(1-2^{-\lambda})$ intersect only at $\lambda = 1$
(which has already been discussed) and $\lambda = 0$. Thus the
only eigenvalue is $\lambda = 0$ with corresponding eigenfunction
$\phi(x) = x - 1$ (since $c_1 = -c_2$).

22a. For positive λ, the general solution of the D.E. is
$y = c_1\sin\sqrt{\lambda}\,x + c_2\cos\sqrt{\lambda}\,x$. The B.C. require that
$\sqrt{\lambda}\,c_1 + \alpha c_2 = 0$, $c_1\sin\sqrt{\lambda} + c_2\cos\sqrt{\lambda} = 0$. Nontrivial
solutions exist if and only if $\sqrt{\lambda}\cos\sqrt{\lambda} - \alpha\sin\sqrt{\lambda} = 0$.
If $\alpha = 0$ this equation is satisfied by the sequence
$\lambda_n = [(2n-1)\pi/2]^2$, $n = 1,2,\ldots$. If $\alpha \neq 0$, λ must
satisfy the equation $\sqrt{\lambda}/\alpha = \tan\sqrt{\lambda}$. A plot of the
graphs of $f(\sqrt{\lambda}) = \sqrt{\lambda}/\alpha$ and $g(\sqrt{\lambda}) = \tan\sqrt{\lambda}$ reveals
that there is an infinite sequence of postive eigenvalues
for $\alpha < 0$ and $\alpha > 0$.

22b. By procedures shown previously, the cases $\lambda < 0$ and
$\lambda = 0$, when $\alpha < 1$, lead to only the trivial solution and
thus by Part(a) all real eigenvalues are positive. For
$0 < \alpha < 1$, the graphs of $f(\sqrt{\lambda})$ and $g(\sqrt{\lambda})$, see Part(a),
intersect once on $0 < \sqrt{\lambda} < \pi/2$. As α approaches 1 from
below, the slope of $f(\sqrt{\lambda})$ decreases and thus the
intersection point approaches zero.

22c. If $\lambda = 0$, then $y(x) = c_1 + c_2 x$ and the B.C. yield
$\alpha c_1 + c_2 = 0$ and $c_1 + c_2 = 0$, which have a non-zero
solution if and only if $\alpha = 1$.

22d. Let $-\lambda = \mu^2$, then $y(x) = c_1\cosh\mu x + c_2\sinh\mu x$ and thus the
     B.C. yield $\alpha c_1 + \mu c_2 = 0$ and $(\cosh\mu)c_1 + (\sinh\mu)c_2 = 0$, which
     have non-zero solutions if and only if $\tanh\mu = \mu/\alpha$. For $\alpha>1$,
     the straight line $y = \mu/\alpha$ intersects the curve $y = \tanh\mu$ in
     one point, which increases as α increases. Thus $\lambda = -\mu^2$
     decreases as α increases.

23. Using the D.E. for ϕ_m and following the hint yields:
     $\int_0^L \phi_m''\phi_n dx + \lambda_m\int_0^L \phi_m\phi_n dx = 0$. Integrating the first term by parts
     yields: $\phi_m'\phi_n\big|_0^L - \int_0^L \phi_m'\phi_n' dx = -\lambda_m\int_0^L \phi_m\phi_n dx$. Upon utilizing he
     B.C. the first term on the left vanishes and thus
     $\int_0^L \phi_n'\phi_m' dx = \lambda_m\int_0^L \phi_m\phi_n dx$. Similarly, the D.E. for ϕ_n yields
     $\int_0^L \phi_m'\phi_n' dx = \lambda_n\int_0^L \phi_n\phi_m dx$ and thus $(\lambda_n - \lambda_m)\int_0^L \phi_m\phi_n dx = 0$. If
     $\lambda_n \neq \lambda_m$ the desired result follows.

24b. The general solution of the D.E. is
     $y = c_1\sin\mu x + c_2\cos\mu x + c_3\sinh\mu x + c_4\cosh\mu x$ where $\lambda = \mu^4$.
     The B.C. require that $c_2 + c_4 = 0$, $-c_2 + c_4 = 0$,
     $c_1\sin\mu L + c_2\cos\mu L + c_3\sinh\mu L + c_4\cosh\mu L = 0$, and
     $c_1\cos\mu L - c_2\sin\mu L + c_3\cosh\mu L + c_4\sinh\mu L = 0$. The first two
     equations yield $c_2 = c_4 = 0$, and the last two have nontrivial
     solutions if and only if $\sin\mu L\cosh\mu L - \cos\mu L\sinh\mu L = 0$. In
     this case the third equation yields $c_3 = -c_1\sin\mu L/\sinh\mu L$ and
     thus the desired eigenfunctions are obtained. The quantity
     μL can be approximated by finding the intersection of
     $f(x) = \tan x$ and $g(x) = \tanh x$, where $x = \mu L$. The first
     intersection is at $x \approx 3.9266$, which gives $\lambda_1 \approx 237.72/L^4$ and
     the second intersection is at $x \approx 7.0686$, which gives
     $\lambda_2 \approx 2,496.5/L^4$.

Section 11.2, Page 714

1.   We have $y(x) = c_1\cos\sqrt{\lambda}\,x + c_2\sin\sqrt{\lambda}\,x$ and thus $y(0) = 0$
     yields $c_1 = 0$ and $y'(1) = 0$ yields $\sqrt{\lambda}\,c_2\cos\sqrt{\lambda} = 0$. $\lambda = 0$
     gives $y(x) = 0$, so is not an eigenvalue. Otherwise
     $\lambda = (2n-1)^2\pi^2/4$ and the eigenfuctions are $\sin[(2n-1)\pi x/2]$,
     $n = 1,2,\ldots$. Thus, by Eq.(20), we must choose k_n so that
     $\int_0^1 \{k_n\sin[(2n-1)\pi x/2]\}^2 dx = 1$, since the weight function
     $r(x) = 1$ (by comparing the D.E. to Eq.(1)). Evaluating the
     integral yields $k_n^2/2 = 1$ and thus $k_n = \sqrt{2}$ and the desired
     normalized eigenfuctions are obtained.

3.   Note here that $\phi_0(x) = 1$, the eigenfunction for $\lambda = 0$, satisifes Eq.(20) and hence it is already normalized.

5.   From Prob. 17 of Sect. 11.1 we have $e^x \sin n\pi x$, $n = 1,2,\ldots$ as the eigenfunctions and thus k_n must be chosen so that $\int_0^1 r(x)k_n^2 e^{2x}\sin^2 n\pi x\, dx = 1$. To determine $r(x)$, we must write the D.E. in the form of Eq.(1). That is, we multiply the D.E. by $r(x)$ to obtain $ry'' - 2ry' + ry + \lambda ry = 0$. Now choose r so that $(ry')' = ry''-2ry'$, which yields $r' = -2r$ or $r(x) = e^{-2x}$. Thus the above integral becomes $\int_0^1 k_n^2 \sin^2 n\pi x\, dx = 1$ and $k_n = \sqrt{2}$. Hence $\phi_n(x) = \sqrt{2}\, e^x \sin n\pi x$ are the normalized eigenfunctions.

7.   Using Eq.(34) with $r(x) = 1$, we find that the coefficients of the series (32) are determined by

$$a_n = (f,\phi_n) = \sqrt{2}\int_0^1 x\sin[(2n-1)\pi x/2]dx$$

$$= (4\sqrt{2}/(2n-1)^2\pi^2)\sin(2n-1)\pi/2. \quad \text{Thus Eq.(32) yields}$$

$$f(x) = \frac{4\sqrt{2}}{\pi^2}\sum_{n=1}^{\infty}\frac{(-1)^{n-1}}{(2n-1)^2}\sqrt{2}\sin[(2n-1)\pi x/2],\ 0 \le x \le 1,$$

which agrees with the expansion using the approach developed in Prob. 39 of Sect. 10.4.

10. In this case $\phi_n(x) = (\sqrt{2}/\alpha_n)\cos\sqrt{\lambda_n}\, x$, where $\alpha_n = (1 + \sin^2\sqrt{\lambda_n})^{1/2}$. Thus Eq.(34) yields $a_n = (\sqrt{2}/\alpha_n)\int_0^1 \cos\sqrt{\lambda_n}\, x dx = \sqrt{2}\sin\sqrt{\lambda_n}/\alpha_n\sqrt{\lambda_n}$.

14. In this case $L[y] = y'' + y' + 2y$. To put this in the form shown in Eq.(3) we must have $(Py')' = y'' + y'$, which is not possible, so this B.V.P. is not self adjoint.

17. In this case $L[y] = [(1+x^2)y']' + y$ and thus the D.E. has the form shown in Eq.(3). However, the B.C. are not separated and thus we must determine by integration whether Eq.(8) is satisfied. Therefore, for u and v satisfying the B.C., integration by parts yields:

$$(L[u],v) = \int_0^1 \{[(1+x^2)u']'+u\}v dx$$

$$= vu'(1+x^2)\Big|_0^1 - \int_0^1 \{(1+x^2)v'u'\}dx + \int_0^1 uv dx$$

$$= vu'(1+x^2)\Big|_0^1 - uv'(1+x^2)\Big|_0^1 + \int_0^1 \{[(1+x^2)v']'+v\}u dx$$

$$= (u,L[v]) \text{ since the integrated terms add to}$$

zero with the given B.C. Thus the B.V.P. is self-adjoint.

21a. Substituting ϕ for y in the D.E., multiplying both sides
     by ϕ, and integrating form 0 to 1 yields
     $\lambda\int_0^1 r\phi^2 dx = \int_0^1\{-[p(x)\phi']'\phi + q(x)\phi^2\}dx$. Integrating the
     first term on the right side once by parts, we obtain
     $\lambda\int_0^1 r\phi^2 dx = -p(1)\phi'(1)\phi(1) + p(0)\phi'(0)\phi(0) + \int_0^1(p\phi'^2+q\phi^2)dx$.
     If $\alpha_2 \neq 0$, $\beta_2 \neq 0$, then $\phi'(1) = -\beta_1\phi(1)/\beta_2$ and
     $\phi'(0) = -\alpha_1\phi(0)/\alpha_2$ and the result follows. If $\alpha_2 = 0$,
     then $\phi(0) = 0$ and the boundary term at 0 will be missing.
     A similar result is obtained if $\beta_2 = 0$.

21b. From the text, $p(x) > 0$ and $r(x) > 0$ for $0 \leq x \leq 1$
     [following Eq.(4)]. If $q(x) \geq 0$ and if β_1/β_2 and $-\alpha_1/\alpha_2$ are
     non-negative, then all terms in the final equation of
     Part(a) are non-negative and thus λ must be non-negative.
     Note that λ could be zero if $q(x) = 0$, $\beta_1 = 0$, $\alpha_1 = 0$ and
     $\phi(x) = 1$.

21c. If either $q(x) \neq 0$ or $\alpha_1 \neq 0$ or $\beta_1 \neq 0$, there is at least
     one positive term on the right and thus λ must be positive.

23a. Using $\phi(x) = U(x) + iV(x)$ in Eq.(4) we have
     $L[\phi] = L[U(x) + iV(x)] = \lambda r(x)[U(x)+iV(x)]$. Using the
     linearity of L and the fact that λ and $r(x)$ are real we
     have $L[U(x)] + iL[V(x)] = \lambda r(x)U(x) + i\lambda r(x)V(x)$.
     Equating the real and imaginary parts shows that both U
     and V satisfy Eq.(1). The B.C. Eq.(2) are also satisfied
     by both U and V, using the same arguments, and thus both
     U and V are eigenfunctions.

23b. By Theorem 11.2.3 each eigenvalue λ has only one linearly
     independent eigenfuction. By Part(a) we have U and V
     being eigenfunctions corresponding to λ and thus U and V
     must be linearly dependent.

23c. From Part(b) we have $V(x) = cU(x)$ and thus
     $\phi(x) = U(x) + icuU(x) = (1+ic)U(x)$.

24. This is an Euler equation, so for $y = x^r$ we have
     $r^2 - (\lambda+1)r + \lambda = 0$ or $(r-1)(r-\lambda) = 0$. If $\lambda = 1$, the general
     solution to the D.E. is $y = c_1 x + c_2 x\ln x$. The B.C. require
     that $c_1 = 0$, $2c_1 + 2(\ln 2)c_2 = 0$ so $c_1 = c_2 = 0$ and $\lambda = 1$ is
     not an eigenvalue. If $\lambda \neq 1$, the general solution to the D.E.
     is $y = c_1 x + c_2 x^\lambda$. The B.C. require that $c_1 + c_2 = 0$ and

$2c_1 + 2^{\lambda}c_2 = 0$. Nontrivial solutions exist if and only if $2^{\lambda} - 2 = 0$. If λ is real, this equation has no solution (other than $\lambda = 1$) and again $y = 0$ is the only solution to the boundary value problem. Suppose that $\lambda = a + bi$ with $b \neq 0$. Then $2^{\lambda} = 2^{a+bi} = 2^a 2^{bi} = 2^a \exp(ib\ln2)$, which upon substitution into $2^{\lambda} = 2$ yields the equation $\exp(ib\ln2) = 2^{1-a}$. Since $e^{ix} = \cos x + i\sin x$, it follows that $\cos(b\ln2) = 2^{1-a}$ and $\sin(b\ln2) = 0$, which yield $a = 1$ and $b(\ln2) = 2n\pi$ or $b = 2n\pi/\ln2$, $n = \pm1, \pm2,\ldots$. Thus the only eigenvalues of the problem are
$\lambda_n = 1 + i(2n\pi/\ln2)$, $n = \pm1, \pm2,\ldots$.

25b. The characteristic equation is $r^2(r^2+\lambda) = 0$. For $\lambda \leq 0$ $y = c_1+c_2x+c_3x^2+c_4x^3$ or $y = c_1+c_2x+c_3e^{\sqrt{\lambda}\,x}+c_4e^{-\sqrt{\lambda}\,x}$. In both cases the B.C. yield zero values for all c's and thus there are no eigenfunctions.

For $\lambda > 0$ $y = c_1 + c_2x + c_3\sin\sqrt{\lambda}\,x + c_4\cos\sqrt{\lambda}\,x$. The B.C. require that
$c_1+c_4 = 0$, $c_4 = 0$, $c_1 + c_2L + c_3\sin\sqrt{\lambda}\,L + c_4\cos\sqrt{\lambda}\,L = 0$, and $c_2 + \sqrt{\lambda}\,c_3\cos\sqrt{\lambda}\,L - \sqrt{\lambda}\,c_4\sin\sqrt{\lambda}\,L = 0$. Thus $c_1 = c_4 = 0$ and for nontrivial solutions to exist λ must satisfy the equation $\sqrt{\lambda}\,L\cos\sqrt{\lambda}\,L - \sin\sqrt{\lambda}\,L = 0$. In this case $c_2 = (-\sqrt{\lambda}\,\cos\sqrt{\lambda}\,L)(c_3)$ and it follows that the eigenfunction ϕ_1 is given by
$\phi_1(x) = \sin(\sqrt{\lambda_1}\,x) - (\sqrt{\lambda_1}\,\cos\sqrt{\lambda_1}\,L)x$ where λ_1 is the smallest positive solution of the equation $\sqrt{\lambda}\,L = \tan\sqrt{\lambda}\,L$. A graphical or numerical estimate of λ_1 reveals that $\lambda_1 \cong (4.4934)^2/L^2$.

25c. Assuming $\lambda \neq 0$ the eigenvalue equation is $2(1-\cos x) = x\sin x$, where $x = \sqrt{\lambda}\,L$. Graphing $f(x) = 2(1-\cos x)$ and $g(x) = x\sin x$ we see that there is an intersection for $6 < x < 7$. Since both $f(x)$ and $g(x)$ are zero for $x = 2\pi$, this then is the precise root and thus $\lambda_1 = (2\pi)^2/L^2$. In addition, it appears there might be an intersection for $0 < x < 1$. Using a Taylor series representation for $f(x)$ and $g(x)$ about $x = 0$, however, shows there is no intersection for $0 < x < 1$. Of course $x = 0$ is also an intersection, which yields $\lambda = 0$, which gives the trivial solution and hence $\lambda = 0$ is not an eigenvalue.

27a. Setting $c(x,t) = X(x)T(t)$ in Eq.(i) we obtain
$$\frac{T'}{DT} = \frac{X'' - (v/D)X'}{X} = -\lambda$$ and thus $X'' - (v/D)X' + \lambda X = 0$,
$X(0) = 0$, $X'(L) = 0$, and $T' + \lambda DT = 0$. To put the first
equation in the Sturm-Liouville form multiply by $p(x)$ and
choose p so that $(pX')' = pX'' - p(v/D)X'$. This gives
$p' = -p(v/D)$ or $p(x) = \exp(-vx/D)$.

27b. The characteristic equation of the first O.D.E. in
Part(a) is $r^2 - (v/D)r + \lambda = 0$, so that
$r_{1,2} = (v/2D) \pm i\sqrt{\lambda - v^2/4D^2}$ and hence
$X(x) = e^{vx/2D}(c_1\cos\mu x + c_2\sin\mu x)$. Now $X(0) = 0 \Rightarrow c_1 = 0$
and $X'(L) = 0$ requires that $(v/2D)\sin\mu L + \mu\cos\mu L = 0$.
Thus μ satisfies $\tan\mu L = -(2D/v)\mu$.

27c. Plot on the same graph $y = \tan Lx$ and $y = -(2D/v)x$. For
large n the straight line will intersect the tangent
curve very close the its vertical asymptotes, which are
$Lx = (2n-1)\pi/2$, or $x = (2n-1)\pi/2L$.

27d. Since $r(x) = \exp(-vx/D)$ we have from Eq.(vi)
$$\int_0^L r(x)X_n^2(x)dx = \int_0^L \sin^2(\mu_n x)dx = L/2 - (1/4\mu_n)\sin 2\mu_n L$$
$$= L/2 + (v/4D\mu_n^2)\sin^2\mu_n L,$$
using the double angle formula for sine and Eq.(vii).

27e. From Part(a) we have $T' + \lambda_n DT = 0$, so that $T(t) = e^{-\lambda_n Dt}$,

where λ_n is given in Part(b). Thus $c(x,t) = \sum_{n=1}^{\infty} a_n T_n(t)X_n(x)$

and the a_n are chosen to satisfy $c(x,0) = \sum_{n=1}^{\infty} a_n X_n(x) = f(x)$.

Multiplying both sides of this last equation by
$e^{-vx/2D}\sin\mu_n x$, integrating from 0 to L, using orthogonality,
and the results of Part(d) gives the desired result.

Section 11.3, Page 727

1.   We must first find the eigenvalues and normalized
     eigenfunctions of the associated homogeneous problem
     $y'' + \lambda y = 0$, $y(0) = 0$, $y(1) = 0$. This problem has the
     solutions $\phi_n(x) = k_n\sin n\pi x$, for $\lambda_n = n^2\pi^2$, $n = 1,2,\ldots$.
     To normalize ϕ_n we choose k_n so that $\int_0^1 \phi_n^2 dx = 1$ and thus
     $k_n = \sqrt{2}$. Hence the solution of the original

nonhomogeneous problem is given by $y = \sum_{n=1}^{\infty} b_n \phi_n(x)$, where

the coefficients b_n are found from Eq.(12), $b_n = c_n/(\lambda_n - 2)$
where c_n is given by $c_n = \sqrt{2} \int_0^1 x \sin n\pi x dx$ (Eq.9). [Note
that the original problem can be written as $-y'' = 2y + x$
and therefore comparison with Eq.(1) yields $r(x) = 1$ and
$f(x) = x$]. Integrating the expression for c_n by parts
yields

$$c_n = \sqrt{2}(-1)^{n+1}/n\pi \text{ and thus } y = \sqrt{2} \sum_{n=1}^{\infty} \frac{\sqrt{2}(-1)^{n+1}}{(n^2\pi^2-2)n\pi} \sin n\pi x.$$

2.   From Prob. 1 of Sect. 11.2 we have
$\phi_n = \sqrt{2} \sin[(2n-1)\pi x/2]$ for $\lambda_n = (2n-1)^2\pi^2/4$ and from
Prob. 7 of that section we have
$c_n = 4\sqrt{2}(-1)^{n+1}/(2n-1)^2\pi^2$. Substituting these values

into $b_n = c_n/(\lambda_n-2)$ and $y = \sum_{n=1}^{\infty} b_n \phi_n$ yields the desired

result.

3.   Referring to Prob. 3 of Sect. 11.2 we have

$y = b_0 + \sum_{n=1}^{\infty} b_n(\sqrt{2} \cos n\pi x)$, where $b_n = c_n/(\lambda_n-2)$ for

$n = 0,1,2,\ldots$. The rest of the calculations follow
those of Prob. 1.

5.   Note that the associated eigenvalue problem is the same
as for Prob. 1 and that $|1-2x| = 1-2x$ for $0 \le x \le 1/2$
while $|1-2x| = 2x-1$ for $1/2 \le x \le 1$.

8.   Writing the D.E. in the form Eq.(1), we have $-y'' = \mu y + f(x)$,
so $r(x) = 1$. The associated eigenvalue problem is
$y'' + \lambda y = 0$, $y'(0) = 0$, $y'(1) = 0$ which has the eigenvalues
$\lambda_n = n^2\pi^2$, $n = 0,1,2\ldots$ and the normalized eigenfunctions
$\phi_0 = 1$, $\phi_n(x) = \sqrt{2} \cos n\pi x$, $n = 1,2\ldots$, as found in Prob. 3,

Sect. 11.2. Now, we assume $y(x) = b_0 + \sum_{n=1}^{\infty} b_n \phi_n(x) = \sum_{n=0}^{\infty} b_n \phi_n(x)$,

Eq.(4), and thus $b_n = \dfrac{c_n}{\lambda_n-\mu}$ [Eq.(12)], where

$c_n = \int_0^1 f(x)\phi_n(x)dx$, $n = 0,1,2\ldots$, Eq.(9). Thus

$$y(x) = \sum_{n=0}^{\infty} \frac{c_n}{\lambda_n - \mu} \phi_n(x) = -c_0/\mu + \sqrt{2} \sum_{n=1}^{\infty} \frac{c_n \cos n\pi x}{\lambda_n - \mu}, \text{ where we}$$

have assumed $\mu \neq \lambda_n$, $n = 0,1,2\ldots$.

10. Since $\mu = \pi^2$ is an eigenvalue of the corresponding
     homogeneous equation, Theorem 11.3.1 tells us that a
     solution will exist only if $-(a+x)$ is orthogonal to the
     corresponding eigenfunction $\sqrt{2} \sin \pi x$. Thus we require
     $\int_0^1 (a+x) \sin n\pi x \, dx = 0$, which yields $a = -1/2$. With $a = -1/2$,
     we find that the particular solution is $Y = (x-1/2)/\pi^2$
     and $y_c = c \sin \pi x + d \cos \pi x$ by methods of Chapt. 3. Setting
     $y = y_c + Y$ and choosing d to satisfy the B.C. we obtain
     the desired family of solutions.

11. Note that in this case $\mu = 4\pi^2$ and $\phi_2 = \sqrt{2} \sin 2\pi x$ are the
     eigenvalue and eigenfunction respectively of the
     corresponding homogeneous equation. However, there is no
     value of a for which $-\sqrt{2} \int_0^1 (a+x) \sin 2\pi x \, dx = 0$, and thus
     there is no solution.

12. In this case a solution will exist only if $-a$ is
     orthogonal to $\sqrt{2} \cos \pi x$., that is if $\int_0^1 a \cos \pi x \, dx = 0$.
     Since this condition is valid for all a, a family of
     solutions exists.

14. Since $\sum_{n=1}^{\infty} c_n \phi_n(x)$ converges to zero we have $\sum_{n=1}^{\infty} c_n \phi_n(x) = 0$.
     Multiplying and integrating as suggested yields
     $\int_0^1 [\sum_{n=1}^{\infty} c_n \phi_n(x)] r(x) \phi_m(x) dx = 0$ or
     $\sum_{n=1}^{\infty} c_n \int_0^1 r(x) \phi_n(x) \phi_m(x) dx = 0$. The integral that multiplies c_n
     is just δ_{nm} [Eq.(22) of Sect. 11.2]. Thus the infinite sum
     becomes c_m, which then yields $c_m = 0$.

18. A twice differentiable function v satisfying the boundary
     conditions can be found by assuming that $v = ax + b$.
     Thus $v(0) = b = 1$ and $v(1) = a + 1$ while $v'(1) = a$.
     Hence $2a + 1 = -2$ or $a = -3/2$ and $v(x) = 1 - 3x/2$.
     Assuming $y = u + v$ we have

$(u+v)'' + 2(u+v) = u'' + 2u + 2(1-3x/2) = 2 - 4x$ or
$u'' + 2u = -x$, $u(0) = 0$, $u(1) + u'(1) = 0$ which is the
same as Ex. 1 of the text.

19. From Eq.(30) we assume $u(x,t) = \sum_{n=1}^{\infty} b_n(t)\phi_n(x)$, where the

ϕ_n are the eigenfunctions of the related eigenvalue
problem $y'' + \lambda y = 0$, $y(0) = 0$, $y'(1) = 0$ and the $b_n(t)$ are
given by Eq.(42). From Prob. 2, we have
$\phi_n = \sqrt{2} \sin[(2n-1)\pi x/2]$ and $\lambda_n = (2n-1)^2\pi^2/4$. To
evaluate Eq.(42) we need to calculate
$B_n = \int_0^1 \sin(\pi x/2)\sqrt{2} \sin[(2n-1)\pi x/2 dx$ [Eq.(41) with

$r(x) = 1$ and $f(x) = \sin(\pi x/2)$], which is zero except for
$n = 1$ in which case $B_1 = \sqrt{2}/2$, and
$\gamma_n = \int_0^1 (-x)\sqrt{2} \sin[(2n-1)\pi x/2]dx$ [Eq.(35) with

$F(x,t) = -x$]. This integral is the negative of the c_n in
Prob. 2 and thus
$\gamma_n = -4\sqrt{2}(-1)^{n+1}/(2n-1)^2\pi^2 = -c_n$, $n = 1,2,\ldots$. Setting
$\gamma_n = -c_n$ in Eq.(42) we then have

$b_1 = \dfrac{\sqrt{2}}{2}e^{-\pi^2 t/4} - c_1\int_0^t e^{-\pi^2(t-s)/4}ds$

$\quad = \dfrac{\sqrt{2}}{2}e^{-\pi^2 t/4} - \dfrac{4c_1}{\pi^2}e^{-\pi^2(t-s)/4}\Big|_0^t$

$\quad = \dfrac{\sqrt{2}}{2}e^{-\pi^2 t/4} - \dfrac{4c_1}{\pi^2} + \dfrac{4c_1}{\pi^2}e^{-\pi^2 t/4}$ and

similarly $b_n = -c_n\int_0^t e^{-\lambda_n(t-s)}ds = -(c_n/\lambda_n)e^{-\lambda_n(t-s)}\Big|_0^t$

$\qquad = -(c_n/\lambda_n)(1-e^{-\lambda_n t})$, where $\lambda_n = (2n-1)^2\pi^2/4$,

$n = 2,3,\ldots$. Substituting these values for b_n along
with $\phi_n = \sqrt{2}\sin[(2n-1)\pi x/2]$ into the series for $u(x,t)$
yields the solution to the given problem.

22. In this case $B_n = 0$ for all n and γ_n is given by

$\gamma_n = \int_0^1 e^{-t}(1-x)\sqrt{2} \sin[(2n-1)\pi x/2]dx$

$\quad = e^{-t}\int_0^1 (1-x)\sqrt{2} \sin[(2n-1)\pi x/2]dx$. This last integral

can be written as the sum of two integrals, each of which
has been evaluated in either Prob. 6 or 7 of Sect. 11.2.
Letting c_n denote the value obtained, we then have

$\gamma_n = c_n e^{-t}$ and thus $b_n = c_n \int_0^t e^{-\lambda_n(t-s)} e^{-s} ds = c_n e^{-\lambda_n t} \int_0^t e^{(\lambda_n - 1)s} ds$

$$= [c_n/(\lambda_n - 1)](e^{-t} e^{-\lambda_n t}), \text{ where}$$

$\lambda_n = (2n-1)^2 \pi^2/4$. Substituting these values into Eq.(30) yields the desired solution.

24. Using the approach of Prob. 23 we find that $v(x)$ satisfies $v'' = 2$, $v(0) = 1$, $v(1) = 0$. Thus $v(x) = x^2 + c_1 x + c_2$ and the B.C. yield $v(0) = c_2 = 1$ and $v(1) = 1 + c_1 + 1 = 0$ or $c_1 = -2$. Hence $v(x) = x^2 - 2x + 1$ and $w(x,t) = u(x,t) - v(x)$ where, from Prob. 23, we have $w_t = w_{xx}$, $w(0,t) = 0$, $w(1,t) = 0$ and $w(x,0) = x^2 - 2x + 2 - v(x) = 1$. This last problem can be solved by methods of this section or by methods of Chapt. 10. Using the approach of this section we have $w(x,t) = \sum_{n=1}^{\infty} b_n(t)\phi_n(x)$ where $\phi_n(x) = \sqrt{2} \sin n\pi x$ [which are the normalized eigenfunctions of the associated eigenvalue problem $y'' + \lambda y = 0$, $y(0) = 0$, $y(1) = 0$] and the b_n are given by Eq.(42). Since the P.D.E. for $w(x,t)$ is homogeneous Eq.(42) reduces to $b_n = B_n e^{-\lambda_n t}$ ($\lambda_n = n^2\pi^2$ from the above eigenvalue problem), where

$B_n = \int_0^1 1 \cdot \sqrt{2} \sin n\pi x \, dx = \sqrt{2}\,[1-(-1)^n]/n\pi$. Thus

$$u(x,t) = x^2 - 2x + 1 + \sum_{n=1}^{\infty} \frac{\sqrt{2}\,[1-(-1)^n]}{n\pi} e^{-n^2\pi^2 t} \sqrt{2} \sin n\pi x,$$

which simplifies to the desired solution.

28a. Since $y_c = c_1 + c_2 x$, we assume that $Y(x) = u_1(x) + x u_2(x)$. Then $Y' = u_2$ since we require $u_1' + x u_2' = 0$. Differentiating again yields $Y'' = u_2'$ and thus $u_2' = -f(x)$ by substitution into the D.E. Hence $u_2(x) = -\int_0^x f(s)ds$, $u_1' = xf(x)$, and $u_1(x) = \int_0^x sf(s)ds$. Therefore $Y = \int_0^x sf(s)ds - x\int_0^x f(s)ds = -\int_0^x (x-s)f(s)ds$ and $\phi(x) = c_1 + c_2 x - \int_0^x (x-s)f(s)ds$.

28b. From Part(a) we have $y(0) = c_1 = 0$. Thus $y(x) = c_2 x - \int_0^x (x-s)f(s)ds$ and hence

$y(1) = c_2 - \int_0^1 (1-s)f(s)ds = 0$, which yields the desired
value of c_2.

28c. From Parts (a) and (b) we have

$$\phi(x) = x\int_0^1 (1-s)f(s)ds - \int_0^x (x-s)f(s)ds$$

$$= \int_0^x x(1-s)f(s)ds + \int_x^1 x(1-s)f(s)ds - \int_0^x (x-s)f(s)ds$$

$$= \int_0^x (x-xs-x+s)f(s)ds + \int_x^1 x(1-s)f(s)ds$$

$$= \int_0^x s(1-x)f(s)ds + \int_x^1 x(1-s)f(s)ds.$$

28d. We have $\phi(x) = \int_0^x s(1-x)ds + \int_x^1 x(1-s)ds$

$$= \int_0^x G(x,s)f(s)ds + \int_x^1 G(x,s)f(s)ds$$

$$= \int_0^1 G(x,s)f(s)ds.$$

30b. In this case $y_1(x) = \sin x$ and $y_2(x) = \sin(1-x)$ [assume
$y_2(x) = c_1\cos x + c_2\sin x$, let $x = 1$, solve for c_2 in terms
of c_1 using $y(1) = 0$ and then let $c_1 = \sin 1$]. Using
these functions for y_1 and y_2 we find $W(y_1,y_2) = -\sin 1$
and thus $G(x,s) = -\sin s \, \sin(1-x)/(-\sin 1)$, since $p(x) = 1$,
for $0 \le s \le x$. Interchanging the x and s verifies $G(x,s)$
for $x \le s \le 1$.

30c. Since $W(y_1,y_2)(x) = y_1(x)y_2'(x) - y_2(x)y_1'(x)$ we find that

$$[p(x)W(y_1,y_2)(x)]' = p'(x)[y_1(x)y_2'(x) - y_2(x)y_1'(x)]$$

$$+ p(x)[y_1'(x)y_2'(x) + y_1(x)y_2''(x) - y_2'(x)y_1'(x) - y_2(x)y_1''(x)]$$

$= y_1[py_2']' - y_2[py_1']' = y_1[q(x)y_2] - y_2[q(x)y_1] = 0$, where
we have used Eq.(i) with $f(x) = 0$.

30d. Let $c = p(x)W(y_1,y_2)(x)$. If $0 \le s \le x$, then
$G(x,s) = -y_1(s)y_2(x)/c$. Since the first argument in
$G(s,x)$ is less than the second argument, the bottom
expression of formula (iv) must be used to determine
$G(s,x)$. Thus, $G(s,x) = -y_1(s)y_2(x)/c$. A similar argument
holds if $x \le s \le 1$.

30e. We have $\phi(x) = \int_0^1 G(x,s)f(s)ds$

$$= -\int_0^x \frac{y_1(s)y_2(x)f(s)}{c} ds - \int_x^1 \frac{y_1(x)y_2(s)f(s)}{c} ds$$

(where $c = p(x)W(y_1,y_2)$ and thus, by Leibnitz's rule,

$$c\phi'(x) = -y_1(x)y_2(x)f(x) - \int_0^x y_1(s)y_2'(x)f(s)ds + y_1(x)y_2(x)f(x)$$

$$-\int_x^1 y_1'(x)y_2(s)f(s)ds. \quad \text{From this we obtain}$$

$$-c(p\phi')' = (py_2')'\int_0^x y_1(s)f(s)ds + py_2'y_1f(x)$$

$$+ (py_1')'\int_x^1 y_2(s)f(s)ds - py_1'y_2f(x). \quad \text{Dividing by}$$

c and adding $q(x)\phi(x)$ we get

$$(-p\phi')'+q\phi = \frac{(py_2')'}{c}\int_0^x y_1(s)f(s)ds - \frac{qy_2}{c}\int_0^x y_1(s)f(s)ds$$

$$+ \frac{(py_1')'}{c}\int_x^1 y_2(s)f(s)ds - \frac{qy_1}{c}\int_x^1 y_2(s)f(s)ds + f(x)$$

$$= \frac{(py_2')'-qy_2}{c}\int_0^x y_1(s)f(s)ds + \frac{(py_1')'-qy_1}{c}\int_x^1 y_2(s)f(s)ds + f(x)$$

$= f(x)$, since y_1 and y_2 satisfy $L[y] = 0$. Using $\phi(x)$ and $\phi'(x)$ as found above, the B.C. are both satisfied since $y_1(x)$ satisfies one B.C. and $y_2(x)$ satisfies the other B.C.

33. In general $y(x) = c_1\cos x + c_2\sin x$. For $y'(0) = 0$ we must choose $c_2 = 0$ and thus $y_1(x) = \cos x$. For $y(1) = 0$ we have $c_1\cos 1 + c_2\sin 1 = 0$, which yields $c_2 = -c_1(\cos 1)/\sin 1$ and thus $y_2(x) = c_1\cos x - c_1(\cos 1)\sin x/\sin 1$

$$= c_1(\sin 1 \cos x - \cos 1 \sin x)/\sin 1$$

$$= \sin(1-x) \text{ [by setting } c_1 = \sin 1].$$

Furthermore, $W(y_1,y_2) = -\cos 1$ and thus

$$G(x,s) = \begin{cases} \dfrac{\cos s \sin(1-x)}{\cos 1} & 0 \le s \le x \\[2mm] \dfrac{\cos x \sin(1-s)}{\cos 1} & x \le s \le 1 \end{cases},$$

and hence

$$\phi(x) = \int_0^x [\cos s \sin(1-x)f(s)/\cos 1]ds$$

$$+ \int_x^1 [\cos x \sin(1-s)f(s)/\cos 1]ds$$

is the solution of the given B.V.P.

Section 11.4, Page 739

2a. The D.E. is the same as Eq.(7) and thus, from Eq.(9), the general solution of the D.E. is $y = c_1 J_0(\sqrt{\lambda}\,x) + c_2 Y_0(\sqrt{\lambda}\,x)$. The B.C. at $x = 0$ requires that $c_2 = 0$, and the B.C. at $x = 1$ requires $c_1\sqrt{\lambda}\, J_0'(\sqrt{\lambda}) = 0$. For $\lambda = 0$ we have $\phi_0(x) = J_0(0) = 1$ and if λ_n is the nth positive root of $J_0'(\sqrt{\lambda}) = 0$ then $\phi_n(x) = J_0(\sqrt{\lambda_n}\,x)$. Note that for $\lambda = 0$ the D.E. becomes $(xy')' = 0$, which has the general solution $y = c_1 \ln x + c_2$. To satisfy the bounded conditions at $x = 0$ we must choose $c_1 = 0$, thus obtaining the same solution as above.

2b. For $n \neq 0$, set $y = J_0(\sqrt{\lambda_n}\,x)$ in the D.E. and integrate from 0 to 1 to obtain $-\int_0^1 (xJ_0')'dx = \lambda_n \int_0^1 xJ_0(\sqrt{\lambda_n}\,x)dx$. Integrating the left side of this equation yields

$$\int_0^1 (xJ_0')'dx = xJ_0'(\sqrt{\lambda_n}\,x)\Big|_0^1 = J_0'(\sqrt{\lambda_n}) - 0 = 0 \text{ since the } \lambda_n$$

are eigenvalues from Part(a). Thus $\int_0^1 xJ_0(\sqrt{\lambda_n}\,x)dx = 0$, which is the desired result for $m = 0$, $n \neq 0$. For other n and m, we let $L[y] = -(xy')'$. Then $L[J_0(\sqrt{\lambda_n}\,x)] = \lambda_n xJ_0(\sqrt{\lambda_n}\,x)$ and $L[J_0(\sqrt{\lambda_m}\,x)] = \lambda_m xJ_0(\sqrt{\lambda_m}\,x)$. Multiply the first equation by $J_0(\sqrt{\lambda_m}\,x)$, the second by $J_0(\sqrt{\lambda_n}\,x)$, subtract the second from the first, and integrate from 0 to 1 to obtain

$$\int_0^1 \{J_0(\sqrt{\lambda_m}\,x)L[J_0(\sqrt{\lambda_n}\,x)] - J_0(\sqrt{\lambda_n}\,x)L[J_0(\sqrt{\lambda_m}\,x)]\}dx =$$

$(\lambda_n - \lambda_m)\int_0^1 xJ_0(\sqrt{\lambda_n}\,x)J_0(\sqrt{\lambda_m}\,x)dx$. Again the left side is zero after each term is integrated by parts once, as was done above. If $\lambda_n \neq \lambda_m$, the result follows with $\phi_n(x) = J_0(\sqrt{\lambda_n}\,x)$.

2c. Since $\lambda = 0$ is an eigenvalue we assume that

$$y = b_0 + \sum_{n=1}^{\infty} b_n J_0(\sqrt{\lambda_n}\,x). \text{ Since}$$

$-[xJ_0'(\sqrt{\lambda_n}\,x)]' = \lambda_n xJ_0(\sqrt{\lambda_n}\,x)$, $n = 0,1,\ldots$, we find that

$-(xy')' = x \sum_{n=1}^{\infty} \lambda_n b_n J_0(\sqrt{\lambda_n}\, x)$ [note that $\lambda_0 = 0$ and b_0 are

missing on the right]. Now assume

$f(x)/x = c_0 + \sum_{n=1}^{\infty} c_n J_0(\sqrt{\lambda_n}\, x)$. Multiplying both sides by

$x J_0(\sqrt{\lambda_m}\, x)$, integrating from 0 to 1 and using the
orthogonality relations of Part(b), we find

$c_n = \int_0^1 f(x) J_0(\sqrt{\lambda_n}\, x)dx / \int_0^1 x J_0^2(\sqrt{\lambda_n}\, x)dx$, $n = 0,1,2,\ldots$.

[Note that $c_0 = 2\int_0^1 f(x)dx$ since the denominator can be

integrated.] Substituting the series for y and $f(x)/x$
into the D.E., using the above result for $-(xy')'$, and
simplifying we find that

$(\mu b_0 + c_0) + \sum_{n=1}^{\infty} [c_n - b_n(\lambda_n - \mu)] J_0(\sqrt{\lambda_n}\, x) = 0$. Thus

$b_0 = -c_0/\mu$ and $b_n = c_n/(\lambda_n - \mu)$, $n = 1,2,\ldots$, where $\sqrt{\lambda_n}$
are obtained from $J_0'(\sqrt{\lambda_n}) = 0$.

4a. Let $L[y] = -[(1-x^2)y']'$. Then $L[\phi_n] = \lambda_n \phi_n$ and
$L[\phi_m] = \lambda_m \phi_m$. Multiply the first equation by ϕ_m, the
second by ϕ_n, subtract the second from the first, and
integrate from 0 to 1 to obtain

$\int_0^1 (\phi_m L[\phi_n] - \phi_n L[\phi_m])dx = (\lambda_n - \lambda_m)\int_0^1 \phi_n \phi_m dx$. The integral
on the left side can be shown to be 0 by integrating each
term once by parts. Since $\lambda_n \neq \lambda_m$ if $m \neq n$, the result
follows. Note that the result may also be written as
$\int_0^1 P_{2m-1}(x)P_{2n-1}(x)dx = 0$, $m \neq n$.

4b. First let $f(x) = \sum_{n=1}^{\infty} c_n \phi_n(x)$, multiply both sides by $\phi_m(x)$,

and integrate term by term from $x = 0$ to $x = 1$. The
orthogonality condition yields

$c_n = \int_0^1 f(x)\phi_n(x)dx / \int_0^1 \phi_n^2(x)dx$, $n = 1,2,\ldots$ where it is

understood that $\phi_n(x) = P_{2n-1}(x)$. Now assume

$y = \sum_{n=1}^{\infty} b_n \phi_n(x)$. As in Prob. 2 and in the text

$-[(1-x^2)y']' = \sum_{n=1}^{\infty} \lambda_n b_n \phi_n$ since the ϕ_n are eigenfunctions.

Thus, substitution of the series for y and f into the D.E. and simplification yields $\sum_{n=1}^{\infty} [b_n(\lambda_n-\mu) - c_n]\phi_n(x) = 0$. Hence $b_n = c_n/(\lambda_n - \mu)$, $n = 1,2,...$ and the desired solution is obtained [after setting $\phi_n(x) = P_{2n-1}(x)$].

Section 11.5, Page 744

1a. Since u(x,0) = 0 we have Y(0) = 0. However, since the other two boundaries are given by y = 2x and y = 2(x-2) we cannot separate x and y dependence and thus neither X nor Y satisfy homogeneous B.C. at both end points.

1b. The line y = 2x is transformed into $\xi = 0$ and y = 2(x-2) is transformed into $\xi = 2$. The lines y = 0 and y = 2 are transformed into $\eta = 0$ and $\eta = 2$ respectively, so the parallelogram is transformed into a square of side 2. From the given equations, we have $x = \xi + \eta/2$ and $y = \eta$. Thus
$u_\xi = u_x x_\xi + u_y y_\xi = u_x$ and
$u_\eta = u_x x_\eta + u_y y_\eta = (1/2)u_x + u_y$. Likewise
$u_{\xi\xi} = u_{xx} \cdot x_\xi + u_{xy}y_\xi = u_{xx}$
$u_{\xi\eta} = u_{xx}x_\eta + u_{xy}y_\eta = (1/2)u_{xx} + u_{xy}$ and
$u_{\eta\eta} = (1/2)u_{xx}x_\eta + (1/2)u_{xy}y_\eta + u_{yx}x_\eta + u_{yy}y_\eta$
$= (1/4)u_{xx} + u_{xy} + u_{yy}$. Therefore,
$(5/4)u_{\xi\xi} - u_{\xi\eta} + u_{\eta\eta} = u_{xx} + u_{yy} = 0$. The B.C. become
$U(\xi,0) = 0$, $U(\xi,2) = f(\xi+1)$ (since $x = \xi + \eta/2$), $U(0,\eta) = 0$, and $U(2,\eta) = 0$.

1c. Substituting $u(\xi,\eta) = U(\xi)V(\eta)$ into the equation of Part(b) yields $5/4U''V - U'V' + UV'' = 0$ or upon dividing by UV
$$\frac{5}{4}\frac{U''}{U} + \frac{V''}{V} = \frac{U'V'}{UV}, \text{ which is not separable.}$$

2. This problem is very similar to the example worked in the text. The fundamental solutions satisfying the P.D.E.(3), the B.C. u(1,t) = 0, $t \geq 0$ and the finiteness condition are given by Eqs.(15) and (16). Thus assume u(r,t) is of the form given by Eq.(17). The I.C. require that $u(r,0) = \sum_{n=1}^{\infty} c_n J_0(\lambda_n r) = 0$, so $c_n = 0$, and

$$u_t(r,0) = \sum_{n=1}^{\infty} \lambda_n ak_n J_0(\lambda_n r) = g(r). \quad \text{From Eq.(26) of}$$

Sect. 11.4 we obtain

$$\lambda_n k_n a = \int_0^1 rg(r)J_0(\lambda_n r)dr \Big/ \int_0^1 rJ_0^2(\lambda_n r)dr, \quad n = 1,2,\ldots .$$

4.   This problem is the same as Prob. 21 of Sect. 10.7. The
     periodicity condition requires that μ of that problem be
     an integer and thus substituting $\mu^2 = n^2$ into the
     previous results yields the given equations.

5a. Substituting $u(r,\theta,z) = R(r)\Theta(\theta)Z(z)$ into Laplace's
     equation yields $R''\Theta Z + R'\Theta Z/r + R\Theta''Z/r^2 + R\Theta Z'' = 0$ or
     equivalently $R''/R + R'/rR + \Theta''/r^2\Theta = -Z''/Z = \sigma$.   In order
     to satisfy arbitrary B.C. it can be shown that σ must be
     negative, so assume $\sigma = -\lambda^2$, and thus $Z'' - \lambda^2 Z = 0$ and,
     after some algebra, it follows that
     $r^2 R''/R + rR'/R + \lambda^2 r^2 = -\Theta''/\Theta = \alpha$.   The periodicity
     condition $\Theta(0) = \Theta(2\pi)$ requires that $\sqrt{\alpha}$ be an integer n
     so $\alpha = n^2$. Thus $r^2 R'' + rR' + (\lambda^2 r^2 - n^2)R = 0$,
     $\Theta'' + n^2\Theta = 0$, and $Z'' - \lambda^2 Z = 0$.

5b. If $u(r,\theta,z)$ is independent of θ, then the $\Theta''/r^2\Theta$ term
     does not appear in the second equation of Part(a) and
     thus $R''/R + R'/rR = -Z''/Z = -\lambda^2$, from which the desired
     result follows.

6.   Assuming that $u(r,z) = R(r)Z(z)$ it follows from Prob. 5
     that $R = c_1 J_0(\lambda r) + c_2 Y_0(\lambda r)$, from Eq.(13), and
     $Z = k_1 e^{-\lambda z} + k_2 e^{\lambda z}$. Since $u(r,z)$ is bounded as $r \to 0$ and
     approaches zero as $z \to \infty$ we require that $c_2 = 0$, $k_2 = 0$.
     The B.C. $u(1,z) = 0$ requires that $J_0(\lambda) = 0$ leading to an
     infinite set of discrete positive eigenvalues
     $\lambda_1, \lambda_2, \ldots \lambda_n \ldots$.   The fundamental solutions of the
     problem are then $u_n(r,z) = J_0(\lambda_n r)e^{-\lambda_n z}$, $n = 1,2,\ldots$.

     Thus assume $u(r,z) = \sum_{n=1}^{\infty} c_n J_0(\lambda_n r)e^{-\lambda_n z}$.   The B.C.

     $u(r,0) = f(r)$, $0 \le r \le 1$ requires that

     $u(r,0) = \sum_{n=1}^{\infty} c_n J_0(\lambda_n r) = f(r)$ so

     $c_n = \int_0^1 rf(r)J_0(\lambda_n r)dr \Big/ \int_0^1 rJ_0^2(\lambda_n r)dr, \quad n = 1,2,\ldots$.

7b. Again, Θ periodic of period 2π implies $\lambda^2 = n^2$. Thus the solutions to the D.E. are $R(r) = c_1 J_n(kr) + c_2 Y_n(kr)$ (note that λ and k here are the reverse of Prob. 3 of Sect. 11.4) and $\Theta(\theta) = d_1 \cos n\theta + d_2 \sin n\theta$, $n = 0,1,2\ldots$. For the solution to remain bounded, $c_2 = 0$ and thus

$$v(r,\theta) = (1/2)c_0 J_0(kr) + \sum_{m=1}^{\infty} J_m(kr)(b_m \sin m\theta + c_m \cos m\theta).$$

Hence $v(c,\theta)$ is then a Fourier Series of period 2π and the coefficients are found as in Sect. 10.2, Eqs.(13), (14) and Prob. 27.

9a. Substituting $u(\rho,\theta,\phi) = P(\rho)\Theta(\theta)\Phi(\phi)$ into Laplace's equation leads to
$\rho^2 P''/P + 2\rho P'/P = -(\csc^2\phi)\Theta''/\Theta - \Phi''/\Phi - (\cot\phi)\Phi'/\Phi = \sigma.$
In order to satisfy arbitrary B.C. it can be shown that σ must be positive, so assume $\sigma = \mu^2$.
Thus $\rho^2 P'' + 2\rho P' - \mu^2 P = 0$. Then we have
$(\sin^2\phi)\Phi''/\Phi + (\sin\phi\cos\phi)\Phi'/\Phi + \mu^2\sin^2\phi = -\Theta''/\Theta = \alpha.$
The periodicity condition $\Theta(0) = \Theta(2\pi)$ requires that $\sqrt{\alpha}$ be an integer λ so $\alpha = \lambda^2$. Hence $\Theta'' + \lambda^2\Theta = 0$ and
$(\sin^2\phi)\Phi'' + (\sin\phi\cos\phi)\Phi' + (\mu^2\sin^2\phi - \lambda^2)\Phi = 0.$

10. Since u is independent of θ, only the first and third of the Eqs. in Prob. 9a hold. The general solution to the Euler equation is
$P = c_1\rho^{r_1} + c_2\rho^{r_2}$ where $r_1 = (-1+\sqrt{1+4\mu^2})/2 > 0$ and
$r_2 = (-1-\sqrt{1+4\mu^2})/2 < 0$. Since we want u to be bounded as $\rho \to 0$, we set $c_2 = 0$. As found in Prob. 22 of Sect. 5.3, the solutions of Legendre's equation, Prob. 9c, are either singular at 1, at -1, or at both unless $\mu^2 = n(n+1)$, where n is an integer. In this case, one of the two linearly independent solutions is a polynomial denoted by P_n (Probs. 23 and 24 of Sect. 5.3). Since $r_1 = (-1 + \sqrt{1+4n(n+1)})/2 = n$, the fundamental solutions of this problem satisfying the finiteness condition are $u_n(\rho,\phi) = \rho^n P_n(s) = \rho^n P_n(\cos\phi)$, $n = 1,2,\ldots$. It can be shown that an arbitrary piecewise continuous function on $[-1,1]$ can be expressed as a linear combination of Legendre polynomials. Hence we assume that

$u(\rho,\phi) = \sum_{n=1}^{\infty} c_n \rho^n P_n(\cos\phi)$. The B.C. $u(1,\phi) = f(\phi)$ requires that $u(1,\phi) = \sum_{n=1}^{\infty} c_n P_n(\cos\phi) = f(\phi)$, $0 \le \phi \le \pi$. From Prob. 28 of Sect. 5.3 we know that $P_n(x)$ are orthogonal. However here we have $P_n(\cos\phi)$ and thus we must rewrite the equation in Prob.9b to find $-[(\sin\phi)\Phi']' = \mu^2(\sin\phi)\Phi$. Thus $P_n(\cos\phi)$ and $P_m(\cos\phi)$ are orthogonal with weight function $\sin\phi$. Thus we must multiply the series expansion for $f(\phi)$ by $\sin\phi P_m(\cos\phi)$ and integrate from 0 to π to obtain

$c_m = \int_0^{\pi} f(\phi)\sin\phi P_m(\cos\phi)d\phi / \int_0^{\pi}\sin\phi P_m^2(\cos\phi)d\phi$. To obtain the answer as given in the text let $s = \cos\phi$.

Section 11.6, Page 753

2a. From Eq.(6), $b_m = \sqrt{2}\int_0^1 x\sin m\pi x\, dx = \sqrt{2}\,(-1)^{m+1}/m\pi$ and thus $S_n = \dfrac{2}{\pi}\sum_{m=1}^{n} \dfrac{(-1)^{m+1}\sin m\pi x}{m}$.

2b. from Eq.(20), $R_n = \int_0^1 [x - S_n(x)]^2 dx$, where $S_n(x)$ is given in Part(a). Using appropriate computer software we find $R_1 = .1307$, $R_2 = .0800$, $R_5 = .0367$, $R_{10} = .0193$, $R_{15} = .0131$ and $R_{19} = .0104$.

2c. As in Part(b), we find $R_{20} = .0099$, and thus $n = 20$ will insure a mean square error less than .01.

4a. Writing $S_n(x)$ as a quotient we have $S_n(x) = \dfrac{n\sqrt{x}}{e^{nx^2/2}}$. Now use L'Hopital's Rule with respect to n to obtain

$\lim_{n\to\infty} S_n(x) = \lim_{n\to\infty} \dfrac{\sqrt{x}}{\dfrac{x^2}{2}e^{nx^2/2}} = 0$ for $x \ne 0$. For $x = 0$,

$S_n(0) = 0$ and thus $\lim_{n\to\infty} S_n(x) = 0$ for all x in $[0,1]$.
$R_n = \int_0^1 [0-S_n(x)]^2 dx = n^2\int_0^1 xe^{-nx^2}dx$

$= -\dfrac{n}{2}e^{-nx^2}\Big|_0^1 = \dfrac{n}{2} - \dfrac{ne^{-n}}{2}$. Since $ne^{-n} \to 0$ as $n \to \infty$, we have that $R_n \to \infty$ as $n \to \infty$.

5. Expanding the integrand of Eq. (6), we get

$$R_n = \int_0^1 r(x)[f(x) - S_n(x)]^2 dx$$

$$= \int_0^1 r(x)f^2(x)dx - 2\sum_{i=1}^n c_i \int_0^1 r(x)f(x)\phi_i(x)dx$$

$$+ \sum_{i=1}^n \sum_{j=1}^n c_i c_j \int_0^1 r(x)\phi_i(x)\phi_j(x)dx,$$

where the last term is obtained by calculating $S_n^2(x)$.
Using Eqs.(1) and (9) this becomes

$$R_n = \int_0^1 r(x)f^2(x)dx - 2\sum_{i=1}^n c_i a_i + \sum_{i=1}^n c_i^2$$

$$= \int_0^1 r(x)f^2(x)dx - \sum_{i=1}^n a_i^2 + \sum_{i=1}^n (c_i - a_i)^2, \text{ by completing}$$

the square. Since all terms involve a real quantity
squared (and $r(x) > 0$) we may conclude R_n is minimized by
choosing $c_i = a_i$. This can also be shown by solving
$\partial R_n / \partial c_i = 0$, which gives $2(c_i - a_i) = 0$.

7b. From Part(a) we have $f_0(x) = 1$ and thus $f_1(x) = c_1 + c_2 x$
must satisfy $(f_0, f_1) = \int_0^1 (c_1 + c_2 x)dx = 0$ and
$(f_1, f_1) = \int_0^1 (c_1 + c_2 x)^2 dx = 1$. Evaluating the integrals
yields $c_1 + c_2/2 = 0$ and $c_1^2 + c_1 c_2 + c_2^2/3 = 1$, which have
the solution $c_1 = \sqrt{3}$, $c_2 = -2\sqrt{3}$ and thus
$f_1(x) = \sqrt{3}(1-2x)$. Notice that $\sqrt{3}(2x - 1)$ also
satisfies the given conditions.

7c. $f_2(x) = c_1 + c_2 x + c_3 x^2$ must satisfy $(f_0, f_2) = 0$,
$(f_1, f_2) = 0$ and $(f_2, f_2) = 1$.

7d. For $g_2(x) = c_1 + c_2 x + c_3 x^2$ we have $(g_0, g_2) = 0$ and
$(g_1, g_2) = 0$, which yield the same ratio of coefficients
as found in Part(c). Thus $g_2(x) = cf_2(x)$, where c may
now be found from $g_2(1) = 1$.

8. This problem follows the pattern of Prob. 7c except now
the limits on the orthogonality integral are from

−1 to 1. That is $(P_i, P_j) = \int_{-1}^{1} P_i(x) P_j(x) dx = 0$, $i \neq j$.
For $i = 0$ we have $P_0(x) = 1$ and for $i = 0$ and $j = 1$ we
have $P_1(x) = c_1 + c_2 x$ and hence

$$(P_0, P_1) = \int_{-1}^{1} (c_1 + c_2 x) dx = (c_1 x + c_2 x^2/2) \Big|_{-1}^{1} = 2c_1 = 0 \text{ and}$$

thus $P_1(1) = 1$ yields $P_1(x) = x$. The others follow in a
similar fashion.

9a. This part has essentially been worked in Prob. 5 by
setting $c_i = a_i$.

9b. Eq.(6) shows that $R_n \geq 0$ since $r(x) \geq 0$ and thus from

Part(a) $\int_{0}^{1} r(x) f^2(x) dx - \sum_{i=1}^{n} a_i^2 \geq 0$. The result follows.

9c. Since f is square integrable [a condition of Thm.11.6.1],
$\int_{0}^{1} r(x) f^2(x) dx = M < \infty$ and therefore the monotone

increasing sequence of partial sums $T_n = \sum_{i=1}^{n} a_i^2$ is bounded

above. Thus $\lim\limits_{n\to\infty} T_n$ exists, which proves the convergence
of the given sum.

9d. This result follows from Part(a) and Part(c).

9e. By definition if $\sum_{i=1}^{\infty} a_i \phi_i(x)$ converges to $f(x)$ in the

mean, then $R_n \to 0$ as $n \to \infty$. Hence $\int_{0}^{1} r(x) f^2(x) dx = \sum_{i=1}^{\infty} a_i^2$.

Conversely, if $\int_{0}^{1} r(x) f^2(x) dx = \sum_{i=1}^{\infty} a_i^2$, $\lim\limits_{n\to\infty} R_n = 0$ and

$\sum_{i=1}^{\infty} a_i \phi_i(x)$ converges to $f(x)$ in the mean.

10. Bessel's inequality implies that $\sum_{i=1}^{\infty} a_i^2$ converges and thus

the nth term $a_n \to 0$ as $n \to \infty$.

12. If the series were the eigenfunction series for a square

integrable function, the series $\sum\limits_{i=1}^{\infty} a_i^2$ would have to

converge. But $a_0 = 1$, $a_1 = 1/\sqrt{2}$, ..., $a_n = 1/\sqrt{n}$, ...,

and $\sum\limits_{n=1}^{\infty} a_n^2 = \sum\limits_{n=1}^{\infty} 1/n$ is the well-known harmonic series which

does not converge.

Chapter 1: Introduction

Definitions:
* Differential Equation Mathematical Model
* Direction (Slope) Field Equilibrium Solution
* Rate (growth) constant
* Initial Condition, Initial Value Problem (IVP)
* General Solution, Integral curves
* Ordinary Differential Equation (ODE),
     Partial Differential Equation (PDE)
* Systems of Differential Equations
* Order, Linear, Nonlinear, Linearization

Important Skills:
* Derive differential equations that mathematically model simple problems. (Ex. 1, p. 2; Also see p. 7)
* Construct a direction field for a first order ODE, and sketch approximate solutions. (Ex. 2, p. 3)
* Graph the integral curves of a general solution. (Ex. 2, p. 13)
* Know what an initial value problem is, and how to show a given function is a solution to one. (Ex. 2, p. 13)
* Know the difference between an ordinary differential equation and partial differential equation. (p. 19)
* Derivation of pendulum differential equation (p. 21)
* Know how to classify differential equations as order, and linearity. (p. 20 - 21)

Chapter 2: First Order Differential Equations

Definitions:
- First Order Ordinary Differential Equation
- Integrating Factor, Integral Curves
- Variation of parameters
- Separable
- Homogeneous differential equations
- Implicit solutions
- Bernoulli Equations
- Logistic equations, intrinsic growth rate
- Existence and Uniqueness of Solutions General Solutions,
- Autonomous, Logistic Growth, Equilibrium Solutions,
- Stable solutions, asymptotically stable solutions, unstable
     equilibrium solution
- Threshold
- Integrating factors, Exact equations
- Critical Points Exact ODE
- Tangent Line Method (Euler's Method)
- First Order Difference Equation
- Method of successive approximations

Theorems:
- Theorem 2.4.1: Existence and uniqueness of solutions to linear
     first order ODE's. (p. 68)
- Theorem 2.4.2: Existence and uniqueness of solutions to first order
     IVP's. (p. 70)
- Theorem 2.6.1: Existence and uniqueness of solutions to exact first
     order ODE's. (p. 95)
- Theorem 2.8.1: Restatement and elaboration of Theorem 2.4.2.
     (p. 112)

Important Skills:
- Be able to determine if a first order differential equation is
linear or nonlinear. Equation (3) on page 32 gives the form for a
linear ODE.
- If the differential equation is linear, compute the integrating
factor, and then the general solution. (Ex. 4, p. 38)
- Be able to graph integral curves for an ODE. (Ex. 4, p. 38)
- If it's nonlinear, is it separable? If it's separable, you will need
to compute two different integrals.
- It is crucial to know integration of basic functions and integral
methods from your calculus course. For Example, various
substitutions, integration by parts, and partial fractions will all be
utilized. (Ex. 2 & 3, p. 45 & 46)
- If the differential equation is not separable, is it exact? If so,
solve it using the method in section 2.6. (Ex. 2, p. 97)
- If it isn't separable or exact, check for substitutions that would
convert it into a linear equation, nonlinear equation that is then

separable. For example, exercises 27 - 31 (Sec. 2.4, p. 77) show how.
• Bernoulli equations can be transformed into linear equations.
• What happens to solutions as time tends to infinity? Understand stability, asymptotic stability and instability.
• These important qualitative classifications are at the heart of dynamical systems. Important with this is the concept of a threshold value. (Sec. 2.5, p. 84 - 88)
• Know how to obtain approximate solutions using Euler's method if an analytical solution cannot be found. (Ex. 2, p. 106)
• Understand the three steps in the process of mathematical modeling: construction of the model, analysis of the model, and comparison with experiment or observation. (Ex. 3, p. 54)
• Determine the existence and uniqueness of solutions to differential equations. (Ex. 2, p. 71)
• Know how to recognize autonomous equations, and utilize the direction field to represent solution to them. Be able to determine asymptotically stable, semi-stable, and unstable equilibrium solutions. (Ex. 1, p. 83)

Relevant Applications:
• Mixing Problems, Compound Interest, Motion in a Gravitational Field, Radioactive Carbon Dating

Chapter 3: Second Order Linear Equations

Definitions:
- Linear and nonlinear
- Homogeneous, Nonhomogeneous
- Characteristic Equation Wronskian
- General Solution, Fundamental Set of Solutions
- Principle of superposition
- Particular Solution
- Method of undetermined solutions
- Period, Natural Frequency, Amplitude, Phase
- Overdamped, Critically Damped, Underdamped
- Resonance
- Transient Solution, Steady-State Solution or Forced Response

Theorems:
- Theorem 3.2.1: Existence and uniqueness of solutions to second
     order linear homogeneous equations. (p. 146)
- Theorem 3.2.2: Principle of Superposition. (p. 147)
- Theorem 3.2.3: Finding solutions to Equation (2) and Equation (3),
     using the Wronskian at the initial conditions. (p. 149)
- Theorem 3.2.4: Representing general solutions to second order
     linear homogeneous GDE's. (p. 149)
- Theorem 3.2.5: Existence of a fundamental set of solutions.(p. 151)
- Theorem 3.2.6: Abel's Theorem. (p. 153)
- Theorem 3.5.1: Relating differences in nonhomogeneous solutions to
     fundamental solutions. (Used to prove the following theorem.)
     (p. 175)
- Theorem 3.5.2: General solutions to linear nonhomogeneous ODE's.
     (p. 175)
- Theorem 3.6.1: General solutions to linear nonhomogeneous ODE's.
     (Using variation of parameters to determine the particular
     solution.) (p. 188)

Important Skills:
- Be able to determine if a second order differential equation is
linear or nonlinear, homogeneous or nonhomogeneous. (If it can be put
into the form given by Equation (3) in page 138, it is linear.)
- Most of the Chapter deals with linear equations. Important
exceptions are two methods given in Section 3.1, Equations (28) - (33)
on page 142, which shows how to solve second order differential
equations missing the dependent variable, and Equations (34) - (36) on
page 143, which show how to solve equations missing the independent
variable.
- Can you recognize a homogeneous equation with constant coefficients,
and derive the characteristic equation? (Ex. 3, p. 149) This
equation will be quadratic, so know the quadratic formula, the types
of solutions one gets: real and distinct, repeated, and complex
conjugate. These three cases will be crucial to the types of solutions

one gets to constant coefficient homogeneous differential equations.
• Be able to write down fundamental solution sets to homogeneous equations. This means find two solutions. (Ex. 3, p. 149).
• Reduction of order is a way to take a known solution and produce a second solution. Know this method. (Ex 3, p. 171)
• What are the fundamental solution sets for each of the three cases of roots when solving constant coefficient equations? The summary is on p. 170. (Ex. 3, p. 149; Ex. 2, p. 169; Ex. 3, p. 162)
• Solutions to second order nonhomogeneous equations have two components. There is the homogeneous solution, and particular or nonhomogeneous solution. (Thm. 3.5.2, p. 175) To find particular solutions you must know the method of undetermined coefficients, and variation of parameters. (Ex. 4, p. 178; Ex. 1, p. 185)
• Mechanical vibrations give excellent examples for utilizing all the techniques in the Chapter. Know the difference between damped and undamped vibrations, forced and unforced situations.
• For the unforced case, if there is no dampening, the motion is sinusoidal. Be able to determine the natural spring frequency. (Ex. 2, p.196) If there is dampening, know the three different kinds: underdamped, critically damped, and overdamped, depending on the roots to the characteristic equation. If underdamped, know the quasi period. (Ex. 3, p. 199) Know how to graph solutions in the three different cases of dampening. For the forced problem, the cases separate into damped or undamped. If undamped, there is the possibility of resonance if the nonhomogeneous forcing term is sinusoidal with the frequency equivalent to the natural spring frequency. (p. 214)
• If there is no resonance, then there will be beats. (p. 214)
• Know how to derive and graph solutions in this case. You will need trigonometric identities in your analyses.
• For the damped case, know how to identify and graph transient and steady state solutions.(p. 212)

Relevant Applications:
• Mechanical Vibrations, Electric Circuits

Chapter 4: **Higher Order Linear Equations**

Definitions:
- **n**th Order Linear ODE
- Fundamental Set of Solutions, General Solution
- Homogeneous and Nonhomogeneous equations
- Linear Dependence and Independence
- Characteristic Polynomial, Characteristic Equation
- Variation of parameters

Theorems:
- Theorem 4.1.1: Existence and uniqueness of solutions to higher
  order linear ODE's. (p. 220)
- Theorem 4.1.2: General solutions to higher order linear ODE's and
  the fundamental set of solutions (p. 221)
- Theorem 4.1.3: Relates linear independence to fundamental sets of
  solutions. (p. 223)

Important Skills:
- The methods for solving higher order linear differential equations
are extremely similar to those in the last chapter. There is simply **n**
times the fun! The general solution to an **n**th order homogeneous linear
differential equation is obtained by linearly combining **n** linearly
independent solutions. (Eq. (5), p. 220)
- The generalization of the Wronskian is given on page 221. It is used
as in the last Chapter to show the linear independence of functions,
and in particular, homogeneous solutions.
- For the situation where there are constant coefficients, you should
be able to derive the characteristic polynomial, and the
characteristic equation, in this case each of **n**th order. Depending
upon the types of roots you get to this equation, you will have
solution sets containing functions similar to those in the second
order case. (Ex. 2 - 4, p. 229 - 231)
- The general solution of the nonhomogeneous problem easily extends to
the **n**th order case. (Eq. (9), p. 227)
- Both variation of parameters and the method of undetermined
coefficients generalize to determine particular solutions in the
higher dimensional situation. (Ex. 3, p. 236; Ex. 1, p. 241)

Relevant Applications:
- Double and multiple spring mass systems

Chapter 5: Series Solutions of Second Order Equations

Definitions:
* Radius of Convergence, Interval of Convergence
* Analytic
* Recurrence Relation
* Ordinary Point, Singular Point
* Regular and Irregular Singular Points
* Euler Equation, Indicial Equation Exponents of Singularity
* Chebyshev equation, Hermite equation, Bessel equation

Theorems:
* Theorem 5.3.1: Existence of series solutions to linear ODE's near ordinary points and their convergence properties. (p. 262)
* Theorem 5.6.1: Series solutions near regular singular points. (p. 289)

Important Skills:
* Review power series, how to shift the index of summation, (Ex. 3, p. 247) and tests for convergence. (Ex. 2, p. 245)
* Know how to find the interval of convergence for a power series. (Ex. 2, p. 245)
* Be able to determine all ordinary and singular points for a differential equation. (p. 250 - 251)
* For all singular points, be able to categorize as either regular or irregular. (Ex. 5 & 6, p. 275) Give the criteria for a regular singular point.
* For ordinary points, Eq. (3) on page 251 gives the form of the solution. Be able to derive the recursion relation, as in Example 1. If the recursion relation can be solved, one obtains the two solutions of the homogenous problem. (Ex. 1, p. 251)
* The method described in the second paragraph on page 244 can be used to find the first several in each of the homogeneous solutions.
* Be able to determine lower bounds on the radius of convergence of the series solutions. (Ex. 4, p. 264)
* Series solutions near regular singular points require the ability to solve Euler equations. Be able to recognize Euler equations, and know how to derive the characteristic equation. Know the general solutions for the Euler Equation (2), page 272, for the three cases of roots to the characteristic equation. (Ex. 2 and 3, p. 270)
* The assumption for the form of the series solution near regular points is given by Equation (7) on page 279.
* Substitution into the differential equations will yield an indicial equation, as well as, a recursion relation. The solutions to the indicial equation are those to the associated Euler problem. (Ex. 1, p. 279) In cases where the roots to the indicial equation are equal or differ by an integer, the method can be slightly modified to obtain solutions, or one can use reduction of order. (p. 288)

• Finally, Bessel equations give good examples of series solutions near regular singular points; several examples are given in section 5.7. Bessel functions are extremely important in applied mathematics, physics, and engineering problems, and seem to arise when there are cylindrical symmetries.

Chapter 6: The Laplace Transformation

Definitions:
- Improper Integral, Piecewise Continuous Function
- Integral Transforms, Kernel
- The Laplace Transform
- Continuous Exponential Order
- Unit Step Function (Heaviside Function)
- Unit Impulse Function, Delta Function
- Convolution
- Transfer Function, Impulse Response

Theorems:
- Theorem 6.1.1:   Comparison Test for Improper Integrals   (p. 307)
- Theorem 6.1.2:   Existence of the Laplace Transform, $F(s)$ (p. 308)
- Theorem 6.2.1:   Laplace Transform of $f'(t)$ (p. 313)
- Corollary 6.2.2: Laplace Transform of $f^{(n)}(t)$ (p. 314)
- Theorem 6.3.1:   Transform of the unit step function, $u_c(t)$, times a shifted function, $f(t - c)$
                      (p. 326)
- Theorem 6.3.2:   First Translation Theorem; Inverse Transforming $F(s - c)$ (p. 328)
- Theorem 6.6.1:   Second Translation Theorem; Convolution Result
     p. 345)

Important Skills:
- The Laplace transformation is defined through an improper integral. You must be comfortable evaluating improper integrals. Hence you should review this topic in any calculus book.
- Be able to calculate the transform of all the basic functions, given in the table on page 317. (Ex. 5, 7, & 8, p. 309 - 310)
- Even more importantly, know how to compute inverse transform functions using manipulative translation methods. You may need to use partial fractions, but you should have already reviewed this for Chapter 2. (Ex. 2 & 3, p. 318 - 319)
- Know how to transform derivatives of functions and linear differential equations. (Thm. 6.2.1 & Cor. 6.2.2, p. 313 - 314; Ex. 2 & 3, p. 318 - 319)
- Understand the unit function, $u_c(t)$, as well as, the unit impulse function, $\delta(t)$, and how to use them in transforming and inverse transforming functions. (Ex. 1, p. 324; Ex. 1, p. 342)

• The process of using the Laplace transform method is as follows:
Given a differential equation, one transforms both sides of the
equation. One will need to input the initial values when transforming
derivatives. Derivatives with respect to **t** transform to polynomials
in **s**. If the differential equation is linear, then the resulting
equation is linear in Y(s). You simply solve the equation for Y(s),
and then use all the methods available to recover y(t). (Ex. 2, p. 318
for continuous forcing; Ex. 1, p. 342 for discontinuous forcing.)

Relevant Applications:
• Mechanical and electrical problems with discontinuous forcing
functions.

Chapter 7: Systems of First Order Linear Equations

Definitions:
- Systems of ODE's
- Linear vs. Nonlinear Systems
- Solution
- Homogenous and Nonhomogeneous Systems Matrix,
- Transpose, Conjugate, Adjoint, Determinant Scalar
- (Inner) Product, Orthogonal
- Nonsingular (Invertible) and Singular (Noninvertible)
- Row Reduction (Gaussian Elimination)
- Linear Systems, Homogeneous, Nonhomogeneous
- Augmented Matrix
- Linear Dependence and Independence
- Eigenvalues, Eigenvectors, Generalized Eigenvectors
- Normalization
- Multiplicity **m**, Simple Multiplicity (**m** = 1)
- Self Adjoint (Hermitian)
- General Solution, Fundamental Set of Solutions
- Phase Plane, Phase Portrait
- Generalized Eigenvector
- Node, Saddle Point, Spiral Point, Improper Node
- Fundamental Matrix
- The matrix exp(At)
- Similarity Transformation, Diagonalizable Matrices

Theorems:
- Theorem 7.1. 1: Existence and uniqueness of solutions for general systems of First Order IVP's (p. 358)
- Theorem 7.1.2. Existence and uniqueness of solutions for linear systems (p. 359)
- Theorem 7.4.1: Superposition of solutions (p. 386)
- Theorem 7.4.2: Theorem Superposition of solutions - general case (p. 387)
- Theorem 7.4.3: Nonvanishing of Wronskian for linearly independent solutions (p. 387)
- Theorem 7.4.4: Existence of fundamental set of solutions (p. 388)

Important Skills:
- Representation of solutions and vectors
- Find the inverse of a matrix. (Ex. 2, p. 369)
- Find the solution to a set of linear algebraic equations. (Ex. 1, p. 375)
- Determine if a set of vectors is linearly independent. (Ex. 3, p. 377)
- Find the eigenvalues and eigenvectors of a matrix. (Ex. 5, p. 381)
- Sketch a direction field for a 2 x 2 system of linear ODE's.(Ex. 2, p. 394)
- Find the general solution of a system of linear ODE's.
- Distinct Eigenvalues (Ex. 3, p. 397)

- Complex Eigenvalues   (Ex. 1, p. 401)
- Repeated Eigenvalues   (Ex. 2, p. 423)
- Find the fund. matrix for a system of linear ODE's.(Ex. 1 & 2,p. 414-415)
- Find the similarity transformation to diagonalize a matrix.(Ex. 3, p. 418)
- Use the method of undetermined coefficients to find the particular solution to a nonhomogeneous linear system of ODE's.   (Ex. 2, p. 435)
- Use the method of variation of parameters to find the particular solution to a nonhomogeneous linear system of ODE's.   (Ex. 3, p. 437)

Relevant Applications:
- Multiple Spring Mass Problems, Multiple Tank Mixture Problems

Chapter 8: Numerical Methods

Definitions:
• Definitions and Algorithms:
• Convergence
• Global Truncation Error
• Local Truncation Error
• Round-off Error
• Euler Method, Backward Euler Method
• Improved Euler Method (Heun formula)
• Modified Euler Formula
• Runge-Kutta Method
• Adaptive Methods
• One-step Method, Multi-step Method, Predictor-Corrector Method, Adams-   Bashforth Formula, Adams-Moulton Formula, Backward Differentiation Formulas
• Stability
• Stiff Problems

Theorems:
• None

Important Skills:
Use a particular method with specified step size to compute approximate solutions to ODE's.
• Euler Method (Ex. 1, p. 445)
• Backward Euler Method (Ex. 2, p. 447)
• Improved Euler Method (Ex. 1, p. 455)
• Runge-Kutta Method (Ex. 1, p. 461)
• Predictor-Corrector Method, Adams-Bashforth Formula, Adams-Moulton Formula (Ex. 1, p. 467)
• Backward Differentiation Method (Ex. 2, p. 468)
• Use numerical methods to find approximate solutions to systems of ODE'S. (Ex. 1, p. 481)
• Observing large errors of approximation for Euler's method (Ex. 1, p. 471)
• Use numerical methods to find approximate solutions to stiff ODE's (Ex. 2, p. 475)

Relevant Applications:
• Any of the applications previously mentioned, where analytical solutions cannot be found, or for which finding analytical solutions are too costly.

Chapter 9: Nonlinear Differential Equations and Stability

Definitions:
• Equilibrium Solutions
• Critical Points
• Trajectory
• Phase Plane
• Phase Portrait
• Node; Nodal Sink, Nodal Source, Saddle Point, Proper Node,
    (Star Point)
  Improper Node, (Degenerate Node), Spiral Sink, Spiral Source
• Autonomous Stable, Unstable Isolated Critical Point
• Locally Linear System
• Basis of Attraction
• Globally Asymptotically Stable
• Region of Asymptotic Stability, Nullclines
• Separatrix
• Liapunov's Second Method
• Positive Definite, Negative Definite, Positive Semidefinite,
    Negative Semidefinite
• Limit Cycle
• Asymptotically Stable
• Strange Attractors
• Chaotic System/Equation

Theorems:
• Theorem 9.3.1: Stability of critical points of linear systems
dependence on eigenvalues. (p. 508)
• Theorem 9.3.2: Stability of critical points of almost linear
systems. (p. 512)
• Theorem 9.6.1: Stability of critical points dependence on negative
definite and negative semidefinite nature of the Liapunov function,
and its derivative. (p. 547)
• Theorem 9.6.2: Conditions on definiteness for an unstable critical
point. (p. 547)
• Theorem 9.6.3: Conditions for positive definite. (p. 549)
• Theorem 9.6.4: Conditions for $V(x, y) = ax^2 + bxy + cy^2$ to be
positive or negative definite. (p.550)
• Theorem 9.7.1: Existence of closed trajectories (p. 557)
• Theorem 9.7.2: Nonexistence of closed trajectories. (p. 557)
• Theorem 9.7.3: Poincare'-Bendixson Theorem. (p.558)

Important Skills:
• Be able to determine the phase plane and phase portraits of a 2 by 2
linear system. The solutions will depend on eigenvalues. Pages 486 -
493 cover the five important cases. Table 9.1.1 on page 494 summarizes
the eigenvalue results.
• Determine the trajectories for a system of ODE's. (Ex. 3 & 4, p.
504 - 505)
• Know how to determine whether a system of ODE's is locally linear.

(Ex. 1 & 2, p.510)
• Be able to determine the linear system associated with the almost linear system.   (Ex. 3, p. 512)
• Relating the ODE system to the possible motions of a pendulum   (Ex. 4, p. 515)
• Sketch phase portraits for competing species.   (Ex. 1 or 2, p.521 and p.524)
• Sketch phase portraits for predator-prey.   (Ex. 1, p.534)
• Use Liapunov's method to determine the stability of a critical point.   (Ex. 1, p.546; Ex. 2, p.548)
• Determine periodic solutions of systems of ODE's.   (Ex. 1, p.555)
• Study the solution of the van der Pol equation.   (Ex. 2, p. 559)

Relevant Applications:
• Population Modeling, Competing Species, Predator-Prey Modeling

Chapter 10: Partial Differential Equations and Fourier Series

Definitions:
- Boundary Conditions, Two-Point Boundary Value Problem (BVP)
- Homogeneous, Nonhomogeneous
- Eigenvalues, Eigenfunctions
- Fourier Series; Periodic, Fundamental Period
- Inner Product, Orthogonal, Mutually Orthogonal
- Piecewise Continuous
- Even and Odd Function
- Fourier Sine Series and Fourier Cosine Series
- Heat or Diffusion Equation
- Thermal diffusivity
- Wave Equation; Natural Frequencies, Natural Mode, Wavelength
- Laplace's Equations (Potential Equation)
- Potential Equation
- Dirichlet and Neumann Problems

Theorems:
- Theorem 10.3.1: Convergence of Fourier Series (p. 596)

Important Skills:
- Be able to solve Boundary Value Problems. (Ex. 1 - 4, p. 579 - 580)
- Know how to compute Fourier Series for functions. (Ex. 1 - 3, p. 588 - 591)
- Understand the convergence properties of Fourier Series and Gibbs Phenomenon. (Ex. 1, p. 597)
- Know the difference between even and odd functions and the ramifications on their Fourier Series. (Ex. 1 & 2, p. 605 - 607)
- Be able to use Separation of Variables to solve heat conduction problems, (Ex. 1, p.616; Ex. 1, p. 623, and Ex. 2, p. 626) as well as wave propagation problems. (Ex. 1, p. 635)
- Know the difference between Dirichlet and Neumann problems. (p. 647)
- Be able to apply Separation of Variables to solve Laplace's equation. (Ex. 1, p. 650)

Relevant Applications:
- Acoustics, Scalar Electromagnetic Propagation, Chemical or Thermal Diffusion

Chapter 11: Boundary Value Problems and Sturm-Liouville Theory

Definitions:
- Eigenvalues and Eigenfunctions
- Separated Boundary Conditions
- Lagrange's Identity
- Othogonality of Eigenfunctions
- Normalized, Orthonormal Set
- Self-Adjoint boundary value problem
- Periodic Boundary Conditions
- Singular Sturm-Liouville Problem
- Continuous Spectrum
- Bessel Equation
- Method of Collocation
- Mean Square sense
- Mean Square Error
- Complete set of functions
- Square Integrable

Theorems:
- Theorem 11.2.1: Eigenvalues of Sturm-Liouville Problems are Real (p. 676)
- Theorem 11.2.2: Orthogonality of Sturm-Liouville Eigen functions (p. 677)
- Theorem 11.2.3: Eigenvalues of Sturm-Liouville Problems are Simple, and Ordered (p. 677)
- Theorem 11.2.4: Convergence of Infinite Sum of Normalized Sturm-Liouville Eigen functions
    (p. 680)
- Theorem 11.3.1: Existence and Uniqueness of Solutions to Nonhomogeneous Sturm-Liouville
    problems (p. 690)
- Theorem 11.3.2: Fredholm Alternative Theorem (p. 690)
- Theorem 11.6.1: Completeness of Sturm-Liouville Eigen functions (p. 720)

Important Skills:
- Be able to compute eigenvalues and eigenfunctions. (Ex. 1, p. 668)
- Know how to normalize a set of eigenfunctions. (Ex. 1 & 2, p. 678 - 679)
- Expand a given function in terms of normalized eigenfunctions. (Ex. 3, p. 681)
- Solve nonhomogeneous PDE's with mixed boundary conditions. (Ex. 1, p. 690)
- Know circular vibration problems, Bessel functions, and how they relate to singular Sturm-Liouville problems. (p. 705 - 706, and Sec. 11.5)

• Discuss the mean convergence of series representations. (Ex. 1, p. 720)

Relevant Applications:
• Any problems where separation of variables leads to a two point boundary value problem

Notes

Notes

Notes

Notes

Notes

Notes

Notes

Notes